Novel Membrane Emulsification

Novel Membrane Emulsification

Principles, Preparation, Processes, and Bioapplications

Guanghui Ma

WILEY-VCH

Author

Prof. Guanghui Ma
Institiute of Process Engineering
Chinese Academy of Sciences
1 Beierjie, Zhongguan Cun
Haidian District
Beijing
China, 100190

Cover Image: © Guanghui Ma

Library of Congress Card No.: applied for

British Library Cataloguing-in-Publication Data
A catalogue record for this book is available from the British Library.

Bibliographic information published by the Deutsche Nationalbibliothek
The Deutsche Nationalbibliothek lists this publication in the Deutsche Nationalbibliografie; detailed bibliographic data are available on the Internet at <http://dnb.d-nb.de>.

Print ISBN: 978-3-527-34881-7
ePDF ISBN: 978-3-527-83083-1
ePub ISBN: 978-3-527-83084-8
oBook ISBN: 978-3-527-83085-5

Typesetting Straive, Chennai, India
Printing and Binding CPI Group (UK) Ltd, Croydon, CR0 4YY

C111876_150523

Co-Authors

Dongxia Hao
Jie Wu
Fangling Gong
Wei Wei
Yufei Xia
Hua Yue
Yi Wei
Weiqing Zhou
Lan Zhao
Xiangming Na
Yanlin Lv
Nan Wu
Yuning Hu

State Key Laboratory of Biochemical Engineering
Institute of Process Engineering
Chinese Academy of Sciences

Contents

Preface

Polymer microspheres and microcapsules have supported the revolutionary development of industry and led to the successful applications of various new techniques. In pharmaceutical and biomedical fields, for example, the dextran microsphere enabled mammalian cell culture in a large bioreactor, which promoted the development of vaccines such as the COVID-19 vaccine and the rabies vaccine. The crosslinked rigid agarose microsphere has enabled the successful scale-up of chromatographic separation and purification of pharmaceutical proteins and vaccines and led the rapid development of biopharmaceutical industry. Polylactide microsphere and microcapsule promoted the development of sustained release formulation of peptide drugs, which changed daily injections to weekly, monthly, and even yearly, improving the life quality of patients apparently.

The size distribution of microspheres is crucial for bio-applications. For example, uniform agarose microsphere can increase the separation resolution, and impurities can be removed from products by fewer steps. Uniform polylactide microcapsules can release the drug constantly, compared with non-uniform particles, and can guarantee the reproducibility of the products. Especially for targeting drug delivery, the size decides the particle distribution in the body; non-uniform particle will result in lower bioavailability and higher side effects because the particle diameter is one of the key factors that control the pharmacokinetics and efficacy of a loaded drug.

Polymer microspheres/microcapsules can be prepared from monomers or preformed polymers. The former includes emulsion/soap-free emulsion polymerization, disperse/precipitate polymerization, microemulsion polymerization, mini-emulsion polymerization, suspension polymerization, and seeded polymerization. By emulsion/soap-free emulsion polymerization and disperse/precipitate polymerization, uniform microspheres with submicron and micron sizes, respectively, can be prepared easily; however, the encapsulation of active components and morphology control are difficult due to their particulate mechanisms. On the other hand, suspension polymerization can encapsulate active components easily, but the size control is difficult and the size distribution is very broad.

The microspheres/microcapsules also can be prepared by using preformed polymer including natural polymer (agarose, etc.) and synthetic polymer (polylactide, etc.) as the materials. Especially for biomedical applications, preformed biocompatible polymers are usually used to prepare microspheres and microcapsules.

Most often used method is the emulsion-solidification method, similar to the above suspension polymerization method. That is, the emulsion is prepared first, and the droplets are solidified into microspheres. Although the active components can be encapsulated easily, the size control is difficult and the size distribution is very broad, which is unfavorite to reproducible industrialization.

Membrane emulsification technique (MET) solved this problem. The uniform droplets can be formed by using monomers or preformed polymer solutions as the dispersed phases. As a result, uniform microspheres can be obtained by polymerizing monomers or solidifying the droplets. Furthermore, the active components can also be encapsulated if they are added to the monomers or preformed polymer solutions before emulsification.

Membrane emulsification has the following advantages, which has been recognized until now:

(1) Uniform droplets can be prepared, so the sizes of droplets and particles solidified from droplets are easy to control by changing the pore sizes of membranes, and the reproducibility is excellent;
(2) W/O, O/W, and W/O/W or O/W/O emulsions have been successfully prepared by controlling the preparation conditions, including interfacial tension of dispersed phase and membrane, transmembrane pressure, flow rate of continuous phase, and so forth. Therefore, hydrophobic, hydrophilic, and composite microspheres and microcapsules can be prepared, respectively, from different emulsion types;
(3) The preparation condition is mild without violent mechanical stirring, so the temperature-sensitive components can be encapsulated without losing their bioactivity;
(4) The coalescence and breakup of droplets during solidification can be avoided, resulting in high stability of the emulsion and high encapsulation efficiency of the microcapsules;
(5) Scale-up is easy in principle because the productivity can be amplified by increasing membrane area;
(6) Because the size is uniform, the effect of the size and morphology of microspheres on application results can be studied systematically and quantitatively, and the optimum size can be selected to obtain the best application results.

However, the MET, especially the direct membrane emulsification process, has the following difficulties to be solved for industrialization:

(1) The preparation condition to obtain uniform droplets has to be optimized for different systems; it needs a huge amount of trial-and-error experiments;
(2) The interfacial tension between the dispersed phase and the membrane should be high, which is unfavorite for polar monomer systems such as hydroxyethyl methacrylate (HEMA);
(3) The control of trans-membrane pressure should be carried out carefully in a narrow window; otherwise, the size distribution will become broader;
(4) The dispersed phase with high viscosity is hard to be emulsified into continuous phase to obtain uniform droplets with high productivity;

(5) If one membrane has a crack, the whole production will fail. And, how to produce pharmaceutical microspheres in the good manufacturing practice of medical products (GMP) is also a big challenge;
(6) The druggability of microcapsules produced by this technique needs to be verified.

In order to overcome these difficulties, we studied systematically from fundamental research to equipment development and industrialization. We developed the mathematical model to predict the experimental conditions (transmembrane pressure, flow rate of continuous phase, etc.) to obtain uniform droplets and particles for different systems with different physicochemical properties (viscosity of dispersed phase, emulsifier type, interfacial tension, etc.) and successfully developed MET process from O/W, W/O, and double emulsion systems. We also developed rapid MET instead of direct MET to prepare smaller particles, especially for viscous systems. In addition, we developed various morphologies of microspheres by using uniform droplets as templates. Furthermore, the applications and their advantages in sustained release formulation of drugs, targeting delivery of anticancer agents, vaccine delivery, and biochemical engineering (cell culture and bio-separation) were studied. In these application studies, it has been proved that the uniform size has superior advantages. In order to industrialize MET, we have designed equipment and GMP production line of uniform microspheres, and some of these products have been approved for clinical trials. It will be expected that more and more products of uniform microspheres by using MET will be industrialized and put on the market. These contents will be introduced in the following chapters.

With the development of science and technology and the need to improve human life quality, more and more uniform microspheres and microcapsules should be designed and used in the biomedical/pharmaceutical fields, and this technique should be extended to more and more systems besides polymer systems to obtain such things as uniform silica gel particle, inorganic particles, and various composite particles for different emerging applications. Furthermore, preparing more smaller particles with several tens of nanometers by MET which are important in targeting drug delivery and mRNA delivery, is still a big challenge in the future.

Membrane has been widely used for filtration in many fields. But in the last 30 years, this hidden talent of membranes for emulsification has been found. We can use MET to prepare uniform emulsions and uniform particles (microspheres and microcapsules) from both monomer and preformed polymer systems, which have never been realized by conventional chemical engineering processes. This new process is becoming universal through a huge amount of R&D work and has been extended to many application fields, especially in bio-applications.

We hope this new process will benefit more researchers in institutes and industries. Furthermore, microspheres/microcapsules used in biomedical area are attracting new attention, such as in vaccine delivery, especially for infectious disease vaccines and personalized tumor therapeutic vaccines. Uniform microspheres have taken and will take very important role in this area.

Therefore, we summarized the advance of MET from principle to practice and applications, which mainly consisted of our research work on MET. We believe this book will benefit the researchers and graduate students in chemical engineering, biomedical and biological engineering, and polymer materials from academia and industry.

Our research achievements in this book were supported and contributed by National Natural Science Foundation of China, Ministry of Science and Technology of China, the Chinese Academy of Sciences, and all colleagues and students in our team, as well as collaborators. I would like to thank their support and cooperation sincerely.

Guanghui Ma
Professor and Director of State Key Laboratory of Biochemical Engineering, Institute of Process Engineering, Chinese Academy of Sciences.

1

Membrane Emulsification Process: Principle and Model

1.1 Introduction

Emulsions or particles prepared from emulsion with size-uniformity and size-controllability present distinctive advantages in industrial applications, such as microcapsules in drug delivery, particulate adjuvants for vaccination, and microspheres media in chromatographic separation and cell culture [1–4]. The early techniques of large-scale producing emulsion mostly depend on externally exerting strong dissipated energy into fluid mixtures, such as the rotor–stator, the high-pressure homogenizers, and the ultrasonic emulsification systems, in which the dissipated energy cannot be controlled homogeneously and the emulsion with broad size-distribution is often obtained. In recent decades, a great amount of work begins to explore the devices with milder and more controllable dissipating techniques to produce more uniform emulsion. Among the emerging devices, membrane emulsification technology (MET) has been addressed as a widely concerned group of uniform emulsification techniques, during which the to-be dispersed phase could grow into uniform droplets at uniform membrane pores with help of gentle and accurately controlled driving force [5]. Due to the controllability of droplet size by uniform membrane pore instead of the turbulent shearing flow, MET offers many advantages not only in narrow size distribution of droplets but also in lower energy requirement and suitability for emulsification of shear or temperature sensitive components.

According to different principles of emulsion preparation, the MET can be divided into two groups, the cross-flow membrane emulsification [6] and the premix membrane emulsification [7], also called as direct membrane emulsification and rapid membrane emulsification. The cross-flow MET is a process of two liquid phase flow. The to-be dispersed phase is pressed through the membrane's inner channel by pressure, generates the droplets at the membrane pores, and then the droplets until to a certain size are carried away by continuous phase fluid, as shown in Figure 1.1a. There are certain similarities existing between droplet breakup in premix and cross-flow emulsification depending upon the operating parameters. Comparatively, premix emulsification begins with a coarse emulsion, and then is extruded and homogenized through a porous membrane under a higher pressure to obtain fine emulsion, as shown in Figure 1.1b. Although the size of

Novel Membrane Emulsification: Principles, Preparation, Processes, and Bioapplications,
First Edition. Guanghui Ma.

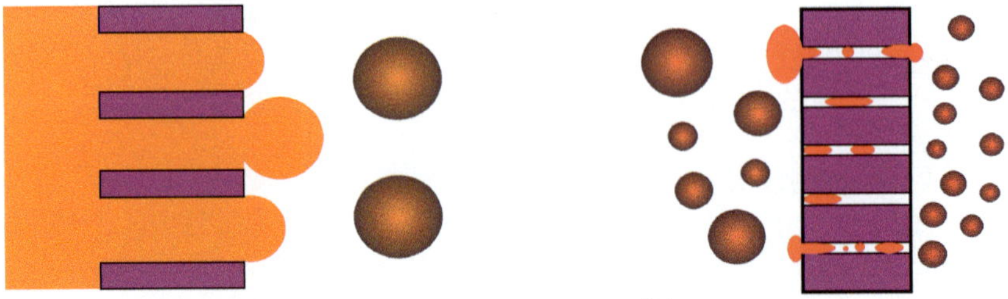

Figure 1.1 Two basic forms of membrane emulsification. (a) Cross-flow MET. (b) Premix MET.

the droplets is not as monodisperse as abovementioned cross-flow emulsification, premix emulsification is still an efficient route to produce emulsions with high dispersed-phase fraction.

The processing parameters of MET controlling emulsion droplet size and distribution have been extensively analyzed both experimentally and theoretically in these years [8–10]. A large number of empirical laws were discovered and a series of microscopic models were established by physics and mathematics language. Here, we would introduce the representative mechanisms of both cross-flow MET and premix MET and their developments over the years from macro to micro scales.

1.2 Cross-Flow Membrane Emulsification

During cross-flow MET, various parameters from different magnitudes would exert combined actions, including the interfacial surface property (diffusion, surface tension, and viscosity of two phases) and the macroscopic operation conditions (disperse phase velocity or pressure and continuous phase velocity) [6]. Generally, the droplet size was experimentally controlled primarily by the choice of membrane, the cross-flow velocity, and the transmembrane pressure. Typically, a factor of 2–6, depending on the properties of dispersed phase and continuous phase and even the structure of membrane is found between pore size and droplet size. Accordingly, the built numerical methodologies also aimed to describe the formation of droplet and predict the droplet size and the size distribution, such as the force and torque balances [11–13], surface free-energy minimization [14, 15], computation fluidic dynamics (CFD) [16], Lattice Boltzmann [17] and phase flow method [18], and so forth. These microscopic descriptions significantly facilitated the prediction of manufacturing throughput of emulsion and particle, optimization of MET operation, and design and scale-up of emulsion modules from laboratory to industrial production. The representative mechanisms and models involving cross-flow MET are intensively discussed as follows.

1.2.1 Mechanism of Droplet Formation

It has been observed that the uniformity of final emulsion mainly depends on the droplet formation behavior in membrane emulsification, except the cases of the emulsion polydispersity caused by instability of the emulsion and wettability of the dispersed phase. Two mechanisms of droplet detachment behavior affecting size distribution were observed in cross-flow MET, i.e. the shear-induced droplet formation (SHE) and the spontaneous transformation-based (STB) droplet formation [19, 20]. The SHE mechanism describes the situation where droplet is detached by uneven shear force from cross-flow or rotary flow of continuous phase. The STB mechanism describes the situation where the droplet breaks off without any additional shearing and just by the variation of interfacial free energy. Droplet detachment driven by mechanism of STB at elongated outlet would result in more uniform emulsion than that driven by mechanism of SHE at circular outlet [21–23]. The Shirasu porous glass (SPG) membrane with a pore geometry composed of tortuous ellipsoidal cylinders has characteristics between circular and elongated outlets. Consequently, both mechanisms were observed via microscope visualization by tailoring the emulsification environments [19].

Here, we compared three typical emulsification environments, respectively, the anionic emulsifier (sodium dodecyl sulfate [SDS]), the nonionic stabilizer (polyvinyl alcohol [PVA]), and the cationic emulsifiers (acetyl trimethyl ammonium bromide [CTAB]) in water continuous phases. Meanwhile, the divinylbenzene was selected as the dispersed phase in O/W emulsion droplet. Figure 1.2 captures a typical spontaneous formation of droplet in continuous phase of 0.2% wt SDS solution. The droplet, respectively, experienced the growth, the staying and the detaching stages. Obviously, the staying stage lasts for the longest time and acted as the speed-limited step in droplet spontaneous formation and detachment.

This spontaneous behavior happened specifically at emulsifiers of nonionic PVA and anionic SDS at high concentrations of emulsifiers. As shown in Figure 1.3a, b, a droplet labeled with the fuzzy circles and arrows is moving out from the pore. Comparatively, the growing droplets tend to adhere to pores and refuse to detach in other emulsification environments, such as pure water and solutions, respectively, of CTAB $\geq 0.05\%$ wt, low concentration of PVA $< 0.5\%$ wt, and SDS $< 0.04\%$ wt. These adhering droplets at adjacent pores tend to coalesce into larger droplets as shown in

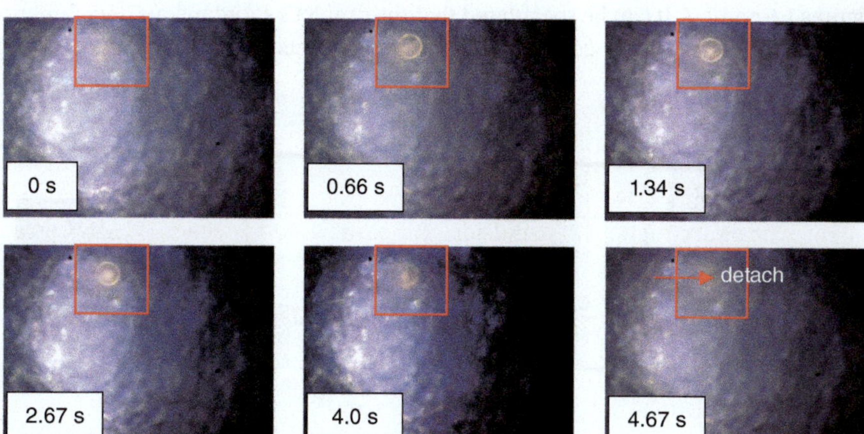

Figure 1.2 Growth of droplets on surface of SPG membrane.

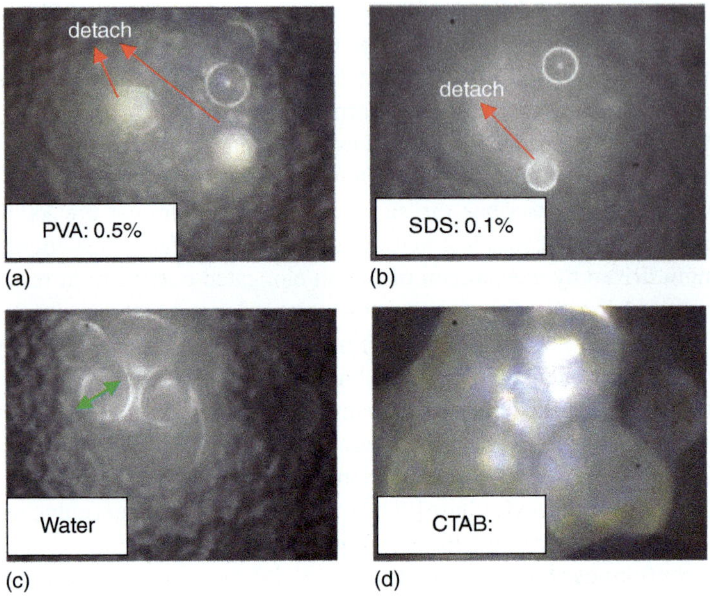

Figure 1.3 Droplets' spontaneous detaching and adhering behaviors.

Figure 1.3c, d, and have to be pulled out at a higher pressure or a strong shearing by continuous phase flow, and finally formed a polydispersed emulsion.

Because the interfacial tension between dispersed phase and continuous phase act as the major adhesion force on droplet, we further investigated the relationship between interfacial tension and droplet detachment behaviors. Figures 1.4 and 1.5 showed that the higher interfacial tension, such as in the case of PVA with concentration below 0.5%, would result in droplets adhesion and final polydispersed emulsion. However, if the interfacial tension could be decreased effectively to the lowest, such as in solution of SDS above 0.2%, the droplet would demonstrate an obvious spontaneous detachment and the uniform emulsion would form finally as shown in Figures 1.6 and 1.7. It can be speculated that the droplet's spontaneous detachments could effectively avoid the coalescence between droplets and nonuniform shearing

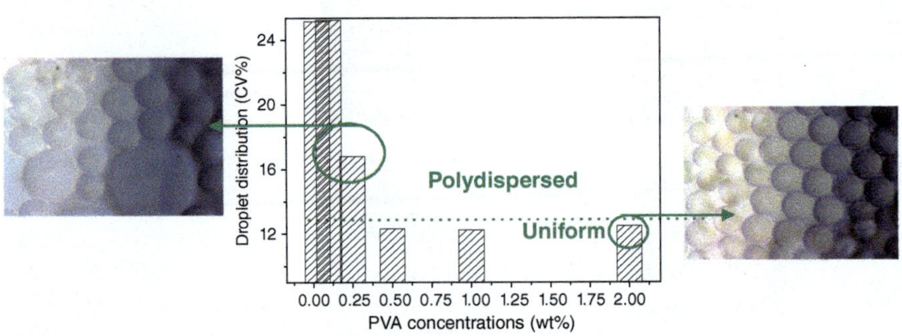

Figure 1.4 Droplets distributions at different PVA concentrations.

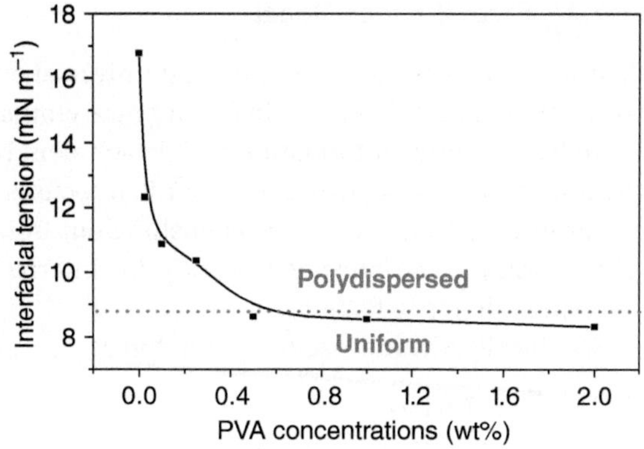

Figure 1.5 Interfacial tension of oil–water phase at different PVA concentrations.

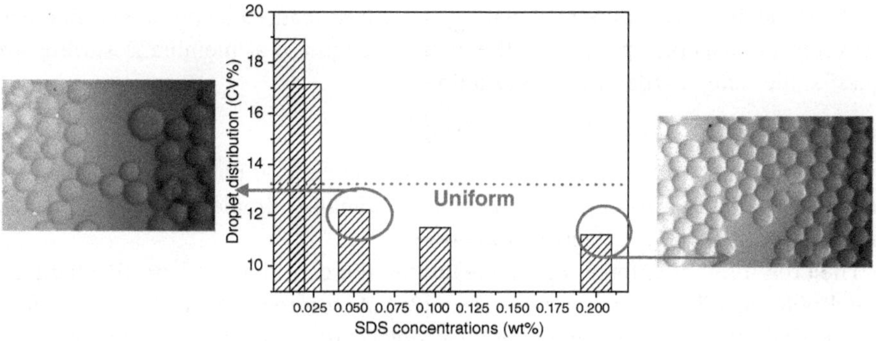

Figure 1.6 Droplets distributions at different SDS concentrations.

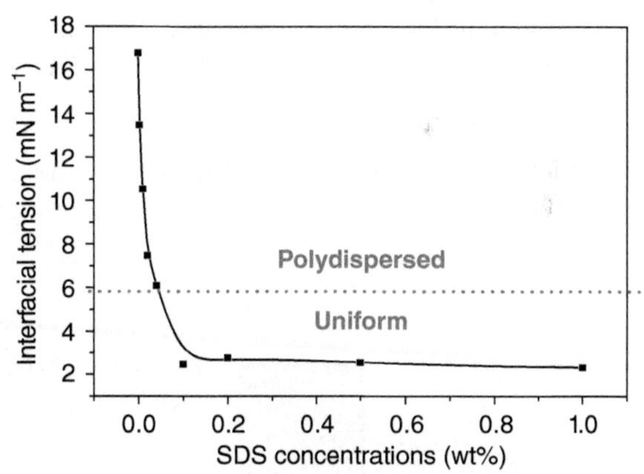

Figure 1.7 Droplets distributions at different SDS concentrations.

field from continuous phase flow, and finally form uniform emulsion as shown in Figures 1.6 and 1.7. Comparatively, SDS showed stronger ability to decrease interfacial tension and presented more significant tendency of spontaneous detachment, and could prepare more uniform droplets. It confirmed again that the spontaneous formation behavior of droplets would facilitate uniform emulsion formation.

1.2.2 Force Balance Model

The force balance model are the most universal explanation for various experimental phenomena and laws during membrane emulsification [12, 24]. The description of individual droplet formation and detachment by forces is very intuitive and easily understood. As shown in Figure 1.8, it focuses on analysis of microscopic forces on droplet at moment of droplet gushing out from membrane pores, including the shear force F_{cf}, the buoyancy force F_B, the interfacial tension force F_γ, and the static pressure difference force F_{sp}.

Specifically, the interfacial tension force,

$$F_\gamma = \pi d_p \gamma(t) \tag{1.1}$$

represents the effects of dispersed-phase adhesion on pore opening as the retaining force for adhesion.

The static pressure difference force F_{sp} is the force caused by pressure difference between the dispersed phase and the continuous phase at membrane surface. In quasi-static state, it is described as as follows:

$$F_{sp} = (P_i - P_o)A_p = \Delta P_\gamma A_p = \pi\gamma(t)d_p^2/d_{dr} \tag{1.2}$$

where A_p is the cross-sectional area of the droplet neck at pore and here approximately assumed as area of pore, and d_{dr} is the dynamic droplet diameter, which is increasing until droplet detaches from pore.

The cross-flow drag force, F_{cf}, is created by the continuous phase flowing past the droplet parallel to membrane surface [25]. According to Stokes' equation in a simple shear flow and assuming that the droplets are formed in laminar sublayer, it is described as as follows:

$$F_{cf} = 3\pi u_c d_{dr} \mu_c \tag{1.3}$$

The dynamic lift force, F_{dl}, results from the asymmetric velocity profile of the continuous phase near droplet and is defined as follows:

$$F_{\text{dl}} = 0.761{\tau_w}^{1.5}{d_{dr}}^3{\rho_c}^{0.5}/\mu_c \tag{1.4}$$

in which, shearing force τ_w is controlled by the continuous phase velocity u_c [22] and is deduced based on the definition of Fanning friction factor as given below:

$$\tau_w = f\frac{\rho_c u_c^2}{2} \tag{1.5}$$

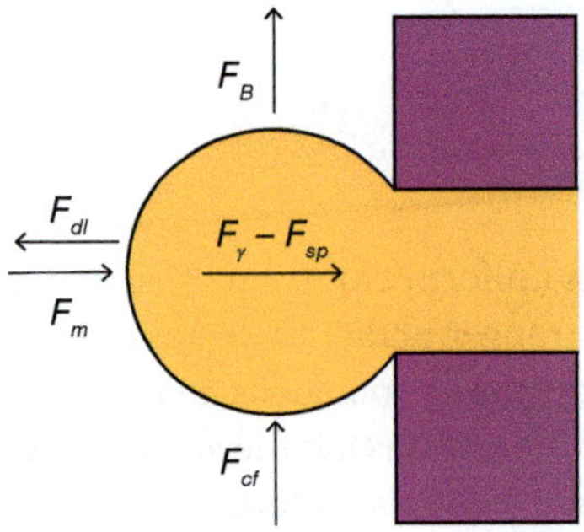

Figure 1.8 Force analysis of droplets on outlet of membrane pore.

where ρ_c is the density of continuous phase, u_c is the flow velocity of continuous phase, and f is a dimensionless factor defined by the Reynolds number Re as follows:

$$f = \begin{cases} 16/\mathrm{Re} & \mathrm{Re} < 2000 \\ 0.00140 + 0.125/\mathrm{Re}^{0.32} & 2 \times 10^3 < \mathrm{Re} < 3 \times 10^6 \end{cases} \tag{1.6}$$

where $\mathrm{Re} = \rho_c u_c D/\mu_c$, D is the inner diameter of the cross-flow channel (membrane channel), and μ_c is the viscosity of continuous phase.

If the density of dispersed phase cannot be neglected, the buoyancy force F_B of a droplet with volume V_d should be amended as the resultant force F_{BG} of buoyancy and gravity F_G, and described as the density difference between two phases.

$$F_{BG} = F_B - F_G = (\rho_c - \rho_d)gV_d \tag{1.7}$$

The linear momentum force F_m is caused by flow movement of a mass of dispersed phase out from the pore outlet.

$$F_m = \int_{A_p} \rho_d {v_d}^2 dA = \frac{\pi}{4}\rho_d u_d^2 d_p^2 \tag{1.8}$$

Among these forces, F_γ is the holding force to make droplets adhere at membrane pore outlet, while F_{dl}, F_{BG}, F_m, and F_{cf} are the detaching forces to drive droplets away from pores, and the inertia forces F_m and buoyancy forces F_{BG} can be neglected when they are approximately from six to eight orders of magnitude smaller than other forces after calculation. If the total adhesion forces are greater than the detaching force, the droplet would continue to grow on outlet of membrane pore. If the adhesion forces are less than the detaching forces, the balance of forces on the droplet is lost, and the droplet begins to deform, elongate, form a neck, detach from the membrane pore, and finally enter the continuous phase. The contributions of various forces depended on droplet size. For smaller micron-size pores, the inertial and buoyancy forces are much smaller than the viscous drag force and surface tension, respectively, and thus can be ignored in balance model; otherwise, they will play vital role for large droplets with hundreds of microns.

The prototype of force balance model was first put forward by Peng and Williams in 1998 [26]. They predicted the size of droplet generated from the capillary with inner diameter of 45 microns and considered that the droplet surface was mainly affected by forces, including the flow drag force (F_{cf}, generated by the continuous phase shearing action parallel to membrane surface), the buoyancy force (F_B, generated by the density difference between two phases), and F_γ (represents the force generated by the interfacial tension). Three forces get balanced as $F_{cf} + F_B = F_\gamma$. Schroder and Wang supplemented this microscopic forces model with more accurate descriptions, including the static pressure difference force (F_{SP}, generated from the pressure difference between the continuous phase inside and outside the droplet), the dynamic lift force (F_{dl}, generated from the asymmetric velocity distribution profile of continuous phase near the droplet), and the inertial force (F_m, related to the momentum of fluid movement flowing out of the membrane orifice) [27, 28]. Further, Xu et al. introduced variables of continuous phase flow velocity in force balance model and attempted to predict the change rule of emulsion droplet size under different continuous phase velocities [29]. They found that the force balance model agreed well with the experimental data. The droplet size and droplet formation time decreased with increase of continuous phase flow rate. Since their emulsification experiment was completed in continuous phase without emulsifier and the droplet size was obtained from microscope observation at membrane pore, the equipment and environment of their prediction were still quite different from the actual membrane emulsification. In addition, the dynamic oil–water interfacial tension of droplet surfaces and the influence of operating conditions of dispersed phase on emulsification process were not considered in both of their experiments and model.

Later, G. De Luca et al. introduced more variables, including the dispersed phase transmembrane pressure and the dynamic interfacial tension into abovementioned force balance model. The coupling effect of these variables on the droplet size was investigated [11]. They found that if droplet formed earlier than the interfacial tension arriving at equilibrium, the droplet size was directly proportional to the transmembrane pressure. Conversely, it was not related to the transmembrane pressure. They developed a modified balance model from another perspective of contact line [30]. It is assumed that the film surface exists as long as the force arrives in balance near the contact line; the droplet will adhere to the edge of membrane pore in an inclined position and deform until it finally detach from pore. In the force balance formula based on the contact line, all the forward contact angles, backward contact angles, and the minimum and maximum volume of droplet growth could be calculated (Figure 1.9). The validation of this model is completed under different conditions, including continuous phase velocity, membrane pore size, and interfacial tension. The linear relationship between droplet size and membrane pore size and the change of interfacial tension during droplet formation are all predicted.

1.2.3 Torque Balance Model

Above mentioned force balance models define the generated droplet as a point and the droplet will peel off immediately after two groups forces arriving balance,

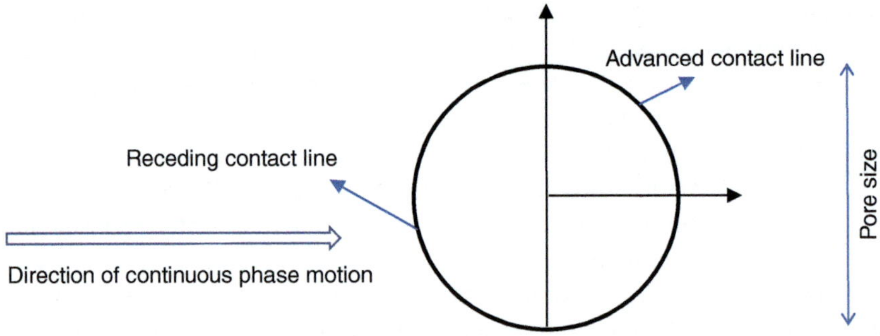

Figure 1.9 Schematic representation of contact line (top view) for a circular shape forming the droplet basis.

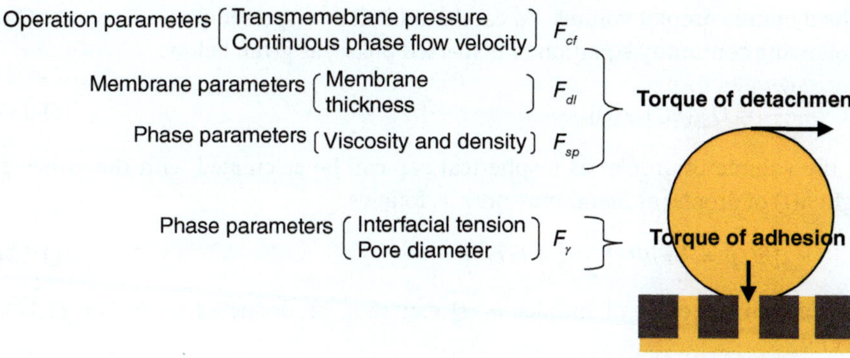

Figure 1.10 Torques and forces on droplet at membrane outlet. Source: Hao et al. [13]/with permission of American Chemical Society.

while the torque model regards the droplet as a sphere, and the droplet will rotate and escape from membrane pore until torques on droplet surface getting balance. The difference between two models lies in their comprehension of emulsion droplet. Torque model provides more accurate description and couples more parameters than force balance model in case the diameter of large droplet during its detachment cannot be ignored. According to the framework of torques on a droplet as shown in Figure 1.10, the torques of adhesion and the torques of detachment can be grouped. It indicates that the rules of torques manipulate the droplet behaviors and can be examined by coupling operation parameters, membrane parameters, and physiochemistry properties of two phases. Therefore, we build a multivariable torque model to obtain the influences of different parameters on droplet detachment mechanism. Before multivariables torque model, two sets of relationships associating the dispersed phase parameters and the continuous phase parameters should be firstly constructed given in the following sections.

1.2.3.1 Associating the Dispersed-Phase Parameters

During droplet growing up at membrane pore, the dispersed phase flow rate Q_{dr} through the membrane pores may be assumed based on Darcy law and Hagen–Poiseuille equation as follows [11]:

$$Q_{dr} = \pi d_p^4 P_{ef}/(128\xi\mu_d L) \tag{1.9}$$

where d_p is membrane pore and here used 5.2 μm, ξ is the pore tortuosity and here approximatively used as 2.1, which was measured and calculated with hydraulic membrane resistance by Vladisavljevic [8], μ_d is the viscosity of dispersed phase, and L is the membrane thickness. P_{ef} is the effective transmembrane pressure to control the dispersed phase and defined as the difference of two pressures,

$$P_{ef} = P_{trm} - P_{\gamma} = P_{trm} - 4\gamma(t)/d_{dr} \tag{1.10}$$

where P_{trm} is the transmembrane pressure applied in emulsification between each side of membrane, and P_{γ} is the capillary pressure.

The dynamic droplet volume V_d can be related to dispersed-phase flow rate Q_{dr} by following continuity equation of dispersed phase, as given below:

$$\frac{dV_{dr}}{dt} = Q_{dr}[h(t), \gamma(t)] \tag{1.11}$$

and the volume of droplet as a spherical cap can be calculated with the growing height $h(t)$ of droplet at membrane pore as follows:

$$V_{dr}[h(t)] = \frac{1}{6}\pi h(t)\left(\frac{3}{4}d_p^2 + h(t)^2\right) \tag{1.12}$$

The dynamic height of droplet $h\ (t)$ can thus be deducted by (1.11), (1.12), and (1.9),

$$\begin{aligned}\frac{dh}{dt} &= \left(\frac{dV_d}{dt}\right) \Big/ \left(\frac{dV_d}{dh}\right) = \left(\frac{dV_d}{dt}\right) \Big/ \left[8\pi\left(d_p^2 + 4h^2\right)\right] \\ &= \frac{d_p^4}{16\mu_d \xi L}\left[\frac{P_{tm}}{d_p^4 + 4h^2} - \frac{16\gamma(t)h}{\left(d_p^2 + 4h^2\right)^2}\right]\end{aligned} \tag{1.13}$$

Droplet diameter $d_{dr}\ (t)$ can also be described by height of droplet $h\ (t)$,

$$\frac{d_{dr}}{2} = \frac{\left(\frac{d_p}{2}\right)^2 + h^2}{2h} \tag{1.14}$$

So, the droplet growth $d_{dr}\ (t)$ can be deducted from Eqs. (1.13) and (1.14):

$$\frac{d(d_{dr})}{dt} = \frac{d(d_{dr})}{dh}\frac{dh}{dt} = \left(1 - \frac{1}{4}\frac{d_p^2}{h^2}\right)\frac{d_p^4}{16\mu_d \xi L}\left[\frac{P_{tm}}{d_p^4 + 4h^2} - \frac{16\gamma(t)h}{\left(d_p^2 + 4h^2\right)^2}\right] \tag{1.15}$$

The initial conditions for Eqs. (1.14) and (1.15) are established according to Laplace equation with initial interfacial tension and critical pressure,

$$d_{dr}(0) = 4\gamma(0)/P_{\gamma}{}^{0} \text{ and } P_{\gamma}^{0} = 16\gamma(0)h(0)/\left[d_p^2 + 4h(0)^2\right]$$

where $h\ (0)$ is also determined by Eq. (1.14).

1.2.3.2 Associating the Continuous Phase Parameters

The shear force τ_w is the function of the continuous phase velocity u_c [29],

$$\tau_w = f\frac{\rho_c u_c^2}{2} \tag{1.16}$$

where ρ_c is the density of continuous phase, u_c is the flow velocity of continuous phase, and f is a dimensionless factor defined by the Reynolds number Re as Equ. 1.6.

1.2.3.3 Torque Balance Model Associating Operation Parameters

Among all the aforementioned forces on droplet, F_γ is the holding force, while F_p, $F_{\gamma m}$, F_{BG}, F_m, and F_τ are detaching forces. The inertia force F_m and the buoyancy force F_{BG}, can be neglected after calculation considering that they are approximately

from six to eight orders of magnitude smaller than other forces. So, all the torques on droplet can be grouped as follows:
the torque of adhesion,

$$T_{\text{adhesion}} = F_\gamma \frac{d_p}{2} = T_{\text{adhesion}}[\gamma(t), d_p] \tag{1.17}$$

and the torque of detachment,

$$T_{\text{detach}} = F_{sp} + F_{dl}\frac{d_p}{2} + F_{cf}h = T_{\text{detach}}[\gamma(t), d_p, h(t), d_{dr}(t), \tau_w(u_c), \rho_c, \mu_c] \tag{1.18}$$

If the adhesion torque (or interfacial tension) was very large at the balance point of two torques, $T_{\text{adhesion}} = T_{\text{detach}}$, the droplet would be difficult to break off and a larger detachment torque (shearing stress) was needed and tend to be dragged by the SHE mechanism instead of the STB mechanism [31]. Therefore, we put forward that the torques at the balance point could act as a comparison criterion to estimate the transformation of the droplet detachment mechanism and further the emulsion uniformity, inspired by that droplet spontaneous formation (STB mechanism) is one of the most important mechanisms to form a uniform emulsion [13]. Since the torque model is constructed by coupling the operating parameters, the membrane parameters, and the physiochemistry properties of the two phases, the disturbances of these parameters on torques can be calculated to predict their influences on emulsion uniformity. Based on above mechanisms and parameters associated in torque models, we further explored the rules of controlling emulsion uniformity by experiment and theoretical approaches [13].

1.2.3.4 Evaluation of Controlling Factors on Droplets Uniformity by Torque Balance Model

Accordingly, the influence of operation conditions, the physical properties of two phases, and the membrane parameters on droplet formation and emulsion uniformity were systemically investigated by evaluating the variable force torques in droplet formation process. The experiment phenomena showed a good coincidence with model prediction. The conditions facilitating the production of uniform droplets were summarized as: (i) the high enough interfacial tension between the dispersed phase and the continuous phase, (ii) low cross-flow velocity of the continuous phase, (iii) low transmembrane pressure, (iv) high viscosity of the dispersed phase, and (v) an emulsifier with great ability and rapid rate to decrease interfacial tension.

1.2.3.4.1 Influence of Continuous Phase Flow on Emulsion Uniformity

The flow rate of continuous phase is a fundamental process parameter to determine membrane emulsification efficiency because the wall shear stress by continuous phase is the major force to drive the droplets detaching from membrane pore. The effects of different continuous phase flow rates on emulsion uniformity were investigated by O/W emulsification experiments. Figure 1.11 shows that the distribution of droplet size (CV%) can keep narrow at a wide range of the continuous phase flow rate from 0.188 to 1.85 m s^{-1}. However, with further increase of continuous phase

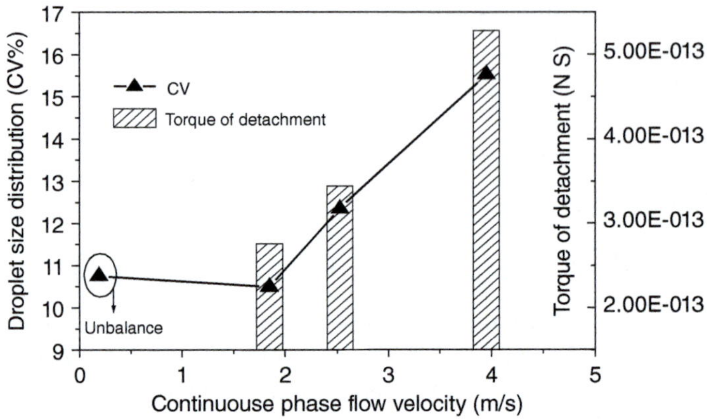

Figure 1.11 Influence of continuous phase flow velocity on droplet size distribution by experiment and detachment torques analysis. Source: Hao et al. [13]/with permission of American Chemical Society.

flow from 1.85 to 3.95 m s^{-1}, the droplet size distribution changed to broad quickly. Figure 1.11 also compared the relationship between droplet size distribution and the related predicted torques at balance. A general tendency was found that the emulsion uniformity (CV%) was spoiled at large detachment torque (at balance point) with continuous flow rate increases.

Figure 1.12 describes the development of two types of torques calculated, the adhesion one and the detachment one. Because the adhesion torque is a function of dynamic interfacial tension, the curve of adhesion torque presents the same change with the dynamic interfacial tension, which decreases sharply at beginning and then comes to equilibrium. The curve of detachment torque also presents a decreasing trend with droplet formation time. Two types of torques get balanced at intersection point of curves. The balance point of torques continues to rise up with continuous

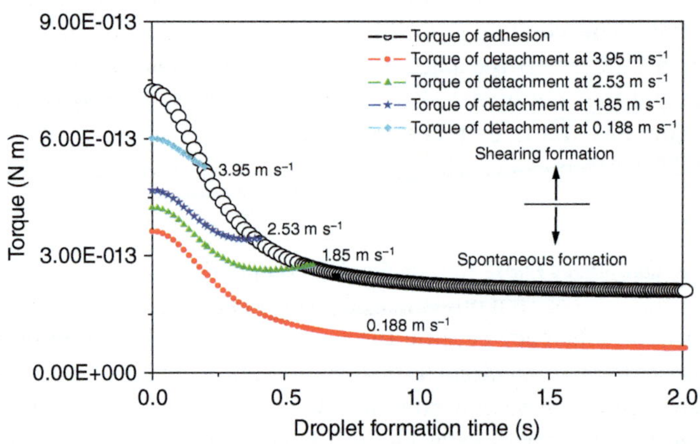

Figure 1.12 The development of torques on global droplet at different continuous phase flow velocity. Source: Hao et al. [13]/with permission of American Chemical Society.

phase flow increasing from 1.85 to 3.95 m s^{-1}, indicating that the droplet detachment tended to be controlled by mechanism of shearing formation and the uniformity of emulsion became worse with increase of continuous phase flow rate. Consequently, based on the different balance points of torques, the droplet detachment behavior can be divided into three zones. (i) At a low continuous phase flow velocity of 0.188 m s^{-1}, the torque of detachment is far below the torque of adhesion and never gets balance. It is speculated that the detachment of droplets is driven by mechanism of STB in this zone. (ii) The torque of detachment could reach to balance point with the torque of adhesion at small level. In this zone, both mechanisms of STB and SHE take effect on droplet, which manipulate the droplet continuously growing up and experiencing a distorting and necking stage before totally breaking off. (iii) The torque of detachment arrived at the balance point with the torque of adhesion at large level. In this zone, the mechanism of SHE is dominant for droplet detachment, which drags the droplets immediately away from the membrane pore and the necking stage can be neglected. Experiments showed that if the operating conditions were controlled in the zones (i) and (ii), the relatively uniform emulsion would be produced.

1.2.3.4.2 *Influence of Transmembrane Pressure on Emulsion Uniformity*

Another parameter strongly affecting the droplet size distribution is the transmembrane pressure, which controls the flux rate of dispersed phase across the membrane channel and thus the development of detachment torque. Figure 1.13 shows that the detachment torque curves rose up with transmembrane pressures increasing. They intersected with adhesion torque curve at the earlier and larger balance point, indicating that the high transmembrane pressure would lead to the detachment mechanism of SHE and the polydisperse emulsion. This speculation was demonstrated by change in droplets' size distribution under different emulsification pressure observed in Figure 1.14. When the dispersed-phase transmembrane pressure increased from critical pressure 8–14 kPa, the CV value increased slightly

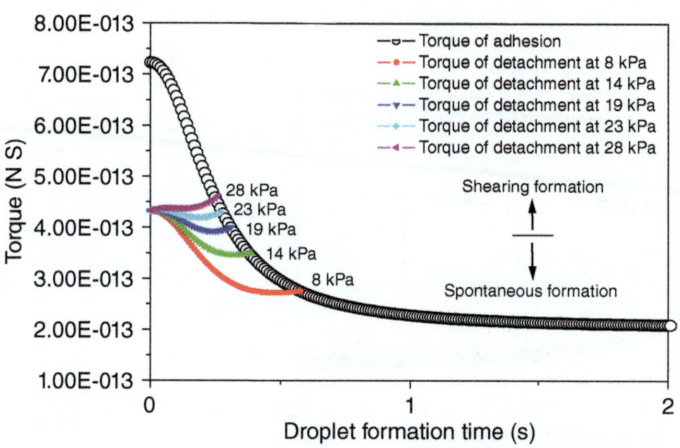

Figure 1.13 The development of torques on global droplet at different transmembrane pressure. Source: Hao et al. [13]/with permission of American Chemical Society.

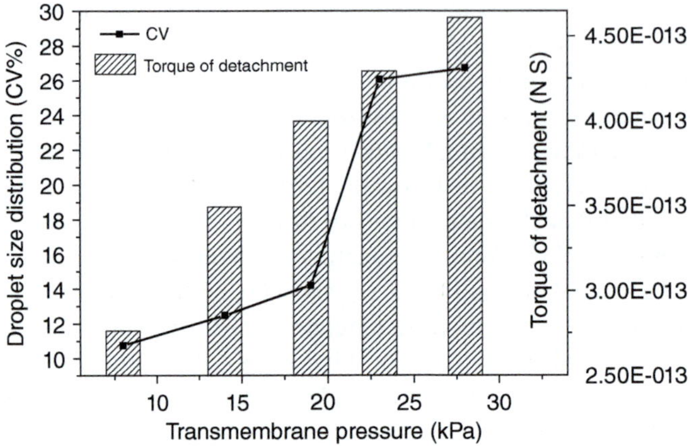

Figure 1.14 Influence of transmembrane pressure on droplet size distribution by experiment and detachment torques prediction. Source: Hao et al. [13]/with permission of American Chemical Society.

from 11% to 14%, and then rise above 26% when transmembrane pressure was increased to 28 kPa.

1.2.3.4.3 Influence of Viscosity of Dispersed Phase on Emulsion Uniformity

To evaluate the influence of viscosity of dispersed phase, two systems of dispersed phase, divinylbenzene and soybean oil, were compared since they have different viscosity but similar dynamic interfacial tension and density. The calculation of torques in Figure 1.15 shows that the torques of dispersed phase with high viscosity would get to balance later, implying that the droplet would easily break from the pores and tend to the spontaneous detachment mechanism and uniform emulsion. Abovementioned speculations were observed by the emulsification experiments in Figure 1.16,

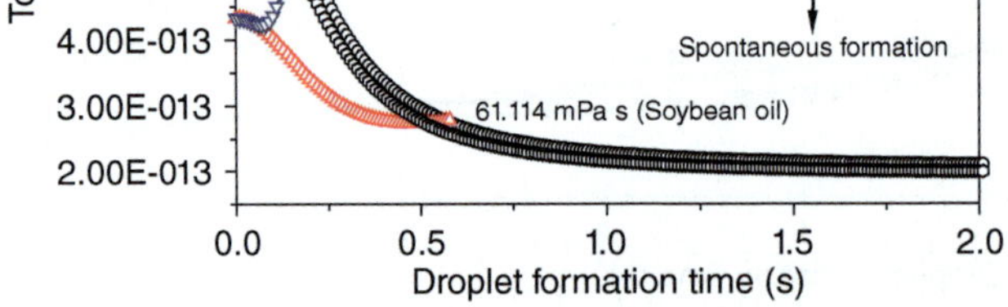

Figure 1.15 The development of torques on global droplet at different viscosity of dispersed phase. Source: Hao et al. [13]/with permission of American Chemical Society.

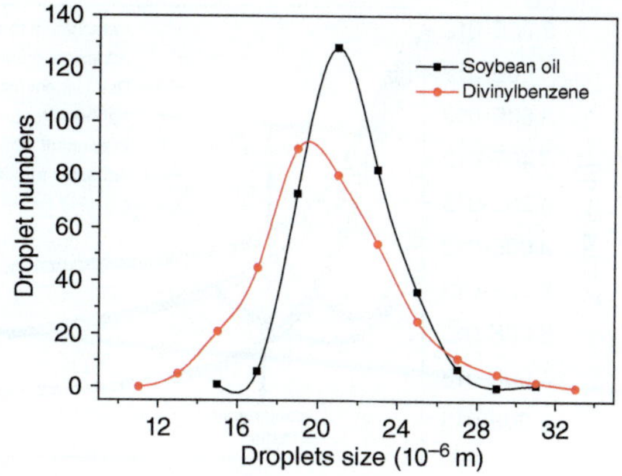

Figure 1.16 Droplet size distribution of dispersed phase with different viscosity. Source: Hao et al. [13]/with permission of American Chemical Society.

where the droplet of soybean oil with high viscosity had a narrower size distribution than that of divinylbenzene with low viscosity.

1.2.3.4.4 Influence of Emulsifiers or Stabilizer on Emulsion Uniformity

The dispersants (such as emulsifiers and stabilizer) dissolved in continuous phase also play critical roles on membrane emulsification [31]. They not only decrease the interfacial tension between dispersed phase and continuous phase by adsorption at newly formed interface of droplet and thus depress the critical transmembrane pressure in emulsification operation but also can restrain the coalescence and aggregation between emulsion droplets. Three typical dispersants, SDS, Tween 20, and PVA often used in membrane emulsification, were evaluated here by torque balance model [28, 31, 32]. SDS with the highest surface activity could rapidly reduce the interfacial tension of dispersed phase to the lowest equilibrium value. Consequently, because the interfacial tension force is a major force holding droplet at membrane pore, SDS with strong ability to lower the interfacial tension would provide the smallest adhesion torque for droplet and a smaller detachment torque for balance (Figure 1.17). It facilitates the droplet easily detaching from the membrane pore at the spontaneous droplet formation (STB mechanism) mode and the uniform emulsion was finally produced (Figure 1.18). Comparatively, the stabilizer PVA cannot effectively reduce the interfacial tension as SDS and Tween 20, and contribute to the larger adhesion torque on droplet in MET (Figure 1.17). It finally drives the droplet detachment with the shearing (SHE) behavior and results in the polydisperse emulsion (Figure 1.18), although it presents the best stabilizing ability for droplets in suspension [32].

Calculation of torque curves with different dispersed phases shows that above tendency also depended on the initial interfacial tension of dispersed phase. When liquid paraffin was used as the dispersed phase, its initial interfacial tension was very large as 46 mN m. It led to a significant raise of adhesion torque curve and larger balance point of torques than the soybean oil (Figure 1.19), which consequently formed the polydispersed emulsion in experiment (Figure 1.20). Comparatively, the soybean oil with low interfacial tension of 17 mN m was calculated to arrive at the lower

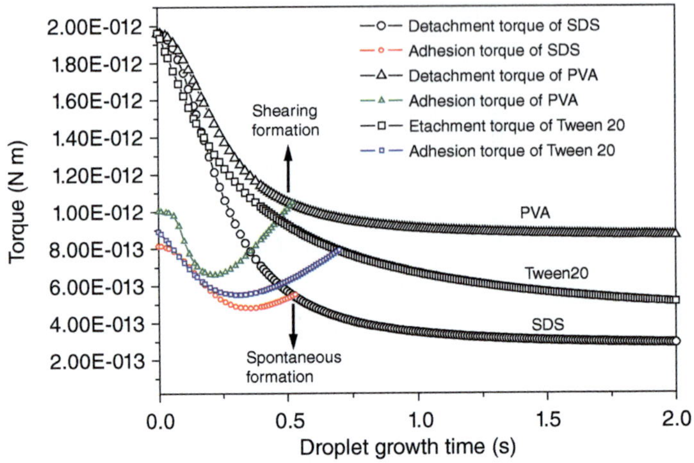

Figure 1.17 The development of torques on global droplet of paraffin by using SDS, Tween 20, and PVA as emulsifiers and stabilizers. Source: Hao et al. [13]/with permission of American Chemical Society.

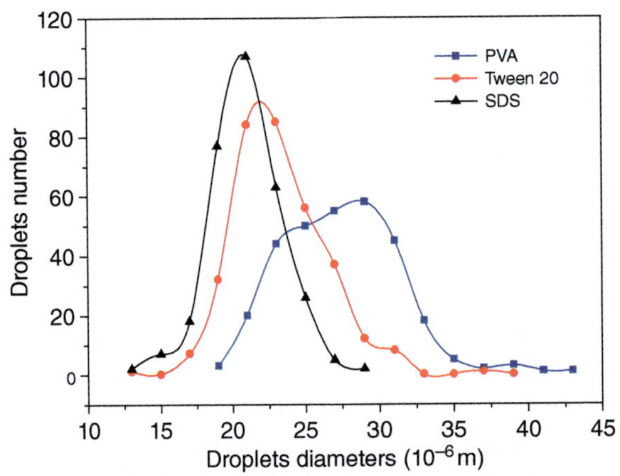

Figure 1.18 The droplet size distribution of paraffin by using SDS, Tween 20, and PVA as emulsifiers and stabilizers [13].

balance point with a spontaneous formation mode and coincide with the uniform emulsion formation in experiment (Figure 1.20).

1.2.4 Computational Fluid Dynamics

Although force and torque models represented sufficient comprehensiveness in incorporating various controlling parameters in membrane emulsification, their macroscopic cognization on emulsification was still limited and their prediction accuracy was much weaker. The microscope models established by CFD provide more dynamic behavior of droplets and are not limited to the droplet size. The basic procedure of CFD in simulation of droplet flow at pores could be described as follows. Firstly, the discrete distribution of flow field around droplet was obtained in continuous region by numerically solving the differential equation controlling

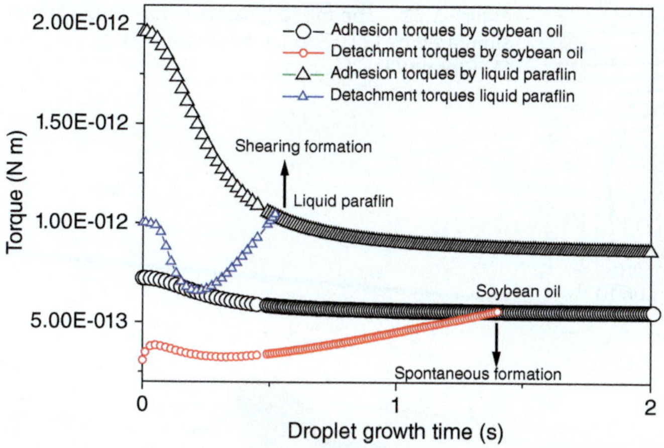

Figure 1.19 The development of torques on global droplet by different dispersed phase by using PVA as a stabilizer. Source: Hao et al. [13]/with permission of American Chemical Society.

Figure 1.20 The droplet size distribution of different dispersed phase with PVA as a stabilizer. Source: Hao et al. [13]/with permission of American Chemical Society.

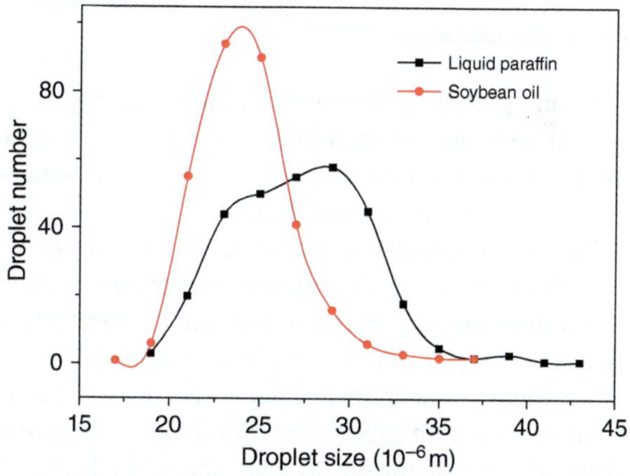

the fluid flow. Then the fluid mechanics of droplet movement was approximately simulated, and meanwhile, the visualization of these calculations was realized. By commercial software CFX4.2 developed based on CFD, A. J. Abrahamse et al. described the shape change of droplet at the membrane pore [33]. The simulation environment and boundary conditions were defined as the three-dimensional multiphase Reynolds flow, 5 μm cylindrical hole, rectangular continuous phase flow channel, 0.5 $m\,s^{-1}$ continuous phase velocity, and 1.3 MPa driving pressure of dispersed phase. Accordingly, a critical distance between pores was identified as the distance between pores must be less than ten times of pore aperture in x-axis direction and seven times in *y*-axis direction, otherwise the droplets at adjacent pores would aggregate with each other. So according to the staggered arrangement of pores, the maximum void fraction without coalescence could be determined as 1.5%. Besides this, CFD also presents a complete visualization of microscale

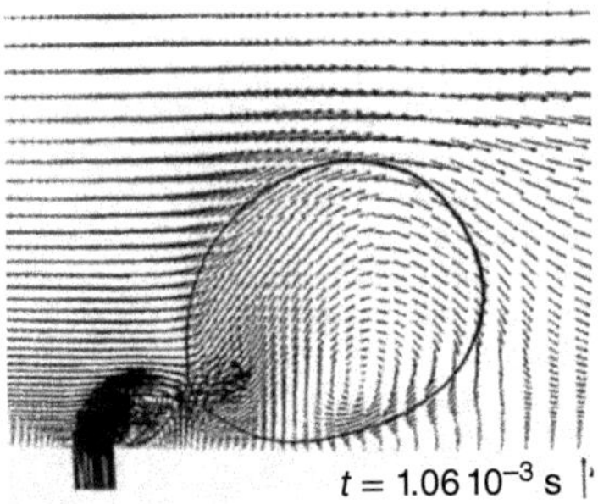

Figure 1.21 The microdynamic behavior of droplet simulated by CFD. Source: Abrahamse et al. [33]/John Wiley & Sons.

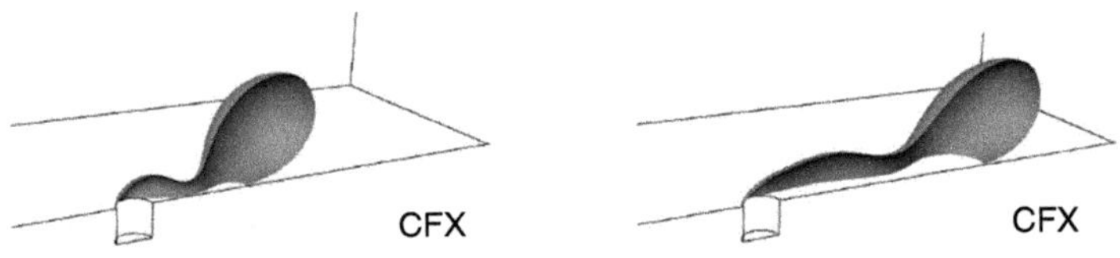

Figure 1.22 Droplet deformation process simulated by CFX: (a) the wall contact angle of 30° and (b) the wall contact angle of 60°. Source: Gijsbertsen Abrahamse et al. [2]/with permission of Elsevier.

dynamic growth of droplets. As shown in Figure 1.21, the droplet firstly formed a long neck during detachment, then the long neck is gradually torn to help the droplet completely detach, and finally the torn neck will stay at the membrane pore to form another new small droplet.

The effect of wall contact angle between droplet and pore surface on formation and shape change of droplet was interpreted by Abrahamse et al. with CFD [2, 34]. Their discovery is consistent with the microscopic observation of membrane emulsification [35]. As shown in Figure 1.22, the large wall contact angle would result in spreading of droplet and polydisperse emulsions by extending the contact edge of droplet on pore solid surface, prolonging their retention and spreading into large droplet. Consequently, a critical value of wall contact angle to avoid droplet spreading could be identified as 60°.

The pore shapes, such as circular and elliptical outlets of pores, were also found to influence the droplet formation by Kobayshi et al. with CFD [16]. As shown in Figure 1.23, the droplet at membrane pore would experience a gradual process of droplet necking, neck elongating, neck fracturing, and droplet detaching. During deformation of droplet, the droplet necking on circular pore will form a complete contact line to cover the whole pore outlet, while the droplet necking on elliptical pore will be permeated by the continuous phase flow into the pore channel and then lead to a rapid increase of fluid pressure and velocity around the necking, the instantaneous cut of neck, and finally the spontaneous formation of smaller droplet.

The advantage of CFD in analysis of membrane emulsification is that it realized both accurate calculation and visualization of microscopic flow field around and inside droplet. It allows for the consideration of more complexities, such as the active pore distribution and the droplets interaction, than the simplified force and torque

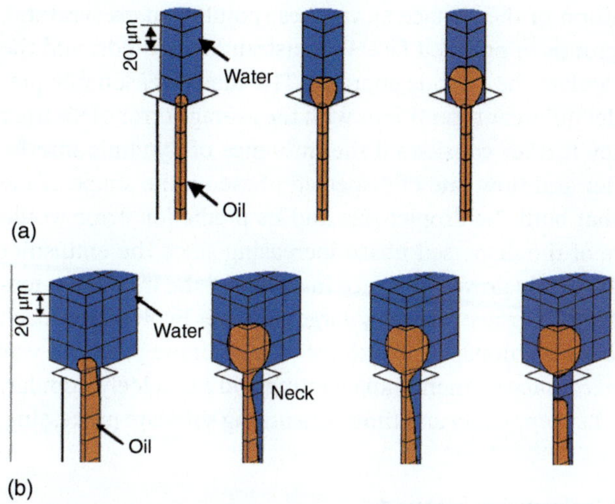

Figure 1.23 Deformation of droplets on surface of different membrane pores: (a) circular pore and (b) elliptical pore. Source: Kobayshi et al. [16]/with permission of American Chemical Society.

balance models. However, CFD still features the inherent limitations, e.g. the reliability relying on the clear definition of sufficient physics data and requirement of commercial software.

1.2.5 Models by Surface Evolver Tool

Surface Evolver tool is a set of interactive finite element software packages developed by the National Center for science and geometry calculation and visualization technology of University of Minnesota. It described the static equilibrium state of liquid and the interfaces shaped by surface potential energy and other energies. Rayner et al. investigated the droplet formation mechanism from the point of view of Gibbs free energy by Surface Evolver (Figure 1.24) [14]. They tracked the droplet shape as it grows and identified the point of instability due to free energy, and thus predict the droplet size. They also compared the prediction with the emulsification experiment with membrane of oblong-shaped pores by Kobayashi [36]. The models were inputted with the pore geometry, oil/aqueous phase interfacial tension, contact

Figure 1.24 Shape of droplets simulated by Surface Evolver. Source: Rayner et al. [14]/with permission of Elsevier.

angle, the geometric definition of the surface as vertices (points), edges (vectors), facets (triangles), bodies (groups of oriented facets), constraints, methods, and the boundary conditions to calculate the various energies. The model reasonably predicted the droplet sizes under quiescent conditions with the average error of 8% from the experimental data. They further considered the influence of dynamic interfacial tension of the emulsifier and flow rate of dispersed phase on the shape of the droplet [15]. They found that both the droplet size and its prediction error would increase with the flow rate of the dispersed phase increasing since the emulsifier cannot be replenished to the newly grown interface in time and the interfacial tension during droplet detaching interface was very large. Surface Evolver approach accounted for the issues about the nonspherical shape of membrane pore, but was limited in the prediction inaccuracy by membrane wetting and not widely popularized as other models due to its complexity and time-consuming software processing.

1.2.6 Models by Lattice Boltzmann Method

Unlike CFD simulating the continuous fluid, Lattice Boltzmann method (LBM) is based on the hypothetical particles (packages of fluid) that move and collide in a lattice according to the kinetic gas theory. It thus can incorporate both physical and thermodynamic components such as surfactants and polymers in membrane emulsification. By LBM, Sandra van der Graaf et al., describe the fluid movement of droplet interface in a T-shaped microchannel emulsification, droplet deformation between two shearing plates, and a sessile drop on a plate with different wetting conditions (Figure 1.25) [17]. They found that the capillary (Ca) number contributes to the growth and detachment stage of droplet. In the case of a higher Ca and a higher dispersed-phase flow rate, the necking volume of droplet will contribute considerably to the final droplet volume, and otherwise the necking volume during detachment stages can be negligible.

As a whole, aforementioned models mostly focused on the moment of single droplet formation and detachment at membrane pore. They realized the prediction of droplet size with different precision preferentially on the physical properties of forces or energy. Besides these, some models from another perspective provided insight into problems of interaction, coalescence, and uniformity between multiple droplets [13, 33, 37–40]. Some others aimed at the flow behavior by introducing the dimensionless groups' number, such as Re, We, and Ca, to describe droplet formation from individual droplet to multiple droplets [41–43]. Considering the

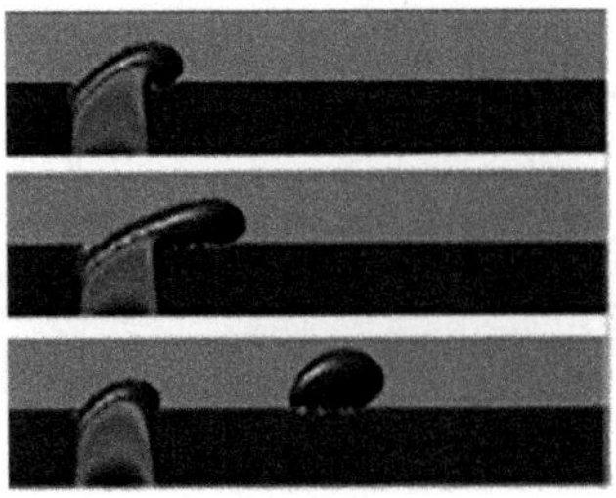

Figure 1.25 Movement of droplets simulated by Lattice Boltzmann method. Source: Van der Graaf et al. [17]/American Chemical Society.

calculation power and time, hardware and software availability, the utilization of them always compromises between accuracy and requirement. Nonetheless, all of these models enriched the microscope information beyond the experiment's visualization from different scales and satisfied the various requirements for membrane emulsification design and optimization.

1.3 Premix Membrane Emulsification

The simplicity of premix emulsification (also named as rapid membrane emulsification), as shown in Figure 1.1b, provided a promising option for large-scale membrane production of emulsions, although many aspects have not been well understood. As mentioned, premix emulsification begins with a coarse emulsion, and then this emulsion is extruded through a membrane under pressure to obtain a fine emulsion. The resulting emulsion is mostly characterized by the productivity and related to the transmembrane flux. The transmembrane flux (J) is defined as follows:

$$J = \frac{Q}{A}$$

where Q is the volumetric flow rate, and A is the cross-sectional area of the membrane.

The actual velocity in pores, $\sigma_{w,p}$, is a function of flux and porosity of membrane as following equation, which is related to local shear forces that are responsible for droplet break up [7].

$$\sigma_{w,p} = \frac{8\eta_c e J \xi}{\varepsilon d_m}$$

where η_c is the continuous phase viscosity, J is the transmembrane flux, and ξ, ε, and d_m are the membrane tortuosity, porosity, and pore diameter, respectively. The relationship between the oil phase and emulsification results can be explained by the flow resistances and the disruption of droplets in pores of membrane. The extent of droplet disruption is related to the wall shear stress inside membrane pores.

In general, it is assumed that shear forces are responsible for droplet breakup during Premix MET. However, it is far from clear how these forces operate, and how they can be related to design of a process. One may expect that more mechanisms operate simultaneously [7]. For example, Van der Zwan et al. microscopically visualized the droplet breakup mechanism in O/W premix emulsification using microfluidic devices and found three factors responsible for breakup, that is the localized shear forces, the interfacial tension effects, and the steric hindrance between droplets [44].

(1) **Localized shear forces**

The breakup mechanism of premix membrane emulsification is due to the shear forces exerted on a droplet coming close to the tip of a channel branching, or due to divergent flow in both legs of a branching, e.g. Y- or T-shaped branching.

So, an expression for critical capillary number (C_{cr}) for breaking a droplet in the T junction can be utilized as follows [45]:

$$C_{cr} = \alpha\varepsilon_0 \left(\frac{1}{\varepsilon_0^{2/3}} - 1 \right)^2$$

where α is a dimensionless constant, which is a function of the viscosity contrast of the two fluids and the geometry of the channel. ε_0 is the droplet's initial extension before entering into the T junction, defined as the ratio of droplet length to its circumference.

(2) **Interfacial tension effects**
When a droplet is squeezed through the neck part in a membrane, the dumbbell-shape of the droplet gives rise to a difference in Laplace pressure between the dispersed phase inside the constraint (ΔP_c) and the dispersed phase before (ΔP_{d1}) and after (ΔP_{d2}) the neck part [45, 46]. So, the snap-off can take place when:

$$\left. \begin{aligned} \Delta P_c > \Delta P_{d1} &\rightarrow \frac{\sigma}{R_{c1}} - \frac{\sigma}{R_{c2}} > \frac{2\sigma}{R_1} \\ \Delta P_c > \Delta P_{d2} &\rightarrow \frac{\sigma}{R_{c1}} - \frac{\sigma}{R_{c2}} > \frac{2\sigma}{R_2} \end{aligned} \right\}$$

where R_{c1} and R_{c2} are the neck radii as shown in Figure 1.26. Further, R_1 and R_2 are the droplet radii before and after the neck part. If R_{c2} is much higher than R_{c1}, snap-off is induced when $2R_{c1} < R_1$ and $2R_{c2} < R_2$. Although shear forces may act simultaneously on the droplet, the lower value of critical capillary number (around 3×10^{-3}) in this case, indicates that the deformation of the droplet inside the neck part already destabilizes the droplets, along the lines of the interfacial tension-induced snap-off mechanism.

(3) **Steric hindrance between droplets**
The dispersed-phase droplets tend to accumulate inside the channels of the membrane. These accumulating droplets would experience collision with each other and thus induce breakup. Breakup in this case is strongly dependent on the interfacial properties. The stable emulsion will resist coalescence, and yield net steric breakup, while a less stable emulsion may coalesce [47].

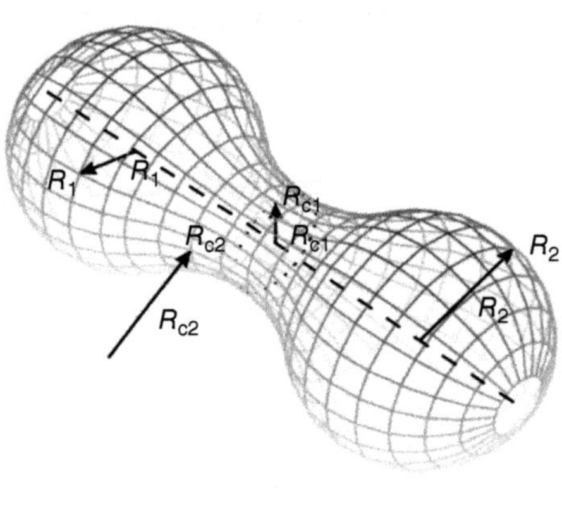

Figure 1.26 Schematic representation of the dumbbell-shaped droplet in a 3D constriction. Source: Link et al. [45]/with permission of Elsevier.

1.4 Summary

The birth of MET created a generation of platform for machinery manufacturing uniform emulsions and particles differing from the past thermodynamics method. It has been rousing great interest in recent decades for high throughput and facility to synthesize microspheres or microcapsules with a controlled size distribution (CV = 10–20%) at high productivity (several tons per hour). Precisely controlling process conditions with enough accuracy and achieving the maximum throughput for industrial production continuously become the general requirement for MET industrial applications. These optimizations and scaling up targeted for industry always rely on deep understanding of MET and decoding the relationship between emulsion droplets size and emulsification parameters, such as phase physical properties (viscosity, interfacial tension, etc.) and operation conditions (transmembrane pressure shear stress, continuous phase flow rate, etc.). Here, we systemically summarize the cognitions and development of principles and models in these years on two basic MET processes, the cross-flow MET and the premix MET. For cross-flow MET, the related mechanism has been thoroughly investigated not only experimentally but also theoretically, including the force and torque balance models, the CFD based models, Surface Evolver and Lattice Boltzmann-based models. Combining the microscope observations, they realize the prediction of both the droplet diameter and the uniformity of emulsion. Comparatively, the knowledge of premix MET is not as clear as cross-flow MET, especially on interaction mechanism between droplets and its influence on the final size of obtained droplets. Nevertheless, some processes that occur in cross-flow emulsification are also of relevance for premix emulsification design and process optimization. These cognitions make MET process more predictable and controllable, and thus provide significant guidance for operation optimization, recipe screening, and pilot scaling up of versatile emulsions.

References

1 Nakashima, T., Shimizu, M., and Kukizaki, M. (2000). Particle control of emulsion by membrane emulsification and its applications. *Advanced Drug Delivery Reviews* 45: 47–56.

2 Gijsbertsen Abrahamse, A.J., van der Padt, A., and Boom, R.M. (2004). Status of cross-flow membrane emulsification and outlook for industrial application. *Journal of Membrane Science* 230: 149–159.

3 Ma, G.H. (2014). Microencapsulation of protein drugs for drug delivery: strategy, preparation, and applications. *Journal of Controlled Release* 193: 324–340.

4 Song, C., Li, F., Wang, S. et al. (2019). Recent advances in particulate adjuvants for cancer vaccination. *Advances in Therapy* 3: 1900115.

5 Lambrich, U. and Schubert, H. (2005). Emulsification using microporous systems. *Journal of Membrane Science* 257: 76–84.

6 Schroder, V. and Schubert, H. (1999). Production of emulsions using microporous, ceramic membranes. *Colloids and Surfaces A: Physicochemical and Engineering Aspects* 152: 103–109.

7 Vladisavljevic, G.T., Shimizu, M., and Nakashima, T. (2004). Preparation of monodisperse multiple emulsions at high production rates by multi-stage premix membrane emulsification. *Journal of Membrane Science* 244: 97–106.

8 Vladisavljevic, G.T. and Schubert, H. (2002). Preparation and analysis of oil-in-water emulsions with a narrow droplet size distribution using Shirasu-porous-glass (SPG) membranes. *Desalination* 144: 167–172.

9 Gijsbertsen-Abrahamse, A.J., van der Padt, A., and Boom, R.M. (2003). Influence of membrane morphology on pore activation in membrane emulsification. *Journal of Membrane Science* 217: 141–150.

10 Vladisavljevic, G.T. and Schubert, H. (2003). Influence of process parameters on droplet size distribution in SPG membrane emulsification and stability of prepared emulsion droplets. *Journal of Membrane Science* 225: 15–23.

11 Luca, G.D., Sindona, A., Giorno, L., and Drioli, E. (2004). Quantitative analysis of coupling effects in cross-flow membrane emulsification. *Journal of Membrane Science* 229: 199–209.

12 Luca, G.D., Renzo, A.D., Maio, F.P.D., and Drioli, E. (2006). Modeling droplet formation in cross-flow membrane emulsification. *Desalination* 199: 177–179.

13 Hao, D.X., Gong, F.L., Hu, G.H. et al. (2008). Controlling factors on droplets uniformity in membrane emulsification: experiment and modeling analysis. *Industrial and Engineering Chemistry Research* 47: 6418–6425.

14 Rayner, M., Trägårdh, G., Trägårdh, C., and Dejmek, P. (2004). Using the surface evolver to model droplet formation processes in membrane emulsification. *Journal of Colloid and Interface Science* 279: 175–185.

15 Rayner, M., Trägårdh, G., and Trägårdh, C. (2005). The impact of mass transfer and interfacial expansion rate on droplet size in membrane emulsification processes. *Colloids and Surfaces A: Physicochemical and Engineering Aspects* 266: 1–17.

16 Kobayshi, I., Mukataka, S., and Nakajuma, M. (2004). CFD simulation and analysis of emulsion droplet formation from straight through microchannels. *Langmuir* 20: 9868–9877.

17 Van der Graaf, S., Nisisako, T., Schroen, C.G.P.H. et al. (2006). Lattice Boltzmann simulations of droplet formation in a T-shaped microchannel. *Langmuir* 22: 4144–4152.

18 Lian, G.P., Jousse, F., Janes, R., and Menech, M.D. (2007: Chapter 7). *Chemical Engineering Research Trends, Modelling the Formation and Flow of Multiphase Droplets in Microfluidics Devices*, 1–22. Nova Science Publishers Inc.

19 Yasuno, M., Nakajima, M., Iwamoto, S. et al. (2002). Visualization and characterization of SPG membrane emulsification. *Journal of Membrane Science* 210: 29–37.

20 Kobayashi, I., Nakajima, M., and Mukataka, S. (2003). Preparation characteristics of oil-in-water emulsions using differently charged surfactants in straight-through microchannel emulsification. *Colloids and Surfaces A: Physicochemical and Engineering Aspects* 229: 33–41.

21 Kobayashi, I. and Nakajima, M. (2002). Silicon array of elongated through-holes for monodisperse emulsion droplets. *AICHE Journal* 48: 1639–1644.

22 Sugiura, S., Nakajima, M., Iwamoto, S., and Seki, M. (2001). Interfacial tension driven monodispersed droplet formation from microfabricated channel array. *Langmuir* 17: 5562–5566.

23 Kobayashi, I., Mukataka, S., and Nakajima, M. (2005). Production of monodisperse oil-in-water emulsions using a large silicon straight-through microchannel plate. *Industrial and Engineering Chemistry Research* 44: 5852–5856.

24 De Luca, G., Maio, F.P.D., Renzo, A.D., and Drioli, E. (2008). Droplet detachment in cross-flow membrane emulsification: comparison among torque and force based models. *Chemical Engineering and Processing* 47: 1150–1158.

25 Wang, Z., Wang, S.C., Schroeder, V., and Schubert, H. (1999). Influence of fluid flow on forces acting on droplet and emulsification results in membrane emulsification process. *Chinese Journal of Chemical Engineering* 50: 505–513.

26 Peng, S.J. and Williams, R.A. (1998). Controlled production of emulsions using a crossflow membrane: part I: droplet formation from a single pore. *Chemical Engineering Research and Design* 76: 894–901.

27 Wang, Z., Wang, S., Schröder, V., and Schubert, H. (2000). Effect of continuous phase viscosity on membrane emulsification. *Chinese Journal of Chemical Engineering* 8: 108–112.

28 Schroder, V., Behrend, O., and Schubert, H. (1998). Effect of dynamic interfacial tension on the emulsification process using microporous ceramic membranes. *Journal of Colloid and Interface Science* 202: 334–340.

29 Xu, J.H., Luo, G.S., Chen, G.G., and Wang, J.D. (2005). Experimental and theoretical approaches on droplet formation from a micrometer screen hole. *Journal of Membrane Science* 266: 121–131.

30 De Luca, G. and Drioli, E. (2006). Force balance conditions for droplet formation in cross-flow membrane emulsifications. *Journal of Colloid and Interface Science* 294: 436–448.

31 Van der Graaf, S., Schroën, C.G.P.H., Van der Sman, R.G.M., and Boom, R.M. (2004). Influence of dynamic interfacial tension on droplet formation during membrane emulsification. *Journal of Colloid and Interface Science* 277: 456–463.

32 Yuyama, H., Watanabe, T., Ma, G.H. et al. (2000). Preparation and analysis of uniform emulsion droplets using SPG membrane emulsification technique. *Colloids and Surfaces A: Physicochemical and Engineering Aspects* 168: 159–174.

33 Abrahamse, A.J., Van der Padt, A., Boom, R.M., and de Heij, W.B.C. (2001). Process fundamentals of membrane emulsification: simulation with CFD. *AICHE Journal* 47: 1285–1291.

34 Abrahamse, A.J. (2003). *Membrane Emulsification: Process Principles.* Netherlands: Wageningen University.

35 Christov, N.C., Ganchev, D.N., Vassileva, N.D. et al. (2002). Capillary mechanisms in membrane emulsification: oil-in-water emulsions stabilized by Tween 20 and milk proteins. *Colloids and Surfaces A: Physicochemical and Engineering Aspects* 209: 83–104.

36 Kobayashi, I. and Nakajima, M. (2002). Effect of emulsifiers on the preparation of food-grade oil-in-water emulsions using a straight-through extrusion filter. *European Journal of Lipid Science and Technology* 104: 720–727.

37 Timgren, A., Tragardh, G., and Tragardh, C. (2009). Effects of pore spacing on drop size during cross-flow membrane emulsification—A numerical study. *Journal of Membrane Science* 337: 232–239.

38 Lepercq Bost, E., Giorgi, M., Isambert, A., and Arnaud, C. (2010). Estimating the risk of coalescence in membrane emulsification. *Journal of Membrane Science* 357: 36–46.

39 Egidi, E., Gasparini, G., Holdich, R.G. et al. (2008). Membrane emulsification using membranes of regular pore spacing: droplet size and uniformity in the presence of surface shear. *Journal of Membrane Science* 323: 414–420.

40 Kosvintsev, S.R., Gasparini, G., and Holdich, R.G. (2008). Membrane emulsification: droplet size and uniformity in the absence of surface shear. *Journal of Membrane Science* 313: 182–189.

41 Lepercq-Bost, E., Giorgi, M., Isambert, A., and Arnaud, C. (2008). Use of capillary number for the prediction of droplet size in membrane emulsification. *Journal of Membrane Science* 314: 76–89.

42 Pathak, M. (2011). Numerical simulation of membrane emulsification: effect of flow properties in the transition from dripping to jetting. *Journal of Membrane Science* 382: 166–176.

43 Sugiura, S., Nakajima, M., Kumazawa, N. et al. (2002). Characterization of spontaneous transformation-based droplet formation during microchannel emulsification. *The Journal of Physical Chemistry* 106: 9405–9409.

44 Zwan, E.V.D., Schroën, K., Dijke, K.V., and Boom, R. (2006). Visualization of droplet break-up in pre-mix membrane emulsification using microfluidic devices. *Colloids and Surfaces A: Physicochemical and Engineering Aspects* 277: 223–229.

45 Link, D.R., Anna, S.L., Weitz, D.A., and Stone, H.A. (2004). Geometrically mediated breakup of drops in microfluidic devices. *Physical Review Letters* 92: 054503.

46 Roof, J.G. (1970). Snap-off of oil droplets in water-wet pores. *Society of Petroleum Engineers Journal* 10: 85–90.

47 Nazir, A., Schroën, K., and Boom, R. (2010). Premix emulsification: a review. *Journal of Membrane Science* 362: 1–11.

2

Preparation of Hydrophobic Microspheres From O/W Emulsion

2.1 Introduction

The preparative techniques for hydrophobic polymer microspheres via a polymerization process from monomers or via a granulation of preformed polymers have been well established in these years, such as suspension polymerization, emulsion polymerization, dispersion polymerization, precipitation polymerization, seeded polymerization, and emulsion–solvent evaporation. For large-scale microsphere production, the most common technique utilized was the rotor–stator system to produce O/W emulsion and then polymerize the suspension–emulsion into microspheres. However, severe classification is necessary to prepare uniform polymeric microspheres and the problem of low yield and high cost is unavoidable. The dispersion polymerization is developed to prepare uniform microspheres above ten microns, but it is difficult to incorporate porogen in polymerization and nonporous microspheres with low specific surface areas are often produced. These limitations are overcome by swelling technique, e.g. multistep swelling and dynamic swelling, by which the monodispersed polymer seed of several hundred nanometers can be swelled with porogen, monomers, and initiators step by step, and then followed by suspension polymerization. Consequently, however, synthesis of microspheres above ten microns with uniform size was not easy and often involved multiple steps with several days and complicated manipulation.

Encouragingly, membrane emulsification technology (MET) provides a one-step and compatible solution, which could fabricate uniform hydrophobic microspheres from ten to dozens of microns just by directly emulsifying the hydrophobic monomer or polymer solution into uniform O/W emulsion and then carrying out suspension polymerization or combining with other techniques such as solvent evaporation of emulsion into target particles [1–4]. Since the porogen and other function materials can also be incorporated in emulsified droplets and various morphologies can be constructed in microspheres, membrane emulsification has roused wide interest in these decades with its facility and high throughput and flexibility to combine other techniques. Now plenty of uniform hydrophobic microspheres with various structures have been successfully synthesized from O/W emulsion by membrane emulsification, including polystyrene (PST) [5, 6], poly(styrene-*co*-methyl methacrylate) [7], polyimide [8], poly(styrene-*co*-*N*,*N*-dimethylaminoethyl methacrylate) [9],

Novel Membrane Emulsification: Principles, Preparation, Processes, and Bioapplications,
First Edition. Guanghui Ma.

poly(glycidyl methacrylate) [10], polylactide [11], polyurethane [12], and magnetic polymer microspheres [13]. These microspheres, with a narrow size distribution of the coefficient of variation (CV) mostly about 10%, were normally prepared by monomer emulsion or the performed polymer emulsion as the initial raw material. In this section, the typical procedures to fabricate these hydrophobic polymer microspheres will be introduced, and the controlling factors on the particle size, size-uniformity, and morphology of microspheres will be, respectively, discussed.

2.2 Preparation from Monomer System

Figure 2.1 demonstrates a typical route to prepare uniform hydrophobic microspheres with monomer systems by membrane emulsification and suspension polymerization. Usually, a special porous glass membrane or other hydrophilic membranes with uniform pores are used. By applying adequate pressure, an oil phase containing a monomer, an initiator, and other additives permeates through the uniform pores of membrane into the aqueous phase to form uniform droplets (O/W emulsion). The stabilizer and surfactant dissolved in the aqueous phase are adsorbed on surface of the droplets to stabilize them. Then, by elevating the temperature to above the decomposition temperature of initiator, the suspended emulsion was polymerized to form uniform particles, during which the monodispersity of particles is maintained if the emulsification and polymerization conditions are adequate.

2.2.1 PST–DVB Microspheres

Highly hydrophobic monomers, such as styrene (ST) and divinyl benzene (DVB), were relatively easily emulsified into uniform O/W droplets during membrane emulsification (ME) process due to their difficult spreading at membrane pore. The MET and direct suspension polymerization are thus an especially perfect route for preparation of polymerized styrene (PST) microspheres with significant uniformity. A standard recipe for emulsification conditions is shown in Table 2.1. The mixture of monomer, porogen dissolving BPO initiator, and lauryl alcohol (LOH) as cosurfactant was used as the dispersed phase. The aqueous solution of poly(vinyl alcohol) (PVA) (stabilizer), sodium lauryl sulfate (SLS) (surfactant), Na_2SO_4 (electrolyte), and hydroquinone (HQ) (inhibitor) was used as

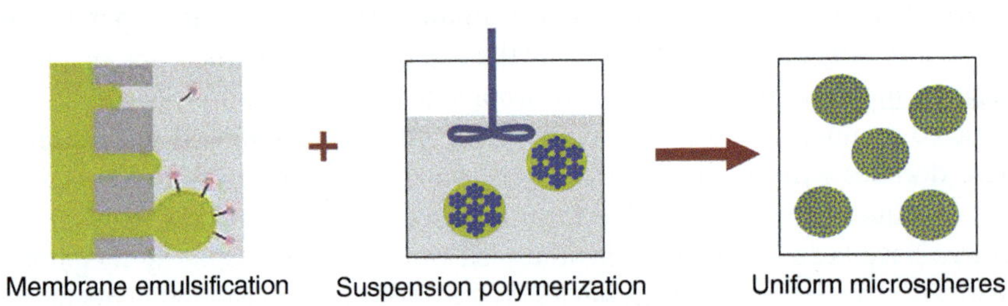

Figure 2.1 Typical preparation procedure of uniform hydrophobic microspheres from O/W emulsion.

Table 2.1 Standard recipe for membrane emulsification.

Oil phase (dispersed phase)	Content (g)	Aqueous phase (continuous phase)	Content (g)
Monomer (DVB + ST)	7.00	Deionized water	300.00
Initiator (BPO)	0.4107	Stabilizer (PVA)	2.7000
Hydrophobic cosurfactant (Lauryl alcohol)	0.70	Emulsifier (SLS)	0.1330
Porogen (Heptane)	7.00	Na_2SO_4	0.0667
		Inhibitor (HQ)	0.0750

Source: Hao et al. [14]/with permission of Elsevier.

Figure 2.2 SEM photos of polyDVB microspheres prepared by heptane porogen. Source: Hao et al. [14]/Elsevier.

the continuous phase [3]. The dispersed phase was pressed through the pores of hydrophilic membrane and formed the droplets into the aqueous phase continuously by nitrogen gas pressure. The droplet size prepared is about two to six times as large as the pore size of membrane [2, 3, 15]. After preparing the uniform droplets composed of monomer, initiator, and hydrophobic cosurfactant, the PST microspheres could be obtained by elevating the temperature above the decomposition point of initiator. Uniform microspheres with an average diameter similar to that of the droplets could be obtained if the droplets were only a nucleation locus. It should be noted that the nucleation in aqueous phase often occurred simultaneously with droplet nucleation since a small amount of monomer and initiator were soluble in aqueous phase to polymerize therein and form the secondary nuclei. Therefore, sometimes a small amount of the water-soluble inhibitor HQ was added in aqueous phase to inhibit the secondary nucleation while maintaining the conversion at a reasonably high value (around 90%). As shown in Figure 2.2, uniform PST microspheres with porous structure can be easily prepared by adding a diluent (porogen) into oil phase before emulsification, where the pore size and the specific area of microspheres can be varied by adjusting the relative amounts of monomer, diluent, and crosslinker [3, 14, 16].

2.2.2 PST-DMAEMA Microspheres

Incorporation of active groups in microspheres and then immobilization of certain ligand on them is always necessary for functional microspheres materials in recognition target molecules in many biologic applications such as bioseparation, bioreactor, and biosensor. For that purpose, the copolymerization between ST and *N,N*-dimethylaminoethyl methacrylate (DMAEMA) was investigated [17, 18]. DMAEMA is a functional monomer, its amino group can be modified into either hydrophobic or hydrophilic properties, and PDMAEMA is a temperature-sensitive polymer in adsorption and desorption of proteins by changing temperature. However, it is quite difficult to incorporate a high amount of DMAEMA into the composite polymerized particles because a majority of DMAEMA will be partitioned in aqueous phase. Furthermore, the secondary nucleation in the aqueous phase will become very serious due to the large amount of DMAEMA dissolved in the aqueous phase. Therefore, it is an important task to prevent the formation of the secondary nucleation when DMAEMA is used in membrane emulsification and polymerization. For the preparation of particles, such as previous PST–DVB microspheres, we usually used benzoyl peroxide (BPO) as an initiator, because the hydrophobicity of BPO is high enough to prevent the secondary nucleation in the aqueous phase. However, BPO was not suitable for the polymerization of DMAEMA, because it lost activity due to its primary radical reacted with DMAEMA.

We prevented the secondary nucleation by changing the type and amount of inhibitor and stabilizer, and amount of initiator and DMAEMA, and also studied the effects of these factors on the fraction of DMAEMA in the copolymer, monomer conversion, and morphology of the particles [17]. Besides this, another effective solution was the addition of hydrophobic substance (hexadecane) in oil phase. The typical recipe was as follows: the mixture of ST and DMAEMA as monomers, hexadecane (HD), and 2,2′-azobis (2,4-dimethylvaleronitdle) (V-65, ADVN) as initiator was used as the oil phase. The aqueous phase dissolved the stabilizer of poly(vinyl pyrrolidone) (PVP), surfactant of SLS and water-soluble inhibitor of diamino phenylene (DAP), or sodium nitrite ($NaNO_2$), was used as a continuous phase. After complete emulsification, the droplets were polymerized at 70 °C. It was found that the diffusion of ADVN into the aqueous phase was a main factor responsible for the secondary nucleation more than that of DMAEMA. The secondary nucleation could be prevented effectively in the presence of water-soluble inhibitor of HQ and DAP when PVP was used as the stabilizer. As a result, the morphology of particles can be manipulated by type of inhibitor. As shown in Figure 2.3, the hollow particles were obtained when $NaNO_2$ was used, while one-hole particles formed in the other cases of DAP and HQ due to phase separation between HD and poly(St-*co*-DMAEMA). By adding cross-linking agent, the hole disappeared and the monomer conversion was improved.

The mechanism of above mentioned structure formation was further verified by measuring the variation of diameter, molecular weight distribution, and monomer conversion by observing morphological changes during the polymerization, as well as by changing the type and amount of hydrophilic monomer, and initiator. It was found that the inert solvent of HD in oil phase played key role in phase separation and morphological changes. The hollow particles formed due to the rapid phase separation between PST and HD, as a consequence that a large amount of monomer diffused into the aqueous phase to form the secondary particles. Rapid phase separation

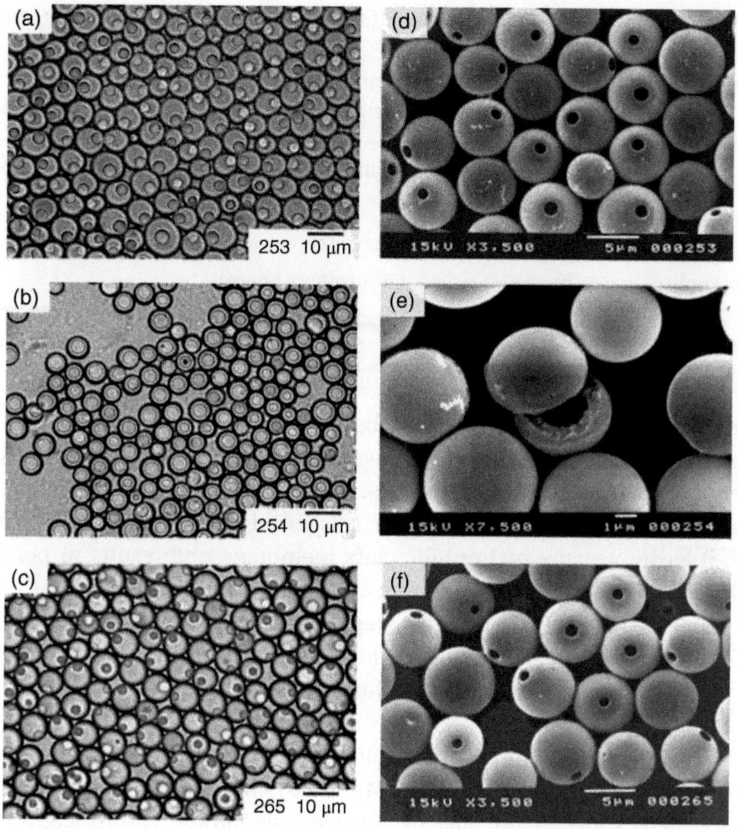

Figure 2.3 Optical microscopy (OM) and scanning electron microscopy (SEM) of polymer particles in the presence of DMAEMA as a function of inhibitor types. (a)–(c): OM; (d)–(f) SEM. Stabilizer: PVP. Inhibitor: (a) (d) DAP; (b) (e) $NaNO_2$; and (c) (f) HQ. Source: Ma et al. [17]/John Wiley & Sons.

confined the HD inside the droplets, resulting in a nonequilibrium morphology. On the other hand, one-hole particles, representing the equilibrium morphology, formed when the phase separation occurred slowly because a lot of monomers existed inside of the droplets to allow mobility of the PST [17, 18].

2.2.3 PGMA Microspheres

Poly(glycidyl methacrylate) (PGMA) microspheres, covered with many epoxy groups, are easily derived to multifunctional material with potential applications in the protein separation and enzyme immobilization. Suspension polymerization is a conventional method to prepare PGMA microsphere but always with the disadvantage of broad size distribution. The conventional seeded polymerization method has been developed to overcome this problem by using small nanoparticle as a seed. However, this method is time-consuming and only effective in the preparation of microspheres with several micrometer diameters and is difficult to

Table 2.2 Standard recipe and results to prepare P(GMA-EGDMA) microsphere (membrane pore size 5.2 mm).

Recipe	Content (g)	Characteristic	Content
GMA	6.00	Droplet average size	26.1 μm
EGDMA	4.0	CV% (droplet)	11.9
Isooctane	3.0	Microspheres average size	25.6
4-Methyl-2-pentanol	0.70	CV% (microspheres)	13.3
		Yield of microspheres	0.0667

Source: Wang et al. [10]/with permission of John Wiley & Sons.

prepare uniform microspheres with diameter above 20 μm, which were commonly used in industrial scale bioseparation by low-pressure chromatography. GMA was a relatively hydrophilic monomer. As described above, when the relatively hydrophilic monomer acts as the oil phase (dispersed phase) in membrane emulsification, it easily wet the hydrophilic glass membrane and results in polydisperse emulsions and microspheres after suspension polymerization. Addition of a nonpolar diluent in oil phase is a one-step solution for this problem since such hydrophobic additive could affect the interfacial tension between oil phase and aqueous phase and alleviate the wettability of the oil phase on membrane surface.

Here, by adding a hydrophobic diluent, isooctane, into the monomer oil phase, we successfully prepared the uniform porous P(GMA-DVB) microsphere and P(GMA-EGDMA) microsphere above 20 μm just by one-step membrane emulsification–polymerization [10]. It was speculated that isooctane was an essential component that worked both as diluent and hydrophobic additive. Its addition not only guaranteed the preparation of uniform monomer droplets and microspheres but also obviously decreased the coagulation between microspheres and increased the yield of microspheres. The standard addition ratio of isooctane/monomer and the characteristic of P(GMA-EGDMA) microsphere can be found in Table 2.2 and Figure 2.4. Although EGDMA was a more hydrophilic cross-linking agent than DVB, uniform P(GMA-EGDMA) microsphere was still prepared by membrane emulsification–polymerization when isooctane was added, illustrating that isooctane was an efficient additive in preparing uniform PGMA microsphere by membrane emulsification.

2.2.4 PST-HEMA Microspheres

2-Hydroxyethyl methacrylate (HEMA) is a frequently utilized monomer in fabricating biocompatible microspheres since it provided the hydrophilic –OH functional group for microspheres surface. However, when such relatively hydrophilic monomer was used as the oil phase in membrane emulsification, the pore wall of membrane was easily wetted and inevitably led to a broad size distribution of emulsion O/W droplets. Therefore, different from hydrophobic monomer as ST,

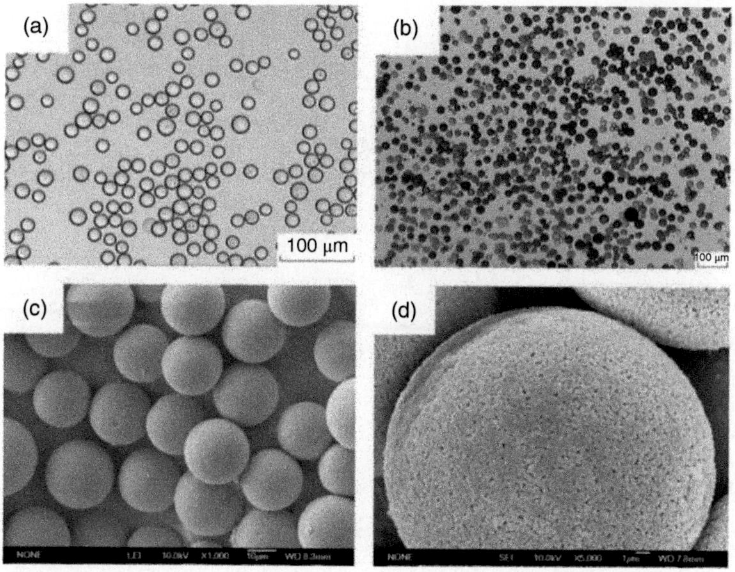

Figure 2.4 Typical OM and SEM photos of P(GMA-EGDMA) microsphere. (a) OM image of droplets in emulsion, (b) OM image of microspheres in serum, (c) SEM of total image, and (d) SEM image of surface. Source: Wang et al. [10]/John Wiley & Sons.

we developed another route combining with membrane emulsification and swelling technique to prepare such polymer particles.

The typical procedure was shown in Figure 2.5. The uniform seed droplets composed of ST, DVB, HD, and BPO were firstly prepared by membrane emulsification as the primary emulsion, and then mixed with the secondary emulsion droplets composed of ST/HEMA or MMA/HEMA prepared by a homogenizer for swelling. Finally, the swollen droplets obtained were polymerized at 75 °C under the nitrogen atmosphere, and the uniform microspheres with a higher content of HEMA were obtained successfully. The effects of the composition of oil phase and the inhibitor in the continuous phase on the incorporated fraction of HEMA, the morphology of particles, and monomer conversion were investigated.

It was found that more HEMA could be incorporated into the polymer when MMA/HEMA was used in preparing the secondary emulsion [19]. Apparently, HEMA shows a higher affinity with MMA than St, thus resulting in more HEMA partitioned into the droplets as MMA was absorbed by the seed droplets. Various microspheres with different morphologies, such as spherical, snow-manlike, and popcornlike, were observed in Figure 2.6, depending on composition of oil phase. It is interesting to note that the microspheres with smooth surfaces were obtained in St/HEMA system, while snowman-like particles were obtained in MMA/HEMA, possibly because more HEMA was incorporated into the particles and leading to phase separation between the crosslinked PST-rich domain and the PHEMA-rich domain. Comparatively, porous P(St-HEMA) particles (Figure 2.7) can be prepared by employing hexanol (HA) or toluene as diluent and ethylene glycol dimethacrylate (EGDMA) as a crosslinker [19–21].

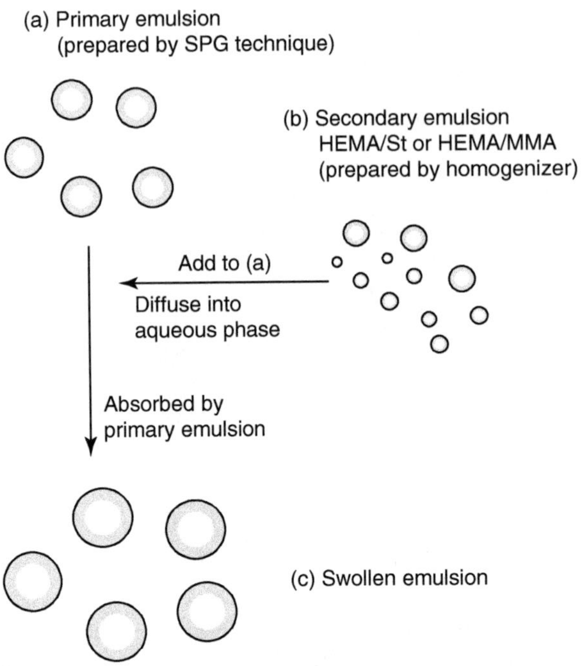

Figure 2.5 Schematic diagram showing the preparation of uniform droplets containing HEMA by the swelling method. Source: Ma et al. [19]/with permission of John Wiley & Sons.

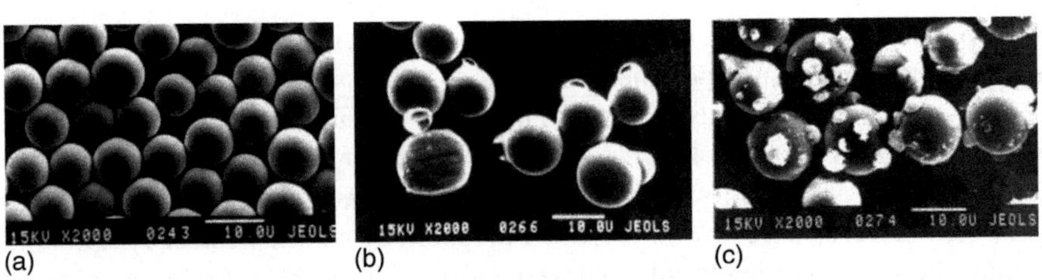

Figure 2.6 Typical SEM micrographs of microspheres prepared by membrane emulsification and swelling methods. (a) Spherical, (b) snow-manlike, and (c) popcornlike. Source: Ma et al. [19]/John Wiley & Sons.

2.2.5 PMMA Microspheres

Uniform micron-scale polymethyl methacrylate (PMMA) microspheres are in great demand commercially, in particular, as a surface modifier, additive, and lubricant for organic pigments, rubbers, cosmetics, and so on. However, MMA monomer was also regarded unsuitable to obtain uniform dispersion by membrane emulsification system. As above is generally believed, the particles with broader size distribution will be obtained when the relative hydrophilic monomers are employed in ME-suspension polymerization since the membrane to prepare such microspheres is hydrophilic and the monomer easily wet the pore wall by generating a jet-like stream of oil phase and lead to nonuniform O/W droplets. This difficulty was once overcome by combining the MET and droplet swelling techniques as described above [22]. The primary uniform emulsion composed of a mixture of hydrophobic diluent (and/or monomer), cosurfactant, and initiator

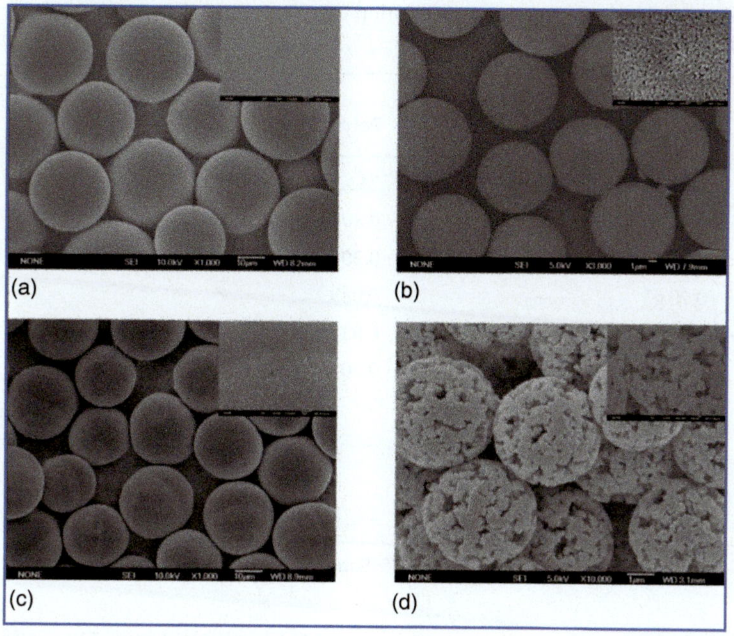

Figure 2.7 SEM micrographs of PHEMA microspheres with different microstructures. The corresponding surface is shown at the top right corner of each micrograph. (a) Nonporous microspheres, (b) porous microspheres, (c) nonporous/porous composite microspheres, and (d) macroporous microspheres. Source: Qu et al. [20]/American Chemical Society.

was prepared with a hydrophilic membrane and was allowed to absorb MMA (and diluent) under the principle of the degradative diffusion process from droplets in the secondary emulsion. By employing such unique two-step emulsion technique, the hydrophobic droplets can be swollen with a hydrophilic mixture of MMA and solvents. Uniform hydrophilic PMMA microspheres of average diameter from 2.5 to 37 μm, can be obtained by SPG membrane of pore diameter from 0.5 to 5.25 μm. Besides, One-eyed spheres, solid and smooth crosslinked spheres, and porous spheres can be, respectively, obtained depending on the polymerization formulation [22].

The standard recipe and procedure to prepare such PMMA microspheres were described as follows. The seed droplets were prepared by direct membrane emulsification (approximately 7 μm using a 1.4 μm pore size membrane) with the inert compound benzene (BZ) as the oil phase, mixed with HD, BPO initiators, etc. The MMA/water secondary emulsion (1–2 μm) was prepared by homogeneous emulsification. In order to make the secondary emulsion unstable, no stabilizer PVA was added and only a small amount of SLS was added. After mixing the two emulsions, the hydrophilic MMA quickly diffused into the water and was continuously absorbed by seed droplets. The secondary droplets disappeared completely within 15 min, and the seed droplets formed a larger swelling liquid droplet. After heating and polymerization, microspheres were obtained. Then the microspheres were precipitated and washed with methanol to remove Bz (Table 2.3). Adding porogens to the

Table 2.3 Standard recipe for two-step emulsification and polymerization to prepare P(MMA) microsphere.

Substance (g)	Primary emulsion	Secondary emulsion	Polymerization
Water	475	110	251.1
PVA-217	3.0	—	0.891
SLS	0.5	0.05	0.20
$NaNO_2$	0.70 g	—	0.089
Bz	23.3	—	6.93
BPO	1.0	—	0.30
HD	1.67	—	—
MMA	—	20	—
EGDMA	—	20	—
Total	504.8	150.1	—

Source: Omi et al. [22]/with permission of John Wiley & Sons.

oil phase before emulsification can prepare porous PMMA microspheres. However, it is important to note that the polar porogens (hexanol, octanol, etc.) should be considered since the hydrophobic porogen will be completely embedded in the polar PMMA microspheres and lead to formation of a dense film on the surface of microspheres.

Later, with the development of novel MET equipment, uniform PMMA microspheres could be prepared by direct membrane emulsification-assisted suspension polymerization [23]. Different from above cross-flow membrane emulsification, a stirred cell-flat at membrane (SCFM) was utilized in emulsification (Figure 2.8a),

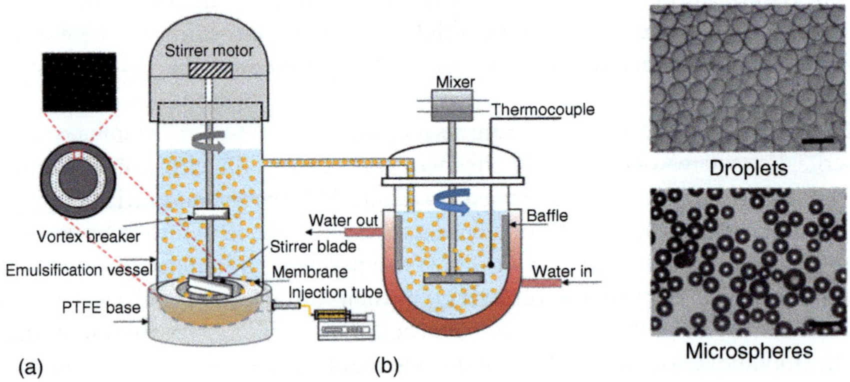

Figure 2.8 Schematic illustration of (a) stirred vessel membrane emulsification, with a micrograph image showing the regular array of pores on the membrane used, and (b) polymerization reactor units used. Source: Alroaithi and Sajjadi [23]/Royal Society of Chemistry.

which stopped the growth of the emerging droplets and resulted in a faster detachment of droplets by increasing the shear stress of impeller and leading to smaller droplets and polymerized microspheres with a higher degree of uniformity (Figure 2.8b) [23]. The uniform droplets were obtained via membrane emulsification at an impeller speed range 500–1500 rpm and flowrates within 1.0–3.0 ml min^{-1}. The emulsification was followed by a shear-controlled suspension polymerization to convert the discrete droplets of monomer to polymer microspheres. The CV of polymer microspheres from assisted-suspension polymerization is smaller than 20%, which shows a significant improvement over that by conventional polymerization of usually greater than 30%.

2.3 Preparation from Performed Polymer System

Another advantage of membrane emulsification process is that uniform polymer microspheres can also be prepared with an organic solution of a preformed polymer as the oil phase. The performed polymer, such as functional polymer, biodegradable polymers, and natural polymers, can be incorporated into emulsion O/W droplets by MET, and then the particles can be obtained after the organic solvent of emulsion was evaporated. We have successfully prepared the PST/PMMA composite particle [4], the polylactide biodegradable particle [11], the highly crosslinked poly(styrene-divinyl benzene-trimethyolpropane trimethacrylate) particle containing a polyimide prepolymer (PIP) [24], the polyurethane particle [12], and the polymer microcapsules containing inorganic material [25].

2.3.1 PST–PMMA Microspheres

The uniform composite microspheres of PST–PMMA can be fabricated using a dichlomethane (DCM) solution of performed PMMA and PST blended as the disperse phase (oil phase), the water containing PVA and SLS as the continuous phase, and followed by membrane emulsification and evaporation of the solvent at room temperature [4, 26, 27]. The morphology of the composite microspheres could be controlled by the PMMA/PST ratio, polymer concentration, and additive in the oil phase such as LOH. When the polymer concentration was low (1 wt%), the core–shell structure was observed (Figure 2.9a). The surface was composed of only PMMA, with PST being completely engulfed in inside of PMMA. However, when the polymer concentration was high (10 wt%), the microphase separation structure was formed (Figure 2.9b), that is, the surface is nonhomogenous, being composed of both PMMA and PST microdomains. This is because the viscosity inside the droplet was too high so the mobility of the polymer became lower. As a result, small domains cannot grow to a single large domain, and the morphology of complete phase separation cannot be attained. It can be concluded that when polymer concentration was low, the core–shell particles were always obtained irrespective of the PMMA/PST ratio in absence of LOH. Comparatively, when polymer concentration was low in presence of LOH, the different morphologies,

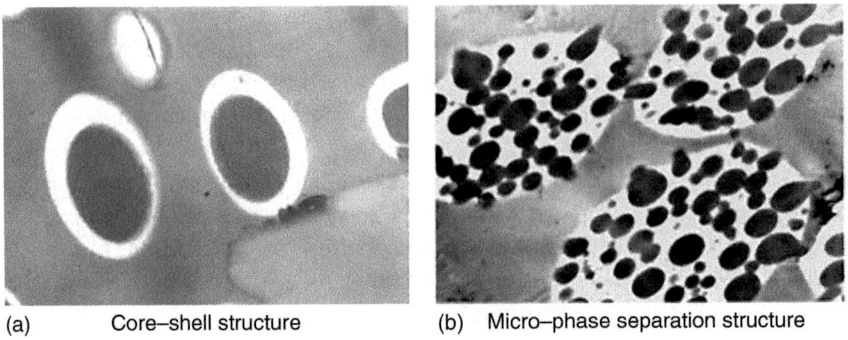

Figure 2.9 Transmission electron microscopy (TEM) photos of PST/PMMA composite microspheres with different structures (microspheres were sliced to c.100 nm thickness with microtome, and then PST dark microdomains were stained by RuO_4). Source: Ma et al. [4]/Elsevier. (a) Core–shell structure and (b) microphase separation structure.

such as microdomain, hemisphere, and inverted core–shell morphologies, could form as the PMMA/PST ratio decreasing from 5/5 to 1/9 (g/g) (Figure 2.10). When polymer concentration was high in presence of LOH, the multiplet and inverted core–core–shell were observed. It consequently could be speculated that the interfacial tension between the LOH and the aqueous phase was lower than

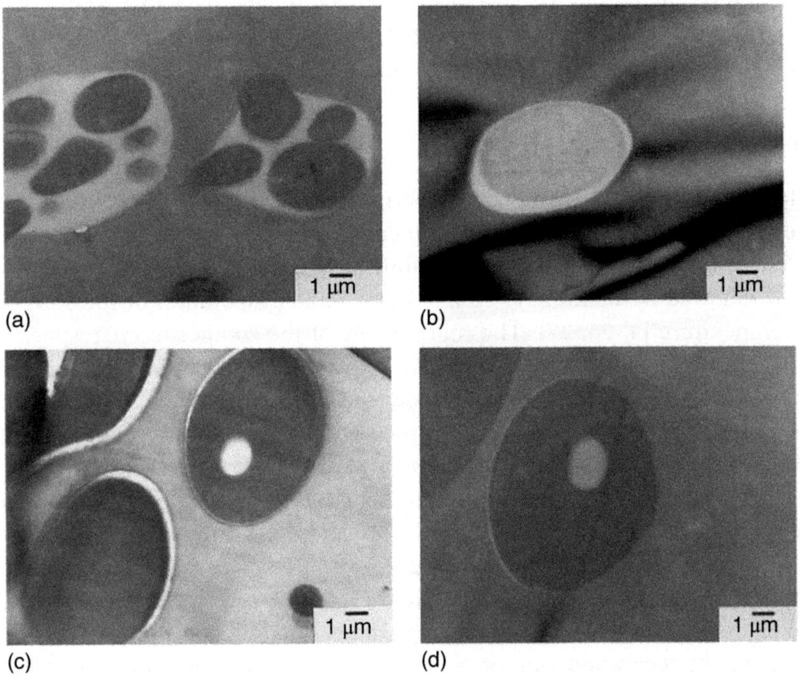

Figure 2.10 TEM of different structure with presence of LOH as PMMA/PST ratio decreased. LOH amount: 2 ml/1.2 g of polymer. Polymer concentration: 2 wt%/v. PMMA/PST (g/g): (a) 5/5, (b) 3/7, (c) 2/8, and (d) 1/9. Source: Ma et al. [4]/Elsevier.

that between polymer and the aqueous phase, and LOH engulfed the composite particles [4]. In this case, the morphology was determined by interfacial tensions of PMMA and PST with LOH, instead of those with the aqueous phase.

2.3.2 Polyurethane Urea Microspheres

Polyurethane is a popular material, since it has intermediate characteristics between rubber and plastics. However, the sizes of polyurethane and hybridized polyurethane particles are limited to the submicron range, no practical method has been established to obtain uniform polyurethane microspheres of micrometer scale. By membrane emulsification and a subsequent chain-extending reaction of prepolymer, a polyurethane urea (PUU) microsphere of 5–20 μm has been successfully prepared by emulsifying a urethane prepolymer (UP) of xylene solution [12]. Firstly, the uniform droplets of 20–40 wt% UP solution of xylene was emulsified and dispersed in an aqueous phase with dissolved stabilizers. Then, the chain extension was allowed at the room temperature with diamines that were added after emulsification. The reaction progressed rapidly with an addition of ethyl acetate to the aqueous phase, promoting the diffusion of diamines into the droplets. The reaction of −NCO groups on polyurethane with water did not hamper the emulsification process, which normally occurred in 1 to 2 h, yielding stable droplets with the CV around 10%. Finally, the solid and spherical PUU particles of 5–20 μm were, respectively, obtained with porous membranes of 1.42, 5.25, and 9.5 μm pore size after removal of the solvent. Here, a small amount of SLS together with the MST-1, an adduct of 2,4-toluene diisocyanate, and poly(ethylene oxide-b-propylene oxide-b-propylene) was used as the stabilizer.

Based on this study, uniform polyurethane urea–vinyl polymer (PUU–VP) microspheres hybridized with a vinyl polymer (polyacrylate or PST) were also prepared by membrane emulsification and a subsequent radical suspension polymerization process. That is, the UP/xylene mixture dissolved in the vinyl monomer (sometimes containing a crosslinker) was emulsified into an aqueous phase by membrane emulsification technique [28]. Then, the prepolymer was chain-extended by adding a chain-extender solution in the aqueous phase. After 2 hours of the chain-extending reaction, the polymerization of the vinyl monomer was carried out by radical suspension polymerization for 24 hours at 70 °C. It was found that hybrid particles of an interpenetrating network structure, with a uniform size and smooth surface, can be obtained by adding a trifunctional crosslinker for the vinyl monomer and a trifunctional chain extender for the UP to the system. The crosslinker was necessary for formation of solid hybrid microspheres with smooth surfaces, high destructive strength, and uniformity without adhering to each other. For example, when the crosslinker was not used and only the chain extender piperazine (PZ) with two functional groups was added, the monomer conversion rate is low. The phase separation between PUU and P(ST–MA) was significant, the surface of microspheres was not smooth (Figure 2.11a), and the dispersion of the particle was unstable since the strong relationship was found between the compatibility of PUU with VP and the stability of the dispersion during polymerization. These

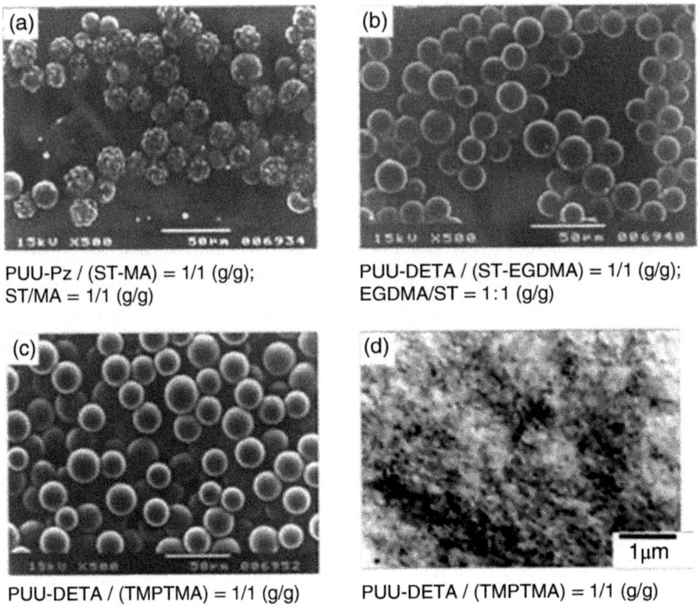

Figure 2.11 Composite microspheres of polyurethane and polystyrene or polyacrylate prepared by membrane emulsification method. Source: Ma et al. [28]/John Wiley & Sons.

disadvantages were overcome by adding an adequate amount of the crosslinker and a trifunctional chain extender of the UP. After using cross-linking agent EGDMA and chain extender diethylenetriamine (DEAE) with three functional groups, the compatibility between two polymers was increased and the phase separation was significantly suppressed (Figure 2.12b). The emulsion became stable and the formation of aggregates decreased. After replacing EGDMA with trimethyol propane trimethacrylate (TMPTMA), a cross-linking agent with three functional groups, the phase separation can be further reduced and the uniformity of particle size can be improved (Figure 2.11c,d). It can be seen that the two polymers are mixed evenly in TEM microscope photos of the microsphere ultrathin film in Figure 2.11d (PUU area is stained with OsO_4).

2.3.3 Polyimide Prepolymer (PIP) Microspheres

The polyimide prepolymer PIP with diphenylmethane-4,4′-bisallylnagiimide (BANI-M) underwent self-curing when it was heated beyond 220 °C and contributed to the adhesion and the mechanical strength. It can be crosslinked and adhered to the glass panels with the heat treatment, and thus could be used as the microsphere spacers for liquid crystal displays (LCDs). Therefore, we prepared a poly(styrene-divinyl benzene) microspheres special for LCD utilization incorporating the PIP-BANI-M, and together adding a trifunctional crosslinker TMPTMA to further improve the destructive strength of such microspheres spacers [24]. Considering that it is difficult to prepare highly crosslinked microspheres and incorporate

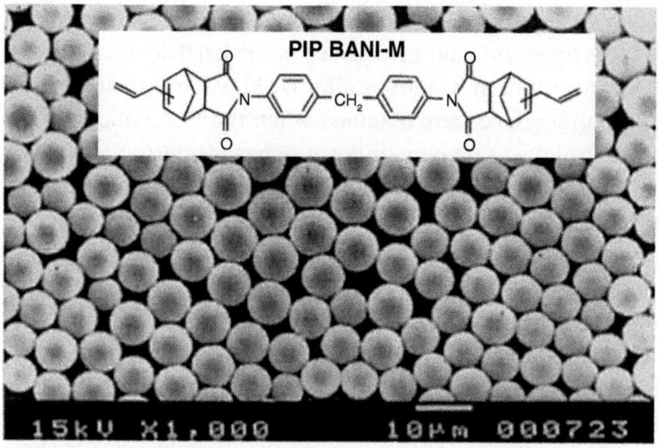

Figure 2.12 The chemical structure of the PIP BANI-M, and the SEM micrograph of microspheres with a narrower size distribution. Source: Ma et al. [24]/John Wiley & Sons.

functional large molecules into the particles by conventional polymerization, such as the activated swelling and nonaqueous dispersion polymerization, we prepared such PIP microspheres by membrane emulsification and a subsequent radical suspension polymerization (Figure 2.12).

The typical procedures are as follows: a mixture of ST, DVB, TMPTMA, and PIP containing an initiator was permeated through the pores of hydrophilic porous glass membrane into an aqueous phase containing a stabilizer to form uniform droplets. Then, the suspension polymerization was carried out at 65 °C for 24 h. The effects of the feed amounts of TMPTMA and PIP on the monomer conversion, size distribution, and destructive strength of the microsphere were investigated. The monomer conversion increased with an increase in TMPTMA. The particle size distribution became narrower as the TMPTMA feed amount increased, but it became broader with an increase in PIP. The destructive strength increased with TMPTMA and PIP increasing. After a heat treatment at 220 °C, the destructive strength decreased because of the heat degradation of the polymer. The addition of TMPTMA suppressed the heat degradation because PIP could undergo self-curing at 220 °C.

2.3.4 Biodegradable Poly(Lactide) Microspheres

A series of biodegradable and biocompatible materials have invoked widely attentions in pharmaceuticals development such as sustained released formulation of drugs, vaccine adjuvant, and bioreactor. Considering that the size-uniformity endows these particles with significant performance, we performed membrane emulsification with a mixture composed of biodegradable poly(1actide) (PLA) polymer, DCM solvent, and LOH cosurfactant as the oil phase, and the PVA and SLS as the continuous aqueous phase [11]. After emulsification, DCM was evaporated at room temperature for 24 hours to obtain the solid particle. It demonstrated that the relatively uniform spherical PLA microsphere could be obtained successfully by using LOH rather than hexadecane HD as a cosurfactant. The phase separation between PLA and LOH was apparent, and the surface of the particle obtained was wrinkled when the PLA concentration was lower after DCM was evaporated, while the particles with the smooth surface were obtained when the PLA concentration was higher. The unique technique to prepare uniform polymer microspheres composed of natural polymers, biodegradable polymers, copolymers or polymer blends, polyesters and those that cannot be polymerized by the radical polymerization.

Besides this, we also prepared the drug-loaded uniform microspheres by membrane emulsification and solvent evaporation of a S/O/W emulsion with a mixture composed of low molecular weight PLGA (PLA/PGA 50:50, average molecular weight MW 30 kDa) and human insulin. The procedure was as follows: the quantitative human insulin powder was added to the 2.5 ml dichloromethane solution containing 0.125 g PLGA directly, homogenized, and emulsified (speed 9800 rpm, time 45 seconds) so that the insulin powder was evenly distributed in the oil phase to form S/O emulsion. Dissolve 0.7 g PVA in 70 ml deionized water, and was used as external water phase. Under the nitrogen pressure, the S/O emulsion was pressed through the membrane to form S/O/W emulsion droplet by MET and then the organic solvent was removed under stirring (rotational speed 100 rpm) for 3 hours. After washing with deionized water by centrifugation three times, the drug-loaded microspheres were obtained by freeze-drying. The addition of PEG 6000 in human insulin could significantly increase the drug encapsulation efficiency to 83.07%. Figure 2.13 shows that the CV value increases with the increase in membrane pore size. When the membrane pore size is 9.6 μm, the CV value reached 19.90%.

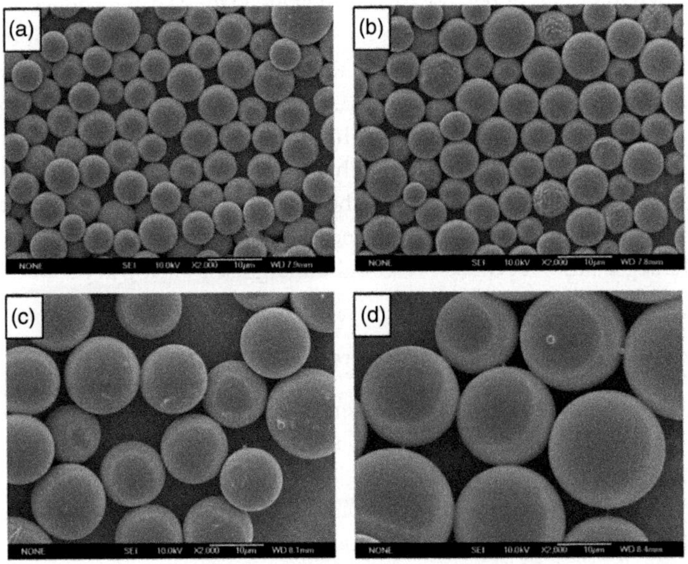

Figure 2.13 SEM photographs of drug-loaded microspheres with different size. Microspheres were prepared with different pore size membranes: (a) 1.4 μm; (b) 2.8 μm; (c) 5.2 μm; and (d) 9.6 μm. The CV value for (a), (b), (c), and (d) was 12.21%, 12.59%, 14.96%, and 19.90%, respectively.

2.3.5 Microcapsules Containing Inorganic Materials

Microcapsule containing inorganic materials, such as the magnetite, has important applications in drug delivery system (DDS) and bioseparation. In DDS, microcapsules could be guided to a target by an external magnetic field and release the drugs they contain. In bioseparation, the magnetic particles carrying the ligand with affinity with cancerous cells could be used to adsorb the cancerous cells. Then, by an external magnetic field, the magnetic microcapsules with adsorbed cancerous cells could be easily separated from other components, which could be returned to the body [16]. By combining the membrane emulsification and evaporation method in the aqueous phase, uniform microcapsules containing inorganic substances can be fabricated. The following is an example of Fe_3O_4 magnetite encapsulated in St-based copolymers. A solution of St-based copolymers, poly(styrene-*co*-acrylic acid) (PST-AA), in a magnetic fluid (25 wt% magnetite dispersed in toluene) was used as the oil phase [29]. An aqueous phase containing dissolved PVA and SLS was used as a continuous phase. After the membrane emulsification, toluene was evaporated under reduced pressure at 50–60 °C. Microcapsules entrapping 30–40 wt% of magnetite were obtained. By comparison, it can be obviously observed that membrane emulsification produced significantly uniform microcapsules with complete encapsulation of magnetite, whereas the microcapsules produced by the conventional mechanical stirring method were polydispersed in size and present magnetite escaping out of microcapsules (Figure 2.14). In the latter case, the droplet size was not uniform due to unstable dispersion, and thus resulted in the breakup of larger droplets and coalescence of smaller droplets among themselves or attachment to larger droplets, and the leakage of magnetite during breakup and coalescence.

2.3.6 Pickering Emulsion

Compared with the conventional emulsions stabilized by surfactants, the particle-stabilized emulsions, also known as Pickering emulsions, are receiving growing attention in both scientific and industrial communities. Pickering emulsion has many advantages, such as the improved stability, low toxicity, and noncontamination, because it uses colloidal particles solely as stabilizers and relies on the irreversible adsorption of the particles at an oil–water interface, which is densely packed to a solid shell the droplets and confer good protection against coalescence. Conventionally, Pickering emulsions are mostly prepared by mixing oil and water using a high-pressure homogenizer. In this device, a pre-emulsion with large

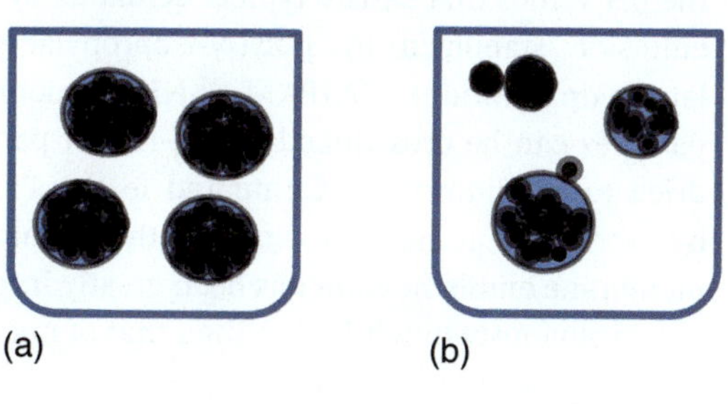

Figure 2.14 Fe_3O_4 polymer particles prepared by different emulsification method. (a) MET technique and (b) homogenizer.

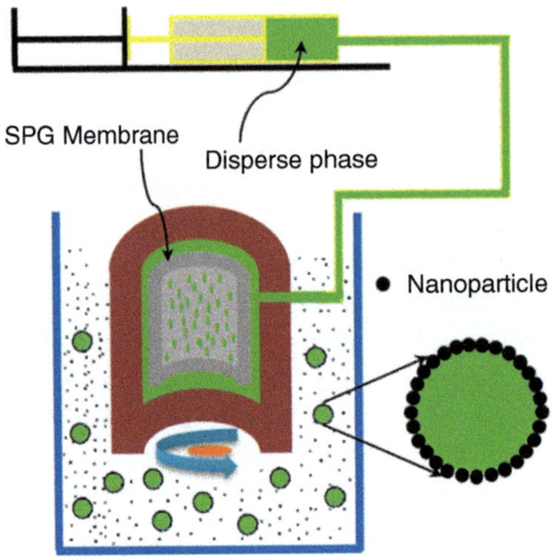

Figure 2.15 Schematic depiction of the Shirasu porous glass (SPG) membrane emulsification process in the production of uniform Pickering emulsions using soft particles as stabilizers. Source: Sun et al. [30]/with permission of American Chemical Society.

droplets is forced through a high shear region in such a way as to promote turbulence and, hence, disrupt large droplets into smaller ones. A drawback of this droplet breakup method is that it is not easy to control the drop size of the emulsion product so that the resultant droplets are usually highly polydisperse, which leads to poor reproducibility and a variable quality of the emulsion product from batch to batch. Membrane emulsification has been demonstrated as a convenient technique to solve the nonuniformity of Pickering emulsions, and a series of preparation routes are designed according to the specific characteristics of colloidal particles [30].

When the soft particles (such as PNIPAM-*co*-MAA microgels and Kollicoat particles) were utilized as stabilizers, the uniform Pickering emulsions were produced by firstly pressing the sunflower oil through the membrane (Figure 2.15), then rapid adsorbing of microgel particles in continuous phase onto droplets [30]. Since the oil–water interface of oil droplet was packed with densely particles and the droplet was protected by an elastic shell, the droplets coalescence was avoided and facilitated formation of uniform emulsions. Consequently, the size of the emulsion droplets ranging from 10–50 μm can be precisely controlled by the size of the membrane pore. The emulsion with uniform droplets was obtained with CV lower than 15% in all of investigated systems. After membrane emulsification, we showed that the Kollicoat particle-stabilized emulsions can be easily destabilized by increasing the pH values that satisfy typical demands by controlled release applications. The emulsion stabilized by poly(*N*-isopropylacrylamide-*co*-2-aminoethyl methacrylate hydrochloride) (PNIPAM-*co*-NH2) microgel particles, interfacial assembled particles can be crosslinked into a robust particle shell and can be subsequently dried to colloidosomes. Compared to the Pickering emulsion droplets prepared by conventional homogenization, the monodispersity of emulsion droplets by membrane emulsification has been greatly improved. The emulsification efficiency is also obviously much higher than that of microfluidic device [30].

2.4 Morphology Control of Microspheres

Membrane emulsification stabilized the structure of emulsion droplet, that is, the adsorption and desorption rates of emulsifier/stabilizer and the diffusion rate of monomer into the aqueous phase are slower. Based on this stability, polymerization could be confined in a droplet when the oil droplet was composed of a monomer, and the mass exchange among the droplets can be neglected. Therefore, MET is more reliable to study the morphology evolution in a microsphere, as compared to the case in which the emulsion droplet is nonuniform and the emulsion structure is unstable because the kinetics and thermodynamic change of adsorbed pattern of emulsifier/stabilizer on droplet and the mass exchange among droplets and the aqueous phase will significantly affect both structures of surface and interior of a microsphere. By combing MET and polymerization process, we summarized the critical factors (e.g. crosslinker and porogen) manipulating the structure formation of porous polymer microspheres and explored their mechanism during morphology evolution.

2.4.1 Effect of Crosslinker on Morphology of Microspheres

Nanostructure of a hydrophobic microsphere prepared from O/W emulsion can be manipulated by crosslinker and the polymerization kinetics during phase separation between porogen and polymer [27]. The formation of nanostructure was presented in Figure 2.16. Porogen, crosslinker, and initiator were firstly dissolved in oil monomer phase, then the O/W emulsion was prepared by membrane emulsification, and finally polymerization was carried out by elevating the temperature. Because the porogen is soluble in monomer but insoluble in polymer, phase separation between porogen and polymer would occur during cross-linking and polymerization of monomer. Since the phase separation structure in presence of cross-linking reaction is nonequilibrium, the final phase separation structure would be governed by compromise between phase separation (diffusion) and cross-linking of the polymer (reaction).

In absence of a crosslinker, the complete phase separation occurred with only one large pore as shown in Figure 2.17. As the amount of crosslinker increased, crosslinked polymer precipitated in porogen more quickly resulting in increased pore number and higher specific surface area. The effect of the amount of crosslinker

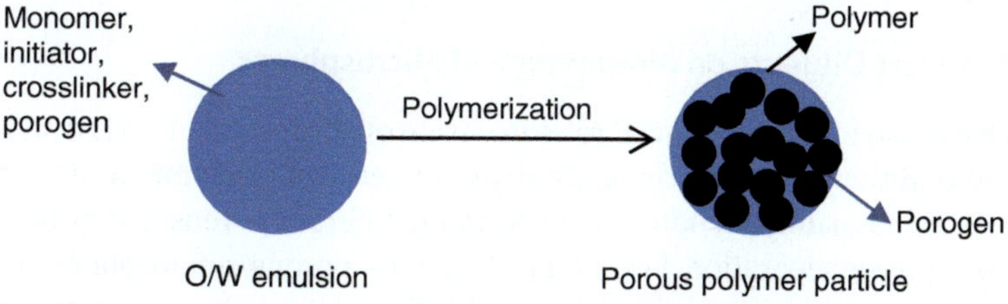

Figure 2.16 Principle of formation of porous polymer particle.

Figure 2.17 Effect of crosslinker amount on phase separation between polymer and porogen.

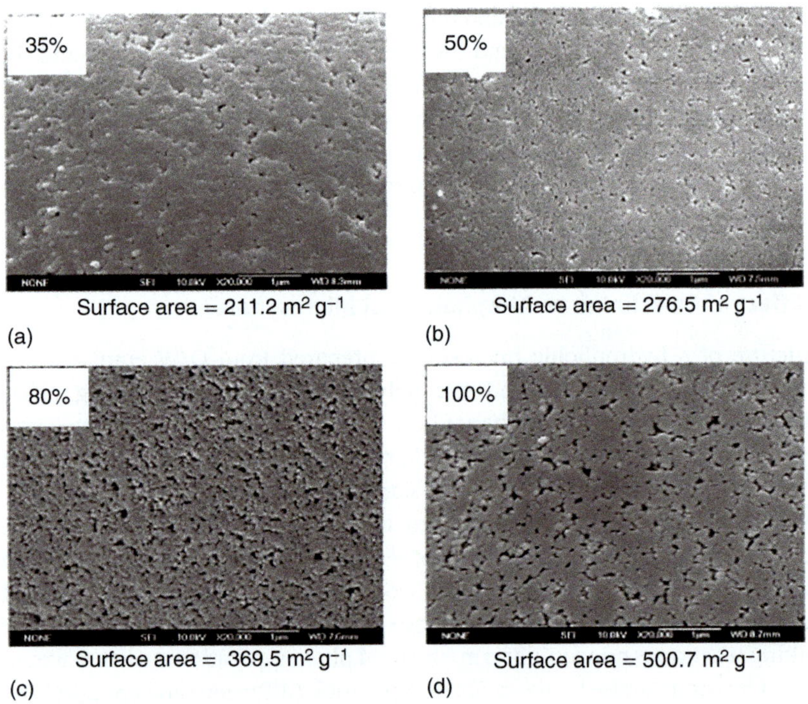

Figure 2.18 SEM showing effect of crosslinker amount on pore size and specific surface area.

on pore size and specific surface area is shown in Figure 2.18, where ST was used as a monomer, DVB as a crosslinker, heptane as a porogen, and BPO was used as an initiator of polymerization.

2.4.2 Effect of Inert Diluents on Morphology of Microspheres

Inert diluents have also presented crucial roles in design of porous structure [20, 21]. It is now known that the thermodynamic affinity of diluent with polymer and thus the phase separation is mainly responsible for forming different porous and hollow structures. Many phase separation behaviors of various porous microspheres in suspension polymerization have been understood quite well by above qualitative theories, but scarcely with quantitative treatments because of the difficulty of in situ monitor phase separation process in one suspended microsphere. Comparatively, for bulk polymerization system, the proceeding of phase separation has been successfully monitored by turbidity change of polymerizing system [31, 32]. The separated phases in polymerization system have different wavelengths of scatter light, which correspondingly give related opacity to the polymer system, the turbidity (transmittance) as a direct result of opacity can thus be used to describe the formation of polymer heterogeneities. If reaction in a suspended monomer droplet was regarded as bulk polymerization, the phase separation behavior in one polymerizing microsphere can also be characterized by monitoring turbidity of bulk polymerization system (monomer–diluent mixture). Therefore, the phase separation behaviors during suspension polymerization in a uniform droplet could be characterized by in situ monitored turbidity (transmittance) of bulk polymerization system including monomer and various diluents, since membrane emulsification can effectively narrow the size distribution of emulsion droplet and thus microspheres, and the volume effect on structure development in suspension polymerization of microspheres was expected to be avoided. Furthermore, because phase's mobility in polymerizing microsphere has been proven important in formation of final structure [33], the gelling point when the polymerization system changes from liquid- to solid-like state was also investigated simultaneously with phase separation monitoring. The following summarized the final structure heterogeneity of uniform particles prepared by two groups of the nonsolvating and the solvating diluents.

2.4.2.1 Nonsolvating Diluent Effects on Microsphere Morphology

A series of nonsolvents with different hydrocarbon chain length were chosen as porogen to fabricate heterogeneous microspheres polymerized by monomer of DVB. Figure 2.19 showed that the heterogeneity of microsphere shifted from porous to hollow structure with increasing chain length of diluents. Specifically, heptane with

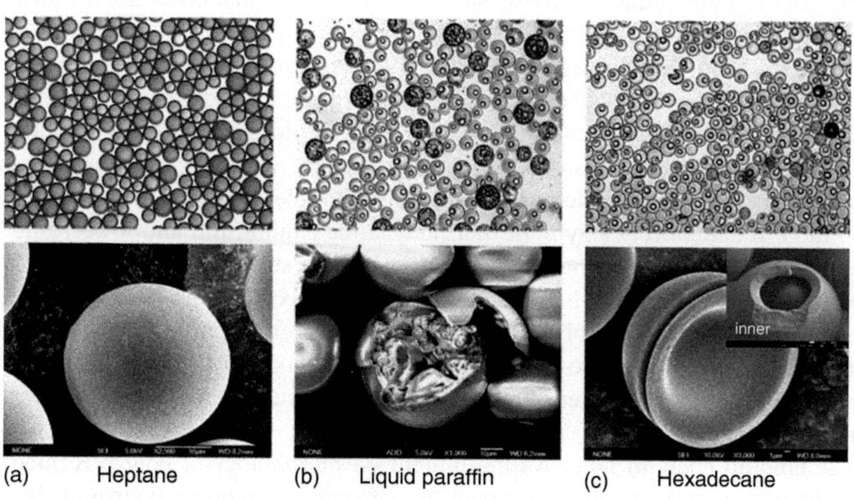

Figure 2.19 OM and SEM micrographs of polydivinylbenzene microspheres. Source: Hao et al. [14]/Elsevier.

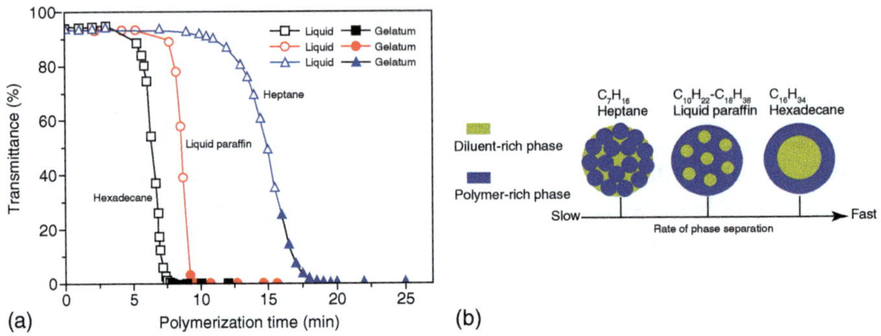

Figure 2.20 Turbidity development with polymerization time (a) and schematic representation of phase separation distribution (b) for three microspheres. Source: Hao et al. [14]/with permission of Elsevier.

shortest chain length formed the porous surface composed of innumerable nuclei, while liquid paraffin formed both multiple microvoids and hollow structures, hexadecane with longest hydrocarbon chain led to hollow void, which can be observed by obvious inner circle and collapsing part of the microspheres.

The phase separations of above three microspheres were investigated by transmittance development of bulk polymerization of their mixture in Figure 2.20. The decrease in transmittance reflected the separation of heterogeneous phases from the homogeneous polymerization solution (monomer–diluent BPO mixture). The hollow dots in transmittance curves denote the observed reaction system still in liquid state, and the solid dots indicate the gelling of reaction system changed to solid-like state (gelatum), which came from the simultaneous observation results of gelling point for polymerization solution. Among three curves, hexadecane induced most rapid decrease of transmittance and late appearance of solid dots at the end of transmittance curves. Based on this, the formation of hollow structures was speculated as follows. Plenty of hexadecane-rich phases and polyDVB-rich phases were quickly separated from polymerization solution at early stage of polymerization with low conversion. The new generated polymer microgels in polymer-rich phases with low molecular weight would make interior of microsphere in liquid and mobile states until to gelling point. Such long-term mobility in reaction environment provided many possibilities for plenty of phases to rearrange into more stable morphology with thermodynamic equilibrium and lowest interfacial free energy. For one polymerizing microsphere suspended in water during suspension polymerization, highly hydrophobic hexadecane-rich phases tended to move into center of microsphere and combined each other into one bulky liquid core. Consequently, the core–shell morphology was constructed as illustrated in Figure 2.20, which speculated the final diverse distributions of phases for above three microspheres.

The amount effect of nonsolvating diluent on morphology of polyDVB microspheres was investigated with heptane as porogen. As heptane increased, both pores and nuclei on microsphere's skeleton were obviously enlarged (Figure 2.21). Correspondingly, Figure 2.21 shows that decline of transmittance curves also went

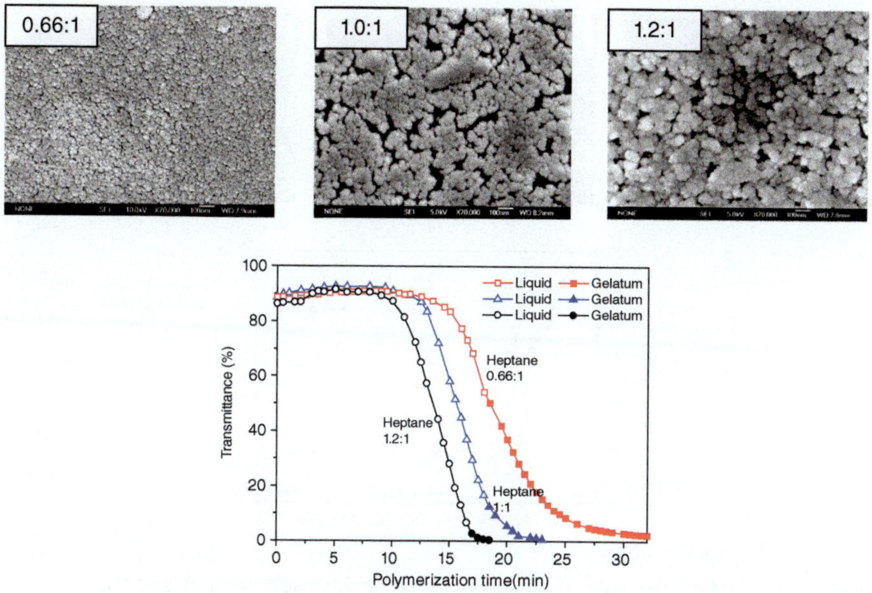

Figure 2.21 SEM (up) of polydivinylbenzene microspheres, and the development (down) of turbidity and gelpoint with polymerization time at different heptane/monomer ratio. Source: Hao et al. [14]/Elsevier.

ahead and the gelling point was postponed with increasing heptane. This tendency demonstrated that increase of nonsolvent porogen accelerated phase separation between porogen and growing polymer because of its poor thermodynamic affinity with polymer network. Since phase separation had occurred early with low conversion when heptane increased, the formed microgels were swollen with a high concentration of monomer, which caused not only fusing of the microgel particles but also significant in-filling of small pores between microgels. Also, because the gelling point was delayed with increase of heptane during the phase separation, the separated microgels gained much time to move and congregated into larger nuclei. Consequently, a polymer matrix with larger nuclei and pores was built up at high addition of heptane. An extreme case can be speculated as follows: if continue to postpone the gelling point at later stage than above, the polymer-rich phases would obtain more mobility and finally arrange into outer shell of hollow microsphere as previous discussion of hexadecane.

2.4.2.2 Solvating Diluent Effects on Microsphere Morphology

Most diluents, which have good solvating for polymer network, generally construct small pores in polymer microspheres, but scarcely construct hollow or macroporous structures if without especially mixing other nonsolvating diluents. Toluene, a typical solvating diluent for polystyrene, was investigated for its effect on porous structure evolution and phase behaviors in preparation of polyDVB microspheres. Figure 2.22 showed that porous feature for toluene can only appear above a high addition (toluene/monomer ratio: 2.0 w/w), and with much smaller pores and

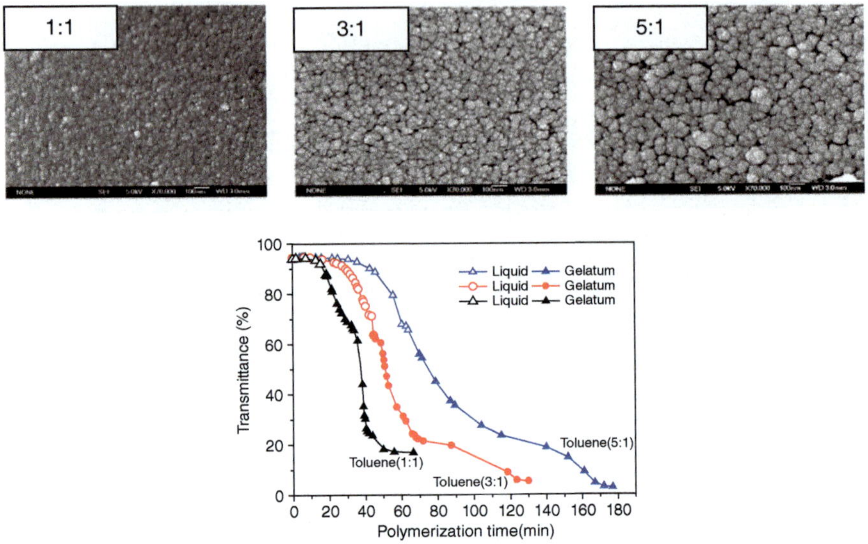

Figure 2.22 SEM micrographs of polydivinylbenzene microspheres (up), and the development (down) of turbidity and gelpoint with polymerization time at different toluene/monomer ratio. Source: Hao et al. [14]/Elsevier.

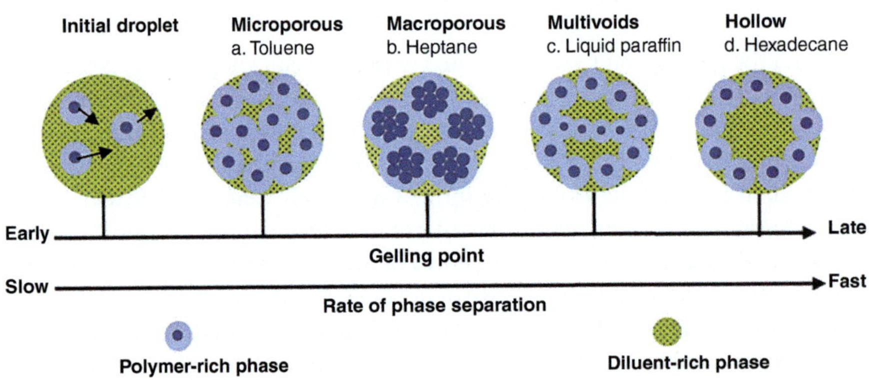

Figure 2.23 Schematic representation of formation of heterogeneous structure in phase separation with different diluents. Source: Hao et al. [14]/with permission of Elsevier.

nuclei than nonsolvating diluent. With further increasing toluene, both pores and nuclei were enlarged. The phase separation behaviors of above microspheres prepared by toluene were reflected in Figure 2.23. Compared with previous nonsolvating diluents, toluene showed much slower decline of transmittance and much earlier appearance of gelling point in descending stage of transmittance curves. It suggested that the proceeding of phase separation was obviously restrained after addition of solvating diluent, which resulted in difficult precipitation of polymer-rich phases. This late phase separation and early occurrence of gelling made the whole reaction mixture solidify into homogenous gel. At such gel environment, the following process including vinyl polymerization, phase separation, and even microgel fusion (and aggregation) has to be in situ finished, because most diffusion and movement of materials have been greatly limited at gelation point as previously discussed. Consequently, the polymerized microgels cannot aggregate into larger nuclei and small pores were constructed in the final polymer network. If continue increasing the amount of toluene, both the transmittance curve and the gelling point were postponed, which means the phase separation was put off with increasing amount of the efficient swelling solvent, and larger pores can be formed because of the late gelling point facilitating formation of larger polymer nuclei. Moreover, all the transmittance curves were not smooth as previous nonsolvent diluents were created. It might result from the formation of many irregular cracks on bulk polymer in spectrophotometer cell, which made the disturbance on the continuous development of transmittance.

According to analysis about turbid (transmission), gelling point, and viscosity in polymerization, we summarized that final heterogeneity of microspheres prepared by membrane emulsification and suspension polymerization mainly depended on the competition of three dynamics in polymerization, i.e. polymerization kinetics, interfacial energy between phases, and mass transport rate governed by gelation. They together manipulated the formation of porous structures during reaction-induced phase separation process [14, 34–36]. We studied the relationship between heterogeneous structures and phase separation in synthesis of uniform polyDVB microspheres with hollow, multivoids, or different porous structures. It shows that the evolution of heterogeneous structures was the result of the diverse phase distribution in final polymer microsphere. Specifically, the final phase distribution in microsphere depended on separation rate between polymer-rich and diluent-rich phases and also gelling point of polymer system. By indirectly tracking the phase separation with real-time transmission of polymer system, it showed that movement of transmittance was influenced by the affinity of inert diluents with polymer. Nonsolvating diluents with poor affinity for polyDVB, such as alkanes, tend to the early decline of transmittance and the fast phase separation, while solvating diluents led to the late declines of transmittance and the slow phase separation. Moreover, the occurrence of gelling points of reaction system during phase separation was demonstrated to contribute much to morphology diversity of microspheres. As shown in Figure 2.24, if the reaction system was solidified at earlier stage in phase separation process, homogenous structure, such as the nonporous or the small pores, was more likely formed. With further postponing occurrence of gelling point, macroporous, microvoids, and hollow structures were, respectively, formed.

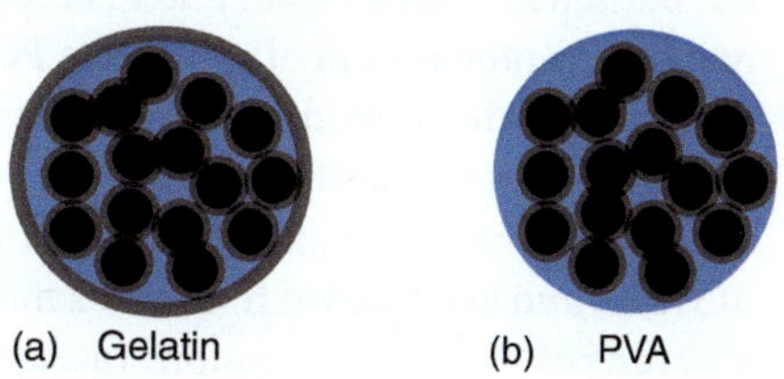

Figure 2.24 PGMA polar microspheres using different stabilizers. (a) gelatin and (b) PVA.

2.4.3 Effect of Emulsifier/Stabilizer on Morphology of Microspheres

Emulsifier or stabilizer also affects the nanostructure of microspheres during preparation of hydrophobic microspheres by membrane emulsification [27]. As described above, the structure in the interior of a microsphere is governed by both diffusion and reaction in its emulsion droplet and depends on the phase separation between porogen and crosslinked polymer. However, the structure on surface is also affected by the interaction between the component of microsphere and aqueous phases. An interesting phenomenon was observed that the surface was usually smooth when the emulsion was stable, whereas the surface was porous when the emulsion was unstable. Figure 2.24 shows such an example a PGMA microsphere, when gelatin was used as a stabilizer in the aqueous phase, producing a very stable dispersion with no coagulation. Although porogen was added, the surface of the microsphere was smooth but the interior was porous. Comparatively, if using PVA as a stabilizer, a porous surface formed as well as the porous interior of microsphere. However, at this case, the dispersed emulsion was not stable and was accompanied by partial coagulation.

A possible explanation is that gelatin preferred to adsorb on surface of polymer more than porogen, resulting in lower interfacial tension of polymer with the aqueous phase and PGMA polymer localizing on surface of microsphere and engulfing the porogen inside. Comparatively, when PVA was used as the stabilizer, the interfacial tension of polymer with the aqueous phase was close to that of porogen with the aqueous phase, so that both porogen and polymer contacted the aqueous phase and led to porous surface. Coagulation of emulsion with stabilizer PVA was also probably due to nonhomogeneous property of surface. Therefore, in order to obtain stable dispersion of microspheres with nonhomogeneous porous surface, it is probably necessary to use a combined emulsifier/stabilizer, of which one can be steadily adsorbed on one domain, while the other on another domain.

2.4.4 Effect of Cosurfactant on Morphology of Composite Microspheres

Addition of cosurfactant was sometimes necessary to prepare uniform O/W droplets when the oil phase was a little hydrophilic since the cosurfactant could increase the hydrophobicity of oil phase and alleviate the wetting of membrane pore wall by oil phase. Besides this, it could prevent the relatively hydrophilic solvent from diffusing into the aqueous phase and thereby maintain the stability of droplets. However, it was found that if the cosurfactant of LOH was too high, the size distribution of the particles became broad due to an apparent phase separation between LOH and polymer in preparation of composite PGMA-ST particles. This phase separation not only spoiled the monodispersity of the particles but also affected the morphologies of composite particles extensively.

This case is shown in Figures 2.25 and 2.26, where the PMMA/PST ratio was 5/5 (g/g) and LOH varied from 0 to 2 ml (per 1.2 g of polymer) [4]. In absence of LOH and the polymer concentration was 10 wt%/v, Figure 2.25a showed that the surface

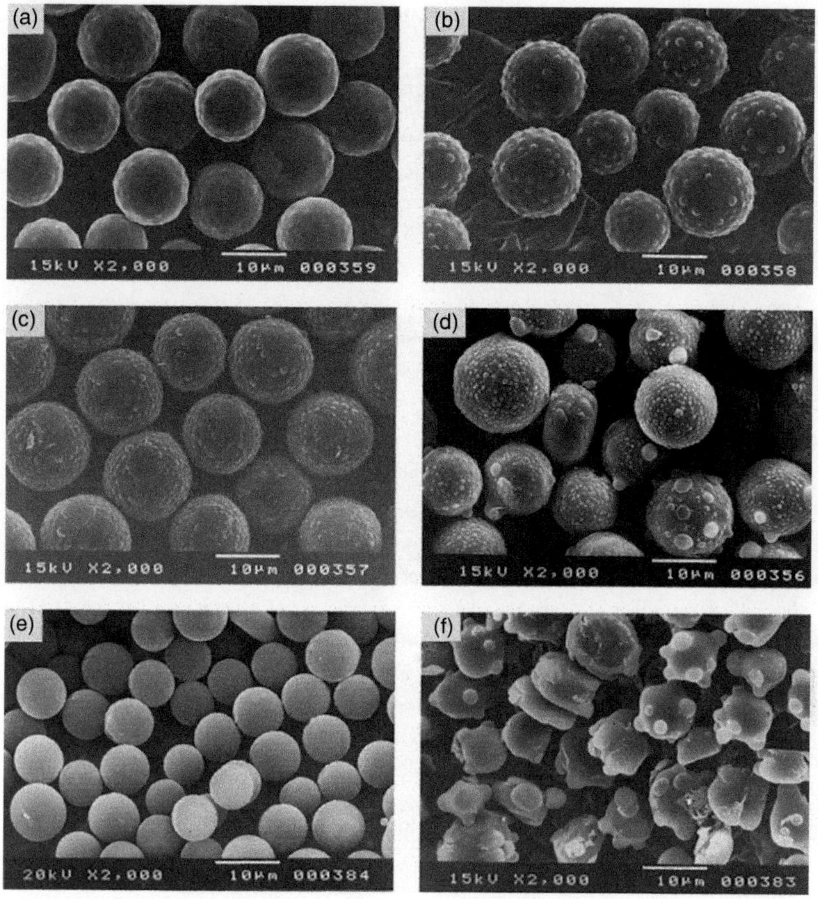

Figure 2.25 SEM micrographs of particles as functions of LOH amount and polymer concentration. PMMA/PST (g/g): 5/5. Polymer concentration (wt%/v): (a–d) 10; (e, f) 2. LOH (ml/1.2 g of polymer): (a, e) 0; (b) 0.5; (c) 1; and (d, f) 2. Source: Ma et al. [4]/Elsevier.

of the microspheres was relatively smooth. In presence of LOH, however, the small PST domain protruded out of surface. The hydrophobic PST and hydrophilic PMMA domains were interwoven on surface. As the polymer concentration was decreased to 2 wt%/v and LOH was not used, the surface was smoother than above (Figure 2.25e), whereas the protruding PST domain became larger when LOH was used. From the corresponding TEM micrographs shown in Figure 2.25, it is clear that the particle showed microdomain morphology for each case when the polymer concentration was 10 wt%/v. It can be confirmed that the protruding domain on surface observed in SEM micrographs was ascribed to the PST domain. The size of the PST domain, especially in vicinity of the surface, became smaller with LOH amount increasing, suggesting that LOH hindered the growth of the PST domain. However, as the polymer concentration was decreased to 2 wt%/v, the core–shell type microsphere was obtained without LOH, and returned to the microdomain type with addition of LOH.

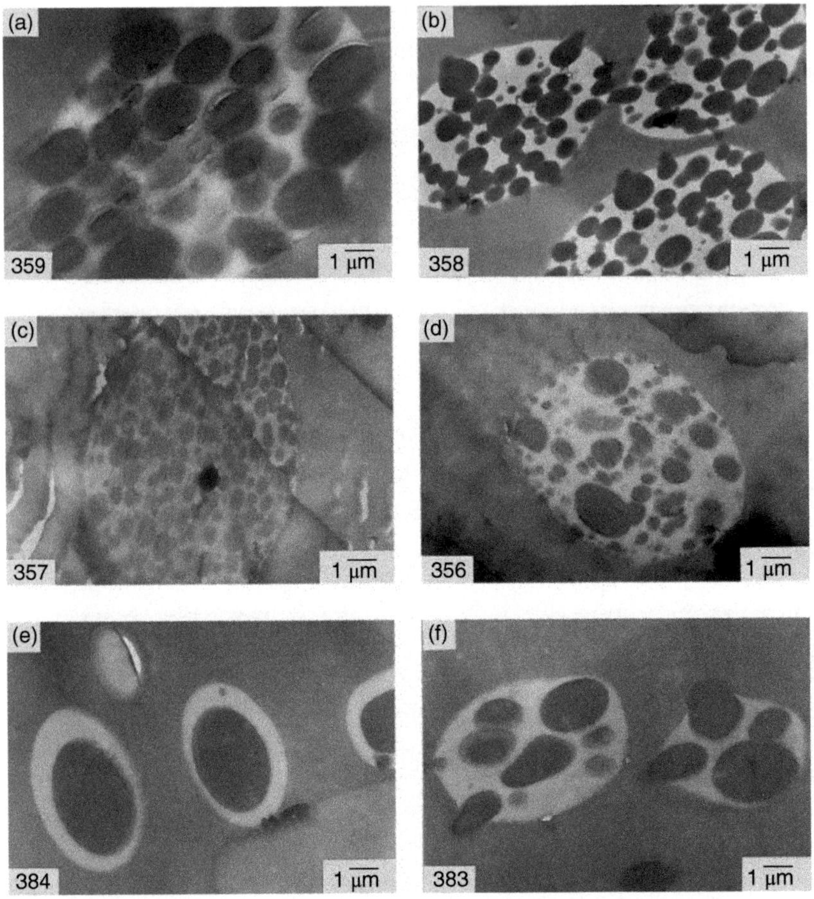

Figure 2.26 TEM micrographs of particles as functions of LOH amount and polymer concentration (PST domain was stained by RuO 4). PMMA/PST (g/g): 5/5.Polymer concentration (wt%/v): (a–d) 10; (e, f) 2. LOH (ml/1.2 g of polymer): (a, e) 0; (b) 0.5; (c) 1; and (d, f) 2. Source: Ma et al. [4]/Elsevier.

If, we compare the cases of 10 and 2 wt.%/v of polymer concentration when LOH was not used (Figures 2.25a,e and Figures 2.26a,e), it can be said that the microdomain type was a nonequilibrium morphology. This was because the viscosity of the PMMA phase was very high when both the polymer concentration and the PMMA/PST ratio were relatively high. As a result, the movement of PST molecules was limited by the high viscosity of the PMMA phase and was also retarded comparing to evaporation of dichloromethane. Consequently, the PST domain cannot grow to one domain due to the high viscosity of the PMMA phase and the large size of the particles. On the other hand, when the polymer concentration was decreased to 2 wt%/v and LOH was not used, the phase separation occurred slowly because a greater amount of solvent was used so the core–shell microsphere of equilibrium phase separation formed. However, when LOH was used, a nonequilibrium morphology was observed again even at very low polymer concentration (2 wt%/v). This result suggested that LOH plays a complicated role in the phase separation mechanism. A majority of LOH was certainly partitioned on the surface of the particles after most of DCM was evaporated because LOH shows a much lower interfacial tension against the aqueous phase. Although the hydrophobicity of PST is higher than that of PMMA, the interfacial tensions of LOH against PST and PMMA are close. Therefore, the PST domain was allowed to protrude out of the surface. Moreover, a small portion of LOH was also partitioned inside the particles to prevent the growth of the PST domain.

Based on the aforementioned fact that LOH was partitioned preferentially on the surface, the theoretical calculation of morphology formation can be carried out [4]. The thermodynamically stable morphology is a system with minimum free energy. From the thermodynamic considerations, a composite microsphere was a mixture of two or more polymers, and its morphology with complete phase separation after solvent evaporation was in the thermodynamically preferred equilibrium. The contribution to the free-energy change is the creation of new interfaces and changes in corresponding interfacial tensions [37, 38]. That is,

$$\Delta G = \sum \gamma_{ij} A_{ij} - \gamma_0 A_0$$

where A_{ij} is the created area of new interfaces between i and j components at the final particle, γ_{ij} is the corresponding interfacial tension, A_0 is the interfacial area of the initial state, and γ_0 is the corresponding interfacial tension. Although the choice of initial state affects the absolute value of ΔG, the determination of the morphological structure that represents the lowest free-energy state is unaffected, since γ_0 is a constant. Based on above, two thermodynamics models to explain many morphology evolutions of composite microspheres were developed as the two-component and the three-component models [34, 35].

It is accepted that a thermodynamically preferred equilibrium morphology with the minimum change of total free-energy is determined by three interfacial tensions for a two-component polymer system, i.e. the interfacial tension between two polymers and that between the aqueous phase (containing surfactant) and the respective polymer. Usually, only core–shell (PST as core, PMMA as shell), inverted core–shell (PMMA as core, PST as shell), and hemispherical particles are stable in a thermodynamic sense.

We utilized three-component model to calculate above morphology of a poly (styrene)–poly(methyl methacrylate) (PST–PMMA) composite microspheres prepared by membrane emulsification, which was tailored extensively by adding cosurfactant of LOH and changing its volume [26]. Agreement was obtained between experimental and calculated results. When 2 ml of LOH was added, LOH could engulf the polymer particles completely. As shown in Figure 2.27, At the high PMMA/PST ratio, a hemicore morphology (HCP1P2P3) was observed with a PST and PMMA, formed hemisphere core inside a LOH shell, while at the low PMMA/PST ratio, a core–shell–shell morphology (CSP1P2P3) was observed with PMMA formed a core, PST formed an inner shell, and LOH formed an outer shell, respectively. Comparatively, when the volume of LOH was below 1 ml (per 1.2 g polymer), LOH could not always engulf the inner polymer particles completely.

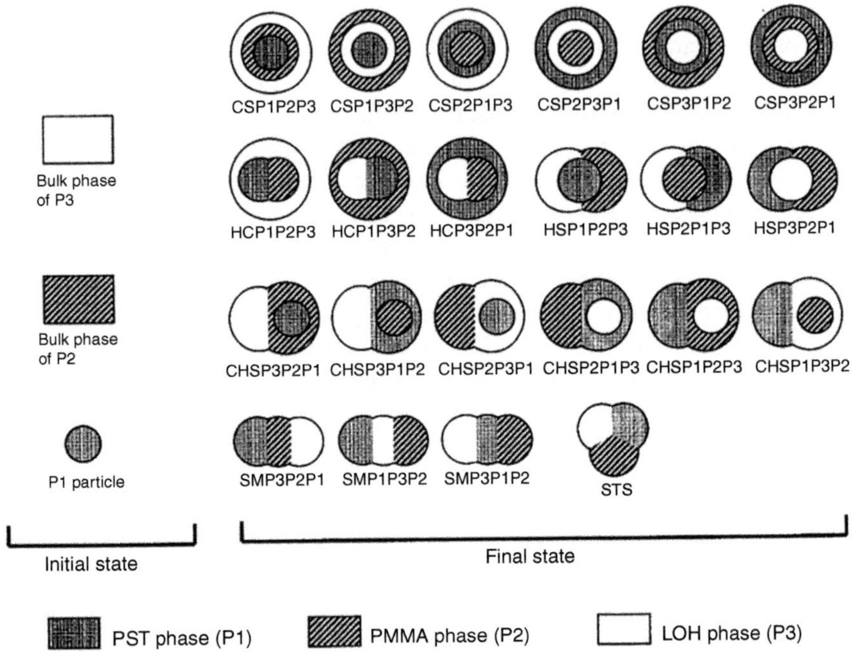

Figure 2.27 Initial and final states for morphology development of a three-component system based on the consideration of Sundberg. Source: Ma et al. [26]/with permission of Elsevier.

The cored hemisphere (CHSP3P2P1), core–shell–shell (CSP2P1P3), and hemicore (HCP1P2P3) morphologies were observed, depending on the PMMA/PST ratio and the volume of LOH.

2.5 Summary

Before the birth of membrane emulsification, many hydrophobic polymer microspheres were prepared by suspension polymerization, during which the hydrophobic monomer was dispersed into aqueous phase and form O/W emulsion by impeller in a stirred tank. Due to the breakage and coalescence of varying droplets in the stirred tank reactors, suspension polymerization results in particles with broad size distributions, and the CV is normally at least as high as 35% [39]. MET approaches, capable of producing droplets with a narrow size distribution, have opened a new chapter for delicate fabrication of a great diversity of polymer microspheres. Here, we summarized a series of typical routes to prepare uniform hydrophobic polymer microspheres by MET, which can be divided into two groups, the ones from the monomer system and the ones from the performed polymer system. For the monomer system, the combination of MET and suspension polymerization was utilized, which began with the emulsification of monomer into uniform O/W droplets by pressing the monomer oil phase through the membrane pores into aqueous phase containing a stabilizer under the controlled pressure, and then initiate the subsequent suspension polymerization of emulsion droplet into uniform microspheres. The key point to guarantee the uniformity of droplets is that the monomer should have enough hydrophobicity and not spread up and wet the membrane pore. If the monomer is mild hydrophilic, the effective solution is to add hydrophobic inert component in monomer phase or combination of seed swelling approach with MET of hydrophobic inert component. For performed polymer system, the functional polymer, biodegradable polymers or natural polymers, etc. all could be dissolved in volatile solvent and then emulsified into uniform O/W droplets by MET, and finally the particles can be obtained by evaporation of organic solvent in oil droplet. Due to the unique size-uniformity of microspheres, the morphology evolution behaviors of microspheres present more regularity than the polydispersed particles, which was demonstrated to depend on the crosslinker, the inert diluents, the emulsifier, the stabilizer, and the cosurfactant by significantly manipulating the phase separation during formation of microspheres. In recent decades, a growing number of functional microspheres with high application value are expected to be size-uniformed by MET. With the development of MET technique and equipment, the uniformity of hydrophobic microspheres become easier to realize and scale up for industry. However, the challenges still exist in: (i) preparation of smaller microspheres and microcapsules, such as tens of nanometers; (ii) expanding MET to the broader polymerization systems, such as daily chemicals, heat storage microcapsules, or electronic information microcapsules; and (iii) establishing the more theoretical guidance and reducing the trial-and-error studies.

References

1 Lambrich, U. and Schubert, H. (2005). Emulsification using microporous systems. *Journal of Membrane Science* 257: 76–84.

2 Omi, S. and Ma, G.H. (2000). Membrane emulsification a versatile tool for the synthesis of polymeric microspheres. *Macromolecular Symposia* 151: 319–330.

3 Hao, D.X., Ma, G.H., and Su, Z.G. (2008). Porogen effects in synthesis of uniform micrometer-sized poly(divinylbenzene) microspheres with high surface areas. *Journal of Colloid and Interface Science* 323: 52–59.

4 Ma, G.H., Nagai, M., and Omi, S. (1999). Study on preparation and morphology of uniform artificial polystyrene–poly(methyl methacrylate) composite microspheres by employing the SPG (shirasu porous glass) membrane emulsification technique. *Journal of Colloid and Interface Science* 214: 264–282.

5 Ma, G.H., Chen, A.Y., Su, Z.G., and Omi, S. (2003). Preparation of uniform hollow polystyrene particles with large voids by a glass membrane emulsification technique and a subsequent suspension polymerization. *Journal of Applied Polymer Science* 87: 244–251.

6 Ma, G.H. (2002). Mechanism of formation of monodisperse polystyrene hollow particles prepared by membrane emulsification technique. effect of hexadecane amount on the formation of hollow particles. *Macromolecular Symposia* 179: 223–240.

7 Nuisin, R., Ma, G.H., Omi, S., and Kiatkamjornnwong, S. (2000). Dependence of morphological changes of polymer particles on hydrophobic/hydrophilic additives. *Journal of Applied Polymer Science* 77: 1013–1028.

8 Omi, S., Matsuda, A., Imamura, K. et al. (1999). Synthesis of monodisperse polymeric microspheres including polyimide prepolymer by using SPG emulsification technique. *Colloids and Surfaces A: Physicochemical and Engineering Aspects* 153: 373–381.

9 Ma, G.H., Su, Z.G., Omi, S. et al. (2003). Microencapsulation of oil with poly (styrene-*N,N*-dimethylaminoethyl methacrylate) by SPG emulsification technique: effects of conversion and composition of oil phase. *Journal of Colloid and Interface Science* 266: 282–294.

10 Wang, R.W., Zhang, Y., Ma, G.H., and Su, Z.G. (2006). Preparation of uniform poly(glycidyl methacrylate) porous microspheres by membrane emulsification polymerization technology. *Journal of Applied Polymer Science* 102: 5018–5027.

11 Ma, G.H., Nagai, M., and Omi, S. (1999). Preparation of uniform poly(lactide) microspheres by employing the Shirasu porous glass emulsification technique. *Colloids and Surfaces A: Physicochemical and Engineering Aspects* 153: 383–394.

12 Yuyama, H., Yamamoto, K., Shirafuji, K. et al. (2000). Preparation of polyurethaneurea (PUU) uniform spheres by SPG membrane emulsification technique. *Journal of Applied Polymer Science* 77: 2237–2245.

13 Yang, J., Hao, D.X., Bi, C.X. et al. (2010). Rapid synthesis of uniform magnetic microspheres by combing premix membrane emulsification and in situ formation techniques. *Industrial and Engineering Chemistry Research* 49: 6047–6053.

14 Hao, D.X., Gong, F.L., Hu, G.H. et al. (2009). The relationship between heterogeneous structures and phase separation in synthesis of uniform microns PolyDVB microspheres. *Polymer* 50: 3188–3195.

15 Omi, S., Katami, K., Yamamoto, A., and Iso, M. (1994). Synthesis of polymeric microspheres employing SPG emulsification technique. *Journal of Applied Polymer Science* 51: 1–11.

16 Ma, G.H. (2003). Control of polymer particle size using porous glass membrane emulsification. *China particuology* 1: 105–114.

17 Ma, G.H., Nagai, M., and Omi, S. (2001). Study on preparation of monodispersed poly(styrene-*co-N,N*-dimethylaminoethylmeth-acrylate) composite microspheres by SPG emulsification technique. *Journal of Applied Polymer Science* 79: 2408–2424.

18 Ma, G.H., Omi, S., Dimonie, V.L. et al. (2002). Study on the preparation and mechanism of formation of hollow monodisperse polystyrene microspheres by SPG emulsification technique. *Journal of Applied Polymer Science* 85: 1530–1543.

19 Ma, G.H., Nagai, M., and Omi, S. (1997). Synthesis of uniform microspheres with higher content of 2-hydroxyethyl methacrylate by employing SPG emulsification technique followed by swelling process of droplets. *Journal of Applied Polymer Science* 66: 1325–1341.

20 Qu, H.H., Gong, F.L., Wang, J. et al. (2008). Strategies for the design of hydrophilic microspheres with controlled size and microstructure. *Industrial and Engineering Chemistry Research* 47: 6399–6404.

21 Qu, H.H., Gong, F.L., Ma, G.H., and Su, Z.G. (Journal of Applied Polymer Science, 2007). Preparation and characterization of large porous poly(HEMA-*co*-EDMA) microspheres with narrow size distribution by modified membrane emulsification method. 105: 1632–1641.

22 Omi, S., Katami, K., Taguchi, T. et al. (1995). Synthesis of uniform PMMA microspheres employing modified SPG (Shirasu porous glass) emulsification technique. *Journal of Applied Polymer Science* 57: 1013–1024.

23 Alroaithi, M. and Sajjadi, S. (2016). Uniform polymer beads by membrane emulsification-assisted suspension polymerisation. *RSC Advances* 6: 79745–79754.

24 Ma, G.H., Fujiwara, J., Su, Z.G., and Omi, S. (2003). Synthesis and characterization of crosslinked uniform polymeric microspheres containing a polyimide prepolymer by a new emulsification process. *Journal of Polymer Science Part A: Polymer Chemistry* s: 2588–2598.

25 Supsakulchai, A., Ma, G.H., Nagai, M., and Omi, S. (2002). Uniform titanium dioxide (TiO_2) microcapsules prepared by glass membrane emulsification with subsequent solvent evaporation. *Journal of Microencapsulation* 19: 425–449.

26 Ma, G.H., Nagai, M., and Omi, S. (1999). Effect of lauryl alcohol on morphology of uniform polystyrene-poly(methyl methacrylate) composite microspheres prepared by porous glass membrane emulsification technique. *Journal of Colloid and Interface Science* 219: 110–128.

27 Ma, G.H., Gong, F.L., Hu, G.H. et al. (2005). Multi-scale structures in emulsion and microsphere complex systems. *China Particuology* 3 (6): 296–303.

28 Ma, G.H., An, C.J., Yuyama, H. et al. (2003). Synthesis and characterization of polyurethaneurea-vinyl polymer (puu–vp) uniform hybrid microspheres by SPG emulsification technique and subsequent suspension polymerization. *Journal of applied polymer science* 89: 163–178.

29 Omi, S., Kanetaka, A., Shimamori, Y. et al. (2001). Magnetite (Fe_3O_4) microcapsules prepared using a glass membrane and solvent removal. *Journal of Microencapsulation* 18: 749–765.

30 Sun, G.Q., Qi, F., Wu, J. et al. (2014). Preparation of uniform particle-stabilized emulsions using SPG membrane emulsification. *Langmuir* 30: 7052–7056.

31 Kara, S., Okay, O., and Pekcan, Ö. (2000). In situ photon transmission technique for monitoring phase separation in real time during gelation. *Polymer Bulletin* 45: 281–285.

32 Serbutoviez, C., Kloosterboer, J.G., Boots, H.M.J., and Touwslager, F.J. (1996). Polymerization-induced phase separation. 2. morphology of polymer-dispersed liquid crystal thin films. *Macromolecules* 29: 7690–7698.

33 Okubo, M., Konishi, Y., Inohara, T. et al. (2003). Size effect of monomer droplets on the production of hollow polymer particles by suspension polymerization. *Colloid & Polymer Science* 281: 302–307.

34 Wang, L., Zhao, J., and Charles, C.H. (2008). Phase separation of polystyrene-*b*-(ethylene-*co*-butylene)-*b*-styrene (SEBS) deposited on polystyrene thin films. *Polymer* 49: 2153–2159.

35 L'Abee, R., Goossens, H., and Duin, M.V. (2008). Thermoplastic vulcanizates obtained by reaction-induced phase separation: Interplay between phase separation dynamics, final morphology and mechanical properties. *Polymer* 49: 2288–2297.

36 Chen, Y.C., Dimonie, Y.L., and El-Aasser, M.S. (1991). Interfacial phenomena controlling particle morphology of composite latexes. *Journal of Applied Polymer Science* 42: 1049–1063.

37 Sundberg, D.C., Casassa, A.P., Pantazopoulos, J. et al. (1990). Morphology development of polymeric microparticles in aqueous dispersions I. thermodynamic considerations. *Journal of Applied Polymer Science* 41: 1425–1442.

38 Sundberg, E.J. and Sundberg, D.C. (1993). Morphology development for three-component emulsion polymers: theory and experiments. *Journal of Applied Polymer Science* 47: 1277–1294.

39 Genwen, Z. and Suzanne, K.M. (1998). Correlation of mean drop size and minimum drop size with the turbulence energy dissipation and the flow in an agitated tank. *Chemical Engineering Science* 53: 2063–2079.

3

Preparation of Hydrophilic Polymer Microspheres from W/O Emulsion

3.1 Introduction

In the past two decades, membrane emulsification has attracted much attention as a novel technology for preparing monodisperse emulsions and microspheres. Compared with other traditional emulsification techniques, membrane emulsification has the advantages of low energy consumption, mild preparation conditions, narrow size distribution of droplets, simple operation conditions, and easy amplification. In recent years, membrane emulsification has been widely applied in many fields, such as cosmetics industry, food industry, pharmaceutical carrier, and separation medium production. In the previous chapter, the development of preparing hydrophobic microspheres by membrane emulsification has been introduced. In this chapter, we will focus on the study of preparing hydrophilic microspheres by direct membrane emulsification.

To prepare hydrophobic microspheres, hydrophilic membranes, such as Shirasu porous glass (SPG) membrane, are usually adopted. SPG membrane is composed of Al_2O_3–SiO_2 and has a large amount of silanol groups on its surface, which has a high degree of hydrophilicity and is suitable for preparing O/W or W/O/W emulsion droplets [1]. However, the necessary condition for preparing uniform-sized emulsion droplets by membrane emulsification is that the interfacial tension between the dispersed phase and the membrane surface is high enough to avoid the spread of dispersed phase on the membrane surface. If SPG membrane is directly used for preparing W/O emulsions or hydrophilic microspheres, due to the wettability of the membrane, hydrophilic substances will easily wet its pore wall and spread on the membrane surface, generating a jetlike stream that leads to nonuniform droplets [2]. In addition, in the oil phase with low dielectric constant, it is difficult for water droplets to be stabilized by the electric repulsion force, which is also the cause of wide size range of droplets [3, 4]. In order to prepare uniform-sized W/O emulsions, Kandori et al. investigated the effect of surfactant [4]. They assumed that by using proper emulsifier, which would form a viscoelastic adsorbed film at the droplet interface and prevent coalescence, stable and monodispersed W/O emulsions could be prepared successfully. They adopted a copolymer surfactant PE-64, which molecular structure is HO $(C_2H_5O)_n$-$(C_3H_7O)_m$-$(C_2H_5O)_n$-H, where the average

Novel Membrane Emulsification: Principles, Preparation, Processes, and Bioapplications,
First Edition. Guanghui Ma.

numbers of *m* and *n* are 30 and 13, respectively. When the surfactant concentration was above 7.5 wt%, the prepared W/O emulsions had excellent dispersion stability. The enhanced stability was attributed to the low interfacial tension and the formation of a viscoelastic adsorbed film of copolymer surfactants on the water droplets. However, the choice scope of surfactants is limited for some applications. And the effect of surfactants on the stability of emulsion is also affected by the type of aqueous and oil phases. Thus, the more common method to prepare monodispersed W/O emulsions was using the hydrophobically modified SPG membranes or hydrophobic polymer membranes directly [5]. Some typical methods are introduced in this chapter.

3.2 Membrane Modification and Preparation

3.2.1 Hydrophobic Modification of the Membrane

The basic principle of hydrophobic modification is using the silicane coupling agent with long hydrophobic carbon chain to react with Si–OH group (Figure 3.1). The modification effect is evaluated by testing the contact angle (CA) between the membrane surface and water. The larger the CA, the better the hydrophobicity of the membrane, indicating that the membrane is well modified.

In the relevant works of litreture, water-soluble silicane coupling agent KP-18C [6], silicone resin polymethylsilsesquioxane (GRT-350) [3, 5], oil-soluble silicane coupling agents – octyltriethoxysilane (ALD) [3] and octadecyltrichlorosilane (OTC) [7] were used as hydrophobic modification reagents. KP-18C is a silane coupling agent with C_{18} hydrophobic chain, which can react with the Si–OH group on the glass membrane [6]. The modification reaction is carried out in the water phase and the process is as follows: the glass membrane was heated in a muffle furnace at 500 °C for 2 hours to remove organic substance, cooled to room temperature, sonicated in deionized water for 0.5 hour, sonicated in a KP-18C solution (−0.07 MPa) for 2 hours, and finally heated again at 120–140 °C for 4 hours to carry out the reaction. In order to make the Si–OH around the membrane pores participate in the reaction effectively, the modification is carried out under reduced pressure and ultrasonic conditions. We used KP-18C of the concentrations 1% (v), 3% (v), and 5% (v) as the modifier, and the modification effect is shown in Figure 3.2 [8]. When the KP-18C concentration was 3% (v), the degree of reaction with the membrane was maximized, and the modification

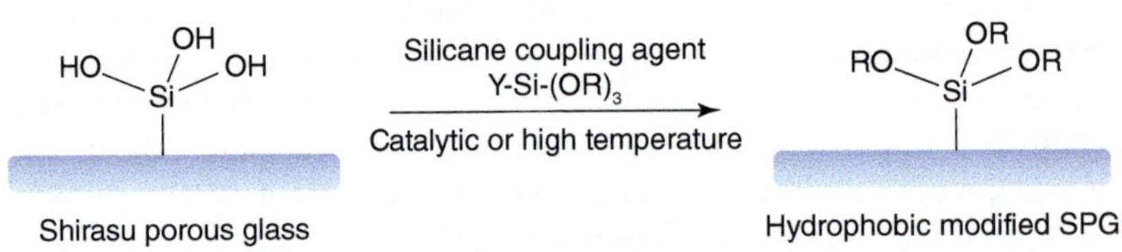

Figure 3.1 Schematic illustration of the principle of hydrophobic modification of SPG membrane.

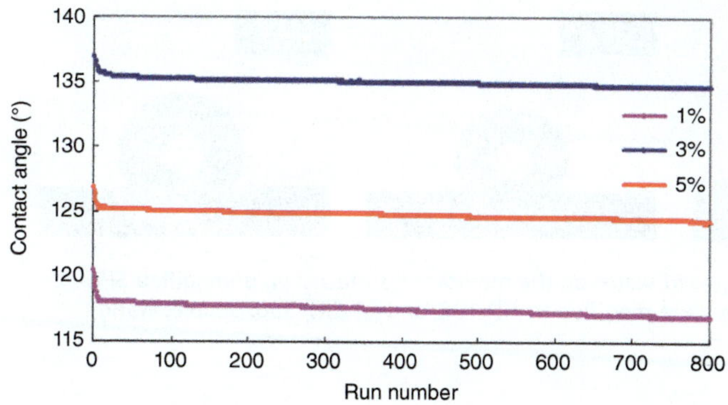

Figure 3.2 The effect of KP-18C concentration on hydrophobicity of SPG membrane after modification. Source: Adapted from Wang [8].

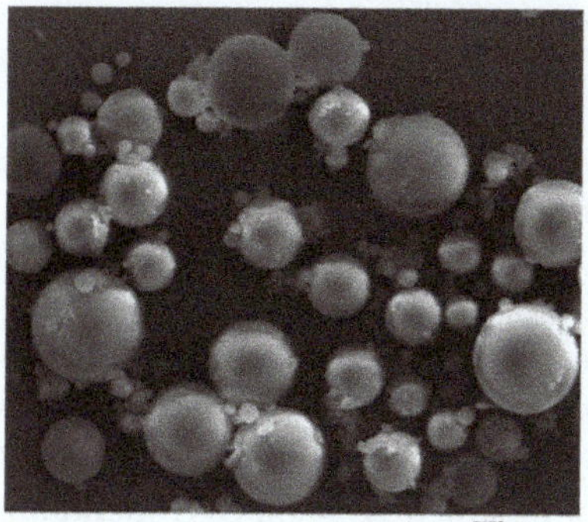

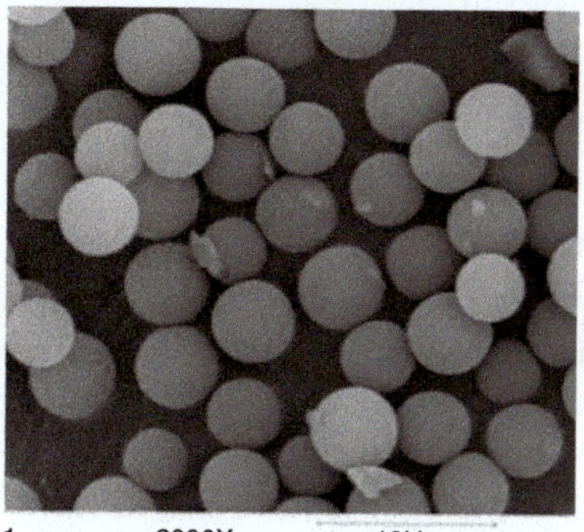

Figure 3.3 Scanning electron microscopy (SEM) photographs of chitosan microspheres prepared by using SPG membrane before (a) and after (b) modification. Source: Wang et al. [6]/Elsevier.

effect was the best. The CA between the membrane surface and water reached 130° or more. And when the concentration of the modifier was increased, the modifier tended to gather and weaken the reaction with the membrane. Figure 3.3 shows the results of preparing chitosan microspheres using the unmodified and modified SPG membrane. The chitosan microspheres prepared with unmodified membranes had a wide particle size distribution. Due to the uneven particle size of the microspheres, many small microspheres adhered to the large microspheres, resulting in the wide size distribution of the microspheres. On the other hand, the size distribution of the microspheres prepared by the modified membrane was very narrow, and the coefficient of variation (CV) value of size distribution was less than 11%, which means the uniform size and good dispersibility of microspheres.

When GRT-350 is used as the hydrophobic modifier, the modification steps are roughly the same as the above KP-18C modification steps, only change the immersed solution from KP-18C aqueous solution to an ethanol solution containing

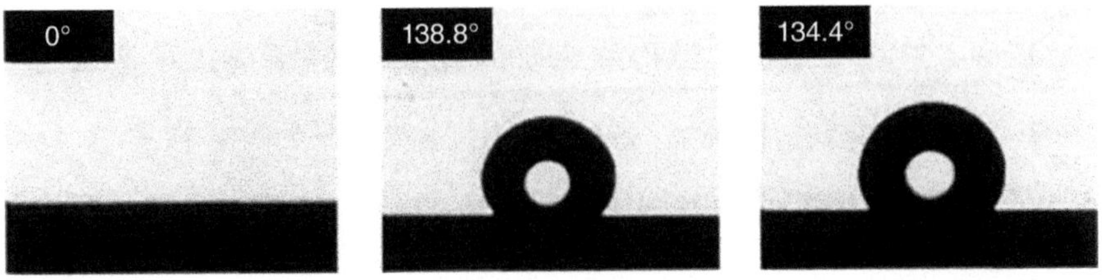

Figure 3.4 Contact angles of water on the membrane surfaces: (a) unmodified SPG membrane, (b) SPG membrane modified by KP-18C, and (c) GRT-350. Source: Wang [8]/Elsevier.

Table 3.1 The effect of the modified layer on the pore size and porosity of SPG membranes.

	Mean pore size (μm)	Porosity (%)
SPG membrane	2.15	61.1
SPG membrane modified by KP-18C	2.09	59.2
SPG membrane modified by GRT-350	2.00	58.5

Source: Adapted from Wang [8].

GRT-350 under reduced pressure and ultrasonic treatment for 24 hours. Our group compared the hydrophobic modification effect of KP-18C and GRT-350 [8]. The difference in the CAs between modified SPG membrane and water is very small (Figure 3.4). The distribution and influence of modified layer on SPG membrane porosity and pore size after modification are shown in Table 3.1. The modified layers formed on the surface of the SPG membranes by KP-18C and GRT-350 reduced the average pore size by 0.06 and 0.15 μm, and the corresponding porosity decreased by 1.9% and 2.6%, respectively. We also compared the changes in CA after the repeated use for emulsification. The result shows that the CA of the SPG membrane modified by GRT-350 decreased sharply after four times of use, while the membrane modified by KP-18C was used repeatedly five times and kept stable.

Compared with the water-soluble KP-18C, ALD and OTC are oil-soluble silicane coupling reagents, which are less used. The modification reaction is carried out in the organic phase of toluene [3] or the mixture of cetane and chloroform [7]. Cheng et al. compared the hydrophobic modification effect of ALD and GRT-350 [3]. SPG membranes treated with either modification reagents possessed good hydrophobicity, and the average CAs are about 120° (Figure 3.5). The molecular structures of the two modification reagents are different. The molecular structure of ALD is $C_8H_{17}Si(OC_2H_5)_3$ and the molecular structure of GRT-350 is $(CH_3SiO_{1.5})_n$, with n is ranging from 300 to 400. During the modification process, only a linear chain of hydrophobic group (C_8H_{17}–) was introduced on the membrane surface for an ALD molecule. When using GRT-350, more than 300 hydrophobic groups were introduced on the SPG membrane surface for single molecule, which is attributed to the intricate spatial network structure of GRT-350. Thus, the required modification

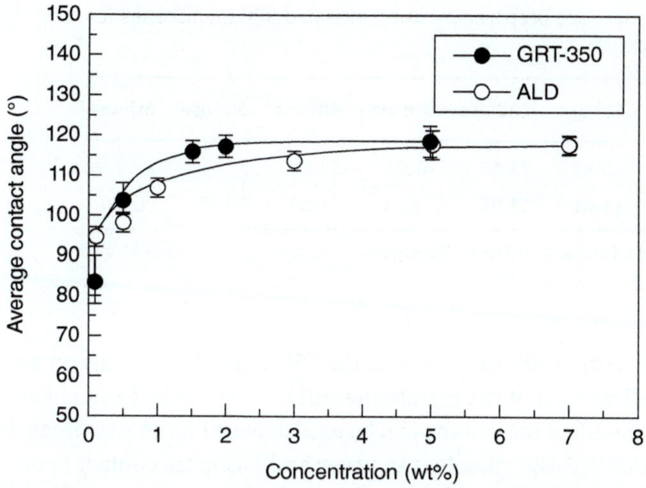

Figure 3.5 Effect of the types and concentrations of the hydrophobic modification reagents on the average contact angles of the SPG membranes surface. Source: Cheng et al. [3]/with permission of Elsevier.

concentration of GRT-350 for obtaining the maximal average CA is lower (2 wt%) than which of ALD (5 wt%) as shown in Figure 3.5. Moreover, the modification method of GRT-350 was simpler and cheaper compared with that using ALD.

The difference between KP-18C- and OTC-modified membranes was also investigated by our group [7]. It can be seen in Figure 3.6 that the CA between the membrane modified by KP-18C and water is larger than that of OTC, which means that using KP-18C can achieve better modification effect. Moreover, as listed in Table 3.2, the CV value of the prepared emulsion droplets showed that when the membrane modified by OTC was used more than three times, the CV value increases significantly up to 60%, indicating that the hydrophobic property of the membrane was poor. When KP-18C was used as the modifier, the size distribution of the emulsion was still very narrow even after five times repeated emulsification, and the CV

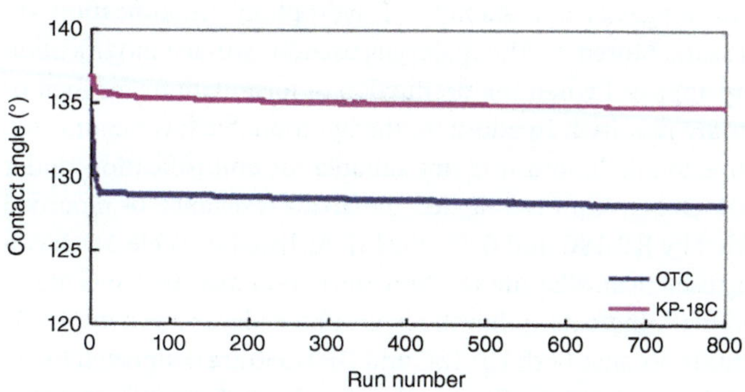

Figure 3.6 The effect type of modifiers on hydrophobicity of SPG membrane after modification. Source: Mi [7]/University of Chinese Academy of Sceinces.

Table 3.2 CV values of emulsions prepared by using modified SPG membranes for several repeated emulsification.

Silicane coupling reagent	1st use	2nd use	3rd use	4th use	5th use	6th use
OTC	15.83	23.43	49.25	73.97	/	/
KP-18C	11.61	13.03	15.23	16.29	16.87	30.51

Source: Mi [7]/University of Chinese Academy of Sceinces.

value was less than 17%. After sixth times of use, the CV value of the emulsion was still lower than 30%, indicating that the membrane still has certain hydrophobicity. Therefore, the KP-18C-modified membrane can be used at least four to five times. It was probably because that KP-18C dissolves in water and has more contact opportunities with the membrane surface, which resulted in the higher degree of modification reaction, and the higher modification density of hydrophobic link to the membrane surface than that of OTC.

Other hydrophobic organosilane compounds have also been reported to modify SPG membranes. Kukuzaki et al. adopted an organosilane compound, 3,3,4,4,5,5,6,6,6-nonafluorohexyltrichlorosilane to modify SPG membranes and investigated the modification effect on the ozone mass transfer [9]. The SPG membranes were modified in the toluene containing 1 wt% of the organosilane compound at 105 °C for 5 hours. Then the membranes were rinsed and dried for use. The measured CAs of the original membranes were between 22° and 27°, while those of the surface-modified membranes were larger than 100°. The change indicates that the SPG membranes were successfully modified.

In summary, the SPG membrane modified by silicane coupling reagents can be used repeatedly, and the corresponding silicane modifier can be selected according to the different modification conditions (oil or water phase). However, the hydrophobically modified methods also have several shortcomings, such as uneven modification, easy to be detached during use, and need to be modified again after several uses. The problems with longevity and stability of hydrophobic modification are crucial at an industrial scale. Moreover, the materials used for surface modification are subject to regulatory approval when the production of formulations for food or medical-grade products are required. In addition, the hydrophobic layer connected by Si–O bond is sensitive to alkali, and it is not suitable for emulsification under alkaline conditions [10]. Our group investigated the alkali resistance of modified SPG membranes modified by KP-18C and GRT-350 [8]. As listed in Table 3.3, it can be found that after being used in an alkaline solution, the CAs of modified SPG membrane are both changed to 0°, indicating that the hydrophobic layer has completely fallen off during use. This is because both KP-18C and GRT-350 are connected to the surface of the SPG membrane through silicon–ether bonds, and the silicon–ether bonds are easily hydrolyzed under alkaline conditions, so the modified membranes are not suitable for alkaline systems.

Table 3.3 The contact angles between water and the surface of the modified SPG membranes before and after used in alkali systems.

	Contact angle	
Sample	**Before used in alkali**	**After used in alkali**
SPG membrane modified by KP-18C	142.2°	0°
SPG membrane modified by GRT-350	135.5°	0°

Source: Adapted from Wang [8].

3.2.2 Preparation of Hydrophobic Membrane

To overcome the shortcomings of membrane modification, many studies have proposed to adopt hydrophobic polymer to prepare porous membranes, such as polytetrafluoroethylene (PTFE), poly(vinylidene fluoride) (PVDF), and polypropylene (PP) [10–14]. Among these polymers, PTFE represents an ideal raw material for manufacturing hydrophobic membranes since it exhibits strong hydrophobicity, stable chemical resistance, and great thermal stability. Huang et al. fabricated a PTFE membrane by sintering electrospun PTFE/poly(vinyl alcohol) (PVA) fiber [15]. The prepared membrane possessed high hydrophobicity with the CA from 138.2 to 147.6° and the continuous fibrous network pore structure as shown in Figure 3.7.

However, this pore structure is not suitable to form uniform emulsion droplets in membrane emulsification process. In order to solve this problem, the group optimized the pore structure by introducing poly(tetrafluoroethylene-hexafluoropropylene) (FEP) particles into the PTFE/PVA precursor nanofiber membrane, as shown in Figure 3.8 [16]. The fibrous network of membrane was transformed to circular pore structure owing to the melting of FEP, which has lower melt temperature and viscosity than PTFE.

The optimized PTFE membrane with uniform circular pore structure was utilized to prepare W/O emulsion. From the optical microscope pictures of prepared W/O emulsions (Figure 3.9a), it can be seen that the emulsion has better uniformity than the one by mechanical stirring method (Figure 3.9b). Moreover, membrane emulsification process needs lower energy consumption than mechanical stirring.

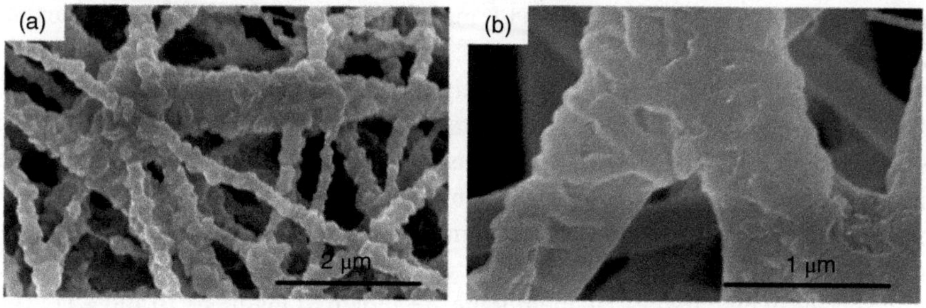

Figure 3.7 Morphologies of PTFE membrane prepared by sintering electrospun (a) 20000 × surface and (b) 40000 × surface. Source: Huang et al. [15]/Elsevier.

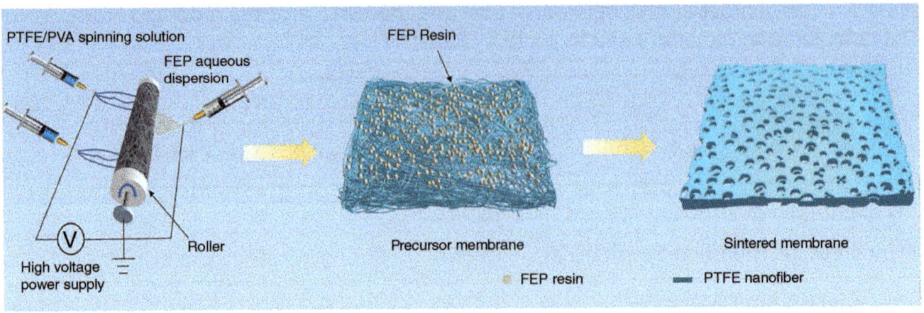

Figure 3.8 Schematic diagram of PTFE nanofiber membrane preparation process. Source: Yu et al. [16]/with permission of Elsevier.

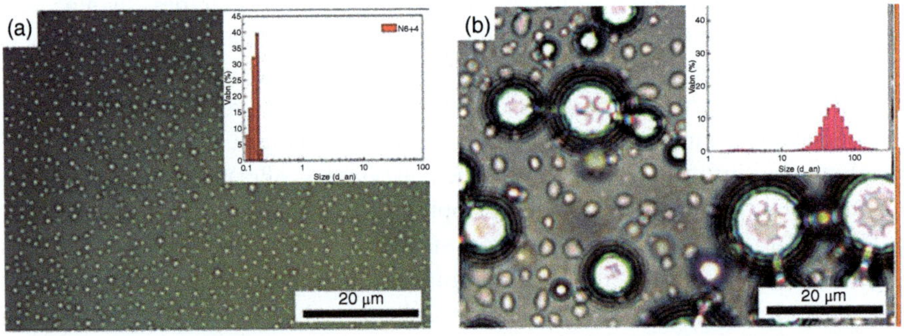

Figure 3.9 The optical microscope pictures and droplet size distributions of W/O emulsion prepared by (a) membrane emulsification process and (b) mechanical stirring method. Source: Yu et al. [16]/Elsevier.

Therefore, the membrane emulsification process showed much higher preparation efficiency.

However, PTFE is difficult to fabricate to specific shape, and commercial PTFE membranes are usually produced through complicate extrusion, rolling, and stretching procedures. This process is not only tedious but also leads to membranes with poor performance stability owing to the creeping deformation of PTFE. There are only few reports about fabricating PTFE membranes by other methods and further study is needed.

PVDF is another hydrophobic polymer used to fabricate porous membranes. Bottino et al. prepared porous PVDF membranes by using a porous glass fiber braid as pore-forming agent [12]. The braid was immersed in polymer solution after a proper period, extracted from the solution, and then immersed in running water to precipitate the polymer and form the membrane. After rinsing, the membrane was kept overnight in demineralized water and dried at 40 °C. The cross-section and outer surface of prepared PVDF membrane are shown in Figures 3.10 and 3.11. The cross-section exhibits a lot of large voids beneath the outer surface layer, and the surface of membrane is smooth with many pores.

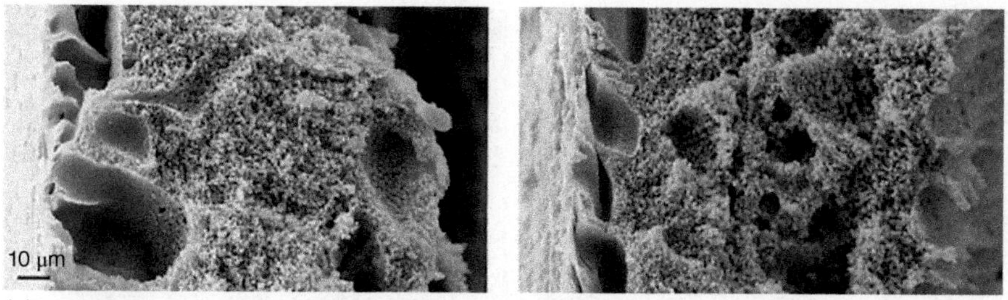

(a) (b)

Figure 3.10 SEM micrographs of the cross-section of PVDF membranes (concentration of the casting solution 15 wt%. (a) Braid immersion time: 20 seconds and Braid extraction speed: 1.2 cm s^{-1}. (b) Braid immersion time: 60 seconds and Braid extraction speed: 1.7 cm s^{-1}). Source: Bottino et al. [12]/Elsevier.

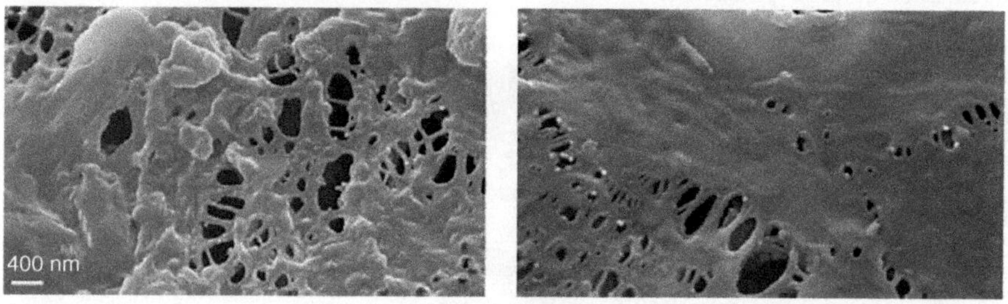

(a) (b)

Figure 3.11 SEM micrographs of the outer surface of PVDF membranes (concentration of the casting solution 15 wt%. (a) Braid immersion time: 20 seconds and Braid extraction speed: 1.2 cm s^{-1}. (b) Braid immersion time: 60 seconds and Braid extraction speed: 1.7 cm s^{-1}). Source: Bottino et al. [12]/Elsevier.

Hydrophobic polypropylene membranes are usually prepared by the molten extrusion technique followed by stretching or thermal phase separation process. Vladisavljević et al. adopted microporous polypropylene hollow membranes to prepare W/O emulsions [17–19]. The obtained W/O emulsions possessed a narrow droplet size distribution and a dispersed phase content up to 25 vol%. Under the optimized conditions, the mean diameter of water droplets ranged between 0.26 and 0.32 μm, which is about one-third below the mean pore size of used membrane.

Polyethylene (PE) is another hydrophobic polymer, which is a thermoplastic resin made by polymerizing ethylene and has good hydrophobicity [14]. Due to its hydrophobic properties, it was prepared into the membrane and applied to the preparation of W/O emulsions by direct or premix membrane emulsification. PE membrane was formed as follows: powder of PE was molded after sieving and then the membrane was formed under sintering at high temperature; the pores of the surface and inside have the same irregular shape (Figure 3.12). Our group firstly prepared agarose W/O emulsion by premix membrane emulsification technique using PE membrane [13]. However, when it was used for direct membrane emulsification, the droplets of W/O emulsion were very large and polydisperse (Figure 3.13). The reason is that the balance between the formation and breaking-off of droplets

(a) (b)

Figure 3.12 SEM images of (a) surface structure and (b) inner structure of PE membrane. Source: Zhou et al. [14]/Elsevier.

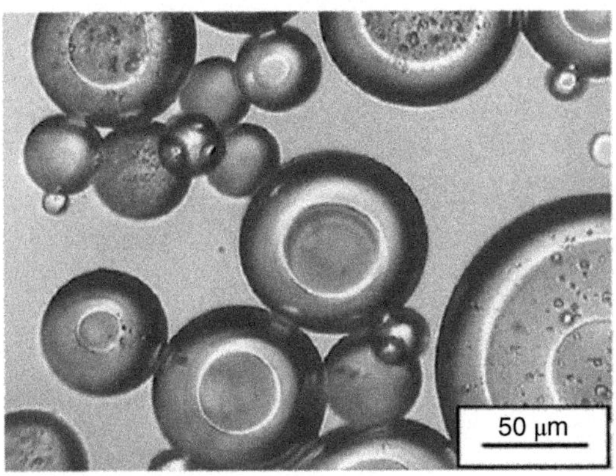

Figure 3.13 Optical micrographs of W/O emulsion prepared by direct membrane emulsification using PE membrane. Source: Zhou et al. [14]/Elsevier.

from the surface relied on the shape of surface pores structure and the time of droplet formation on the surface relied on the balance time. Thus, the irregular pores shape of PE membrane greatly broadened the formed droplets' size.

Due to the uneven distribution of pore size of PE membrane, it is difficult to use PE membrane in direct membrane emulsification. Therefore, some studies have proposed the epoxy-based polymer (EP) membrane, which has important applications in many fields due to its good properties, such as low density, good thermal and electric insulation, high specific surface area, and special pore structure. The feature of special pore structure is taken advantage of for producing W/O emulsion [13, 20, 21]. Our group explored the effect of the solvents, ratio of epoxy group and amine, and curing temperature on the reaction dynamics and final monoliths morphology. The specific preparation process is as follows: diglycidyl ether of bisphenol A (DGEBA) is used as the reactive monomer, 4,4′-diaminodicyclohexylmethane (DDCM) is used as the cross-linking agent, polyethylene glycol 200 (PEG 200) and *N*,*N*-Dimethylformamide (DMF) are used as poor solvent and good solvent, respectively. Mix a certain proportion of DGEBA and PEG 200 (or a certain proportion of PEG 200 and DMF mixture) evenly. Then add the cross-linking agent, mix again,

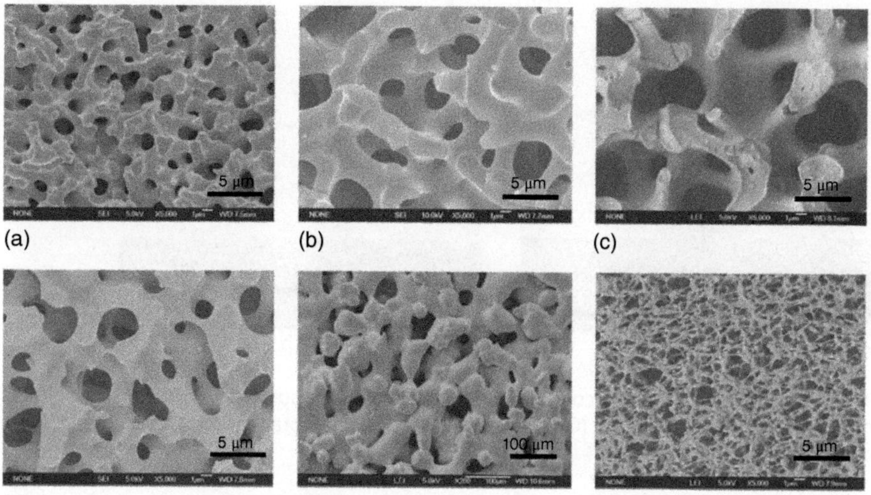

(a) (b) (c) (d) (e) (f)

Figure 3.14 SEM photographs of membranes: (a) 1.22 μm EP membrane, (b) 2.77 μm EP membrane, (c) 4.67 μm EP membrane, (d) 2.39 μm SPG membrane, (e) 25.60 μm PE membrane, and (f) 0.22 μm PTFE membrane. Source: Mi et al. [21]/Elsevier.

and pour the blend solution into a closed mold and cure at 80 °C for 24 hours. After curing, the porogen in the material is removed by ethanol extraction. Place the sample under −70 °C for 2 hours, and after it is completely frozen, place it in a freeze dryer for 24 hours to obtain a porous epoxy resin material. After optimization, EP materials with three-dimensional bicontinuous structure and pore size range of 1.22–25.6 μm are successfully prepared (Figure 3.14). The prepared EP membranes have narrow pore size distribution (CV = 19.9%), high porosity (57.91%, v/v), and three-dimensional bicontinuous structure, which is similar to that of SPG membrane. This special structure is conducive to the formation and detachment of droplets on the membrane surface during the emulsification process. Compared with SPG membrane, EP membranes also have strong alkali resistance (tolerance to 1 M NaOH) and hydrophobic surface (contact angle CA = 130°) (Figure 3.15), which can be directly used in W/O emulsion preparation without modification.

After the successful fabrication of EP membrane, we used it to prepare uniform agarose microspheres [21]. Compared with unmodified SPG membrane, silicane coupling agent modified SPG membrane and EP membrane, agarose microspheres prepared by EP membrane had better dispersion and uniform size distribution. The CV value (11.8%) was better than that prepared by unmodified SPG membrane (56.3%) and modified SPG membrane (12.2%) (Figure 3.16). This indicates that EP membrane can be directly used to prepare uniform hydrophilic microspheres, avoiding the problems encountered in membrane modification.

In addition to the modified SPG membrane and hydrophobic polymer membrane, which are mentioned above, there are also a few studies about preparing W/O emulsions using hydrophobic modified metal membranes, such as stainless steel, anodic porous alumina, and nickel membranes [22–25]. Yanagishita et al. prepared several

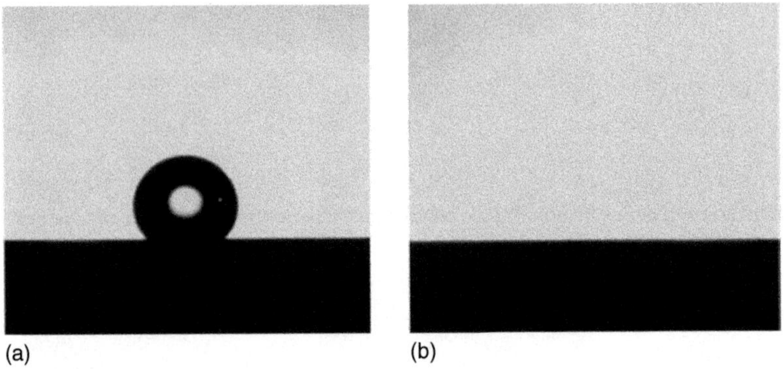

Figure 3.15 Photographs of contact angles of the water droplets on the surface of membranes: (a) EP membrane (CA = 130°) and (b) SPG membrane (CA = 0°). Source: Mi et al. [21]/Elsevier.

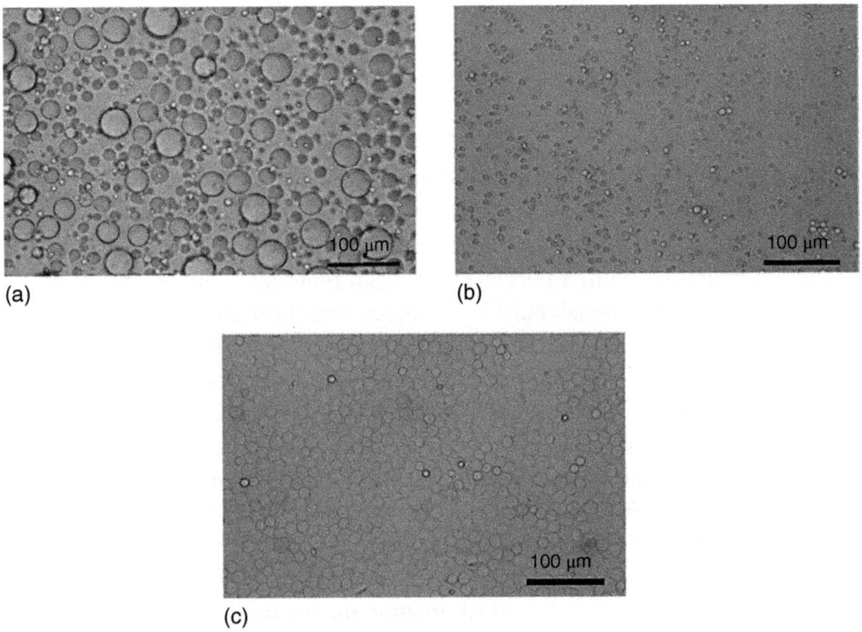

Figure 3.16 Optical micrographs of agarose microspheres prepared by (a) unmodified SPG membrane, (b) hydrophobically modified SPG membrane, and (c) EP membrane. Source: Mi et al. [21]/Elsevier.

metal oxide nanoparticles by using anodic porous alumina membrane [22, 23]. The anodic porous alumina membrane is formed by anodization of Al in acidic solution and possessed highly ordered hole array structure composed of straight cylindrical holes. This alumina membrane was treated with octadecyltrichlorosilane to give the surface hydrophobicity and used to prepare various types of metal oxide nanoparticles with narrow size distribution. The sol solutions containing primary nanoparticles of metal oxides were used as the dispersed phase, and the continuous phase was kerosene containing 3 wt% Span 80. Then the formed emulsion droplets were heated at 70 °C for one hour to be solidified.

The metal membranes take the advantage of high mechanical resistance, good chemical resistance, easy to clean, and narrow pore size distribution. But they also have some disadvantages, such as complex fabrication processes and less variety. And the metal membranes also need to be hydrophobically modified. For example, to obtain hydrophobized nickel membrane, the membrane needs to be placed in an undecane solution of 0.01M 1H, 1H, 2H, 2H-Perfluorododecyltriethosysilane for a minimum of 4 hours at 45 °C [24]. The examples of some hydrophobically modified membranes and hydrophobic membranes are given in Table 3.4.

3.3 Preparation Microparticles from Monomer System

After hydrophobic modification or preparation of hydrophobic membranes, researchers have carried out the preparation of various hydrophilic particles, such as particles polymerized from monomers and from natural polymers. Hydrophilic particles have important application value in many fields, but their size control is difficult. And some researchers adopted membrane emulsification technique to achieve successful size control.

3.3.1 Preparation of Poly(*N*-isopropylacrylamide) (PNIPAM) Microspheres and Microcapsules

Poly(*N*-isopropylacrylamide) (PNIPAM) is a well-known thermosensitive polymer with a sharp volume transition, which occurs near the low critical solution temperature (LCST) [26]. When the ambient temperature is lower than the LCST, PNIPAM will absorb a large amount of water and present a swollen and hydrophilic state. Above the LCST, the backbone of PNIPAM folds onto itself and shields the hydrophobic groups from water, then free water is discharged from the polymer network and the polymer shows sharp volume shrinkage and becomes hydrophobic (Figure 3.17) [27]. Due to this remarkable property of PNIPAM, it has many applications in biomedicine and biotechnology, including thermosensitive controlled drug delivery systems [28], artificial organs [29], and enzyme and protein immobilized carriers [30].

A simple two-step method has been used to prepare monodisperse PNIPAM microspheres and hollow microcapsules by SPG membrane emulsification and UV-induced free radical polymerization as shown in Figure 3.18 [31, 32]. First, monodisperse monomer-contained W/O emulsions were prepared by direct membrane emulsification. Then, polymerization of monomer dissolved in the aqueous droplets was initiated by initiator ammonium persulfate (APS) dissolved in the disperse phase to obtain microspheres or by initiator 2,2-dimethoxy-2-phenylaceto-phenone (BDK) dissolved in the continuous phase to form hollow microcapsules under the UV irradiation at 20 °C. The size of PNIPAM microspheres and microcapsules are respectively, 16.64 and 15.83 μm, and both of

Table 3.4 Summaries of hydrophobically modified membranes and hydrophobic membranes.

Membrane	Concentration of silicane coupling agent	Pore size	Particle size	Emulsion formulation	References
KP-18C-modified SPG membranes	3% (v)	4.7 μm	Chitosan microsphere 3.92 μm	DP: 1.5 wt% chitosan in 1 wt% acetic acid aqueous solution containing 0.9 wt% NaCl CP: Liquid paraffin and petroleum ether in a volume ratio of 7 : 5, and added PO-500	[6]
GRT-350-modified SPG membranes	2.5 wt%	4.5 μm	Agarose microsphere 10.19 μm	DP: 4 wt% agarose and 3.0 wt% NaCl aqueous solution 10 g CP: 100 g liquid paraffin containing 5 wt% PO-500	[13]
OTC-modified SPG membranes	1% (v)	4.7 μm	Emulsion N/A	DP: 1.5 wt% chitosan in 1 wt% acetic acid aqueous solution containing 0.9 wt% NaCl CP: Liquid paraffin and petroleum ether in a volume ratio of 7 : 5, and added PO-500	[8]
PE membranes	N/A	N/A	Agarose beads N/A	DP: 10 wt% agarose solution CP: liquid paraffin and petroleum ether (V/V = 7 : 5) containing 4 wt% PO-500	[14]
EP membranes	N/A	0.6–15 μm	Agarose microsphere 16.53 μm	DP: 4 wt% agarose and 3.0 wt% NaCl aqueous solution 10 g CP: 100 g liquid paraffin containing 5 wt% PO-500	[13]
Nickel membranes	N/A	15 μm	Spherical silica particles 65–240 μm	DP: 10–15 wt% sodium silicate solution CP: 5 wt% Span 80 in low-odor kerosene	[25]

DP: Disperse Phase; CP: Continuous Phase.

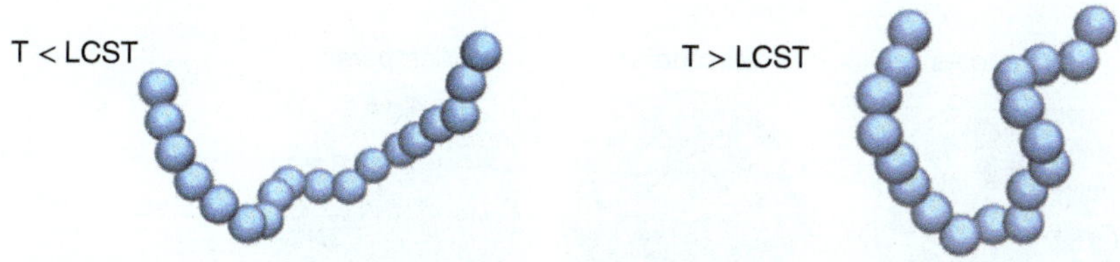

Figure 3.17 Schematic illustration of PNIPAM temperature-sensitive mechanism (the blue beads represent the NIPAM monomers). Source: Abbott et al. [27]/with permission of American Chemical Society.

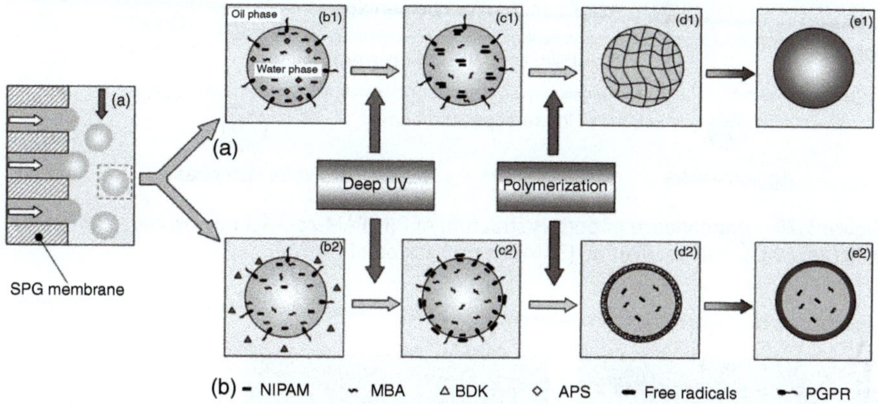

Figure 3.18 Schematic illustration of preparation of monodisperse PNIPAM microspheres (a) and microcapsules (b). (N,N_0-methylenebisacrylamide (MBA), 2,2-dimethoxy-2-phenylacetophenone (BDK), ammonium persulfate (APS), and polyglycerolpolyricinoleate (PGPR)). Source: Cheng et al. [31]/with permission of Springer Nature.

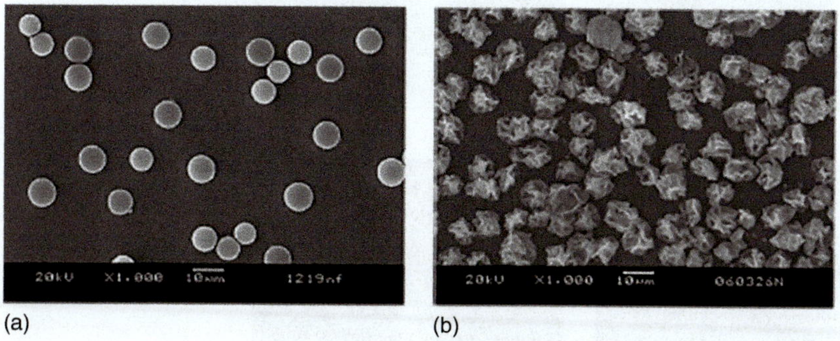

Figure 3.19 SEM image of air-dried PNIPAM microspheres (a) and microcapsules (b). Source: Cheng et al. [31]/Elsevier.

them showed good monodispersity (Figure 3.19) and excellent thermal sensitivity. The microspheres had solid structures (Figure 3.19a), and the hollow microcapsules collapsed and possessed crinkled structures after dying (Figure 3.19b).

Our group also utilized membrane emulsification technique to prepare poly(*N*-Isopropyl acrylamide-*co*-acrylic acid) (P(NIPAM-*co*-AAc)) microspheres with different membrane pore sizes [33]. Due to the thermo-sensitivity of NIPAM

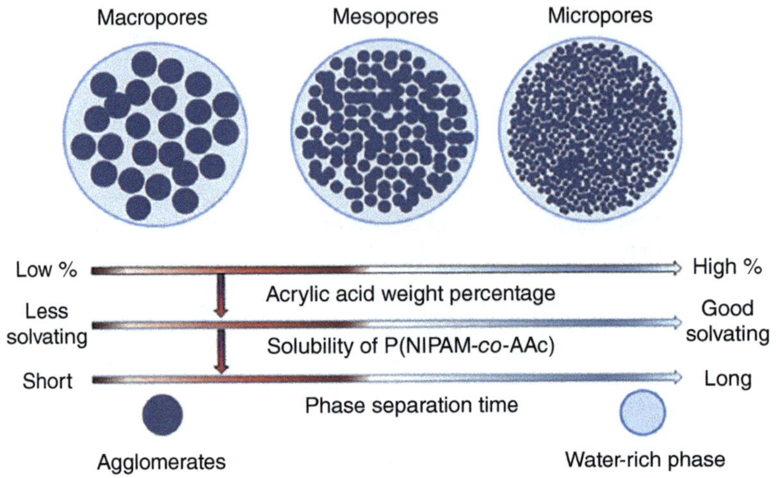

Figure 3.20 Dependence of porous structure of P(NIPAM-*co*-AAc) microspheres on acrylic acid (AAc) wt%. Source: Si et al. [33]/with permission of Elsevier.

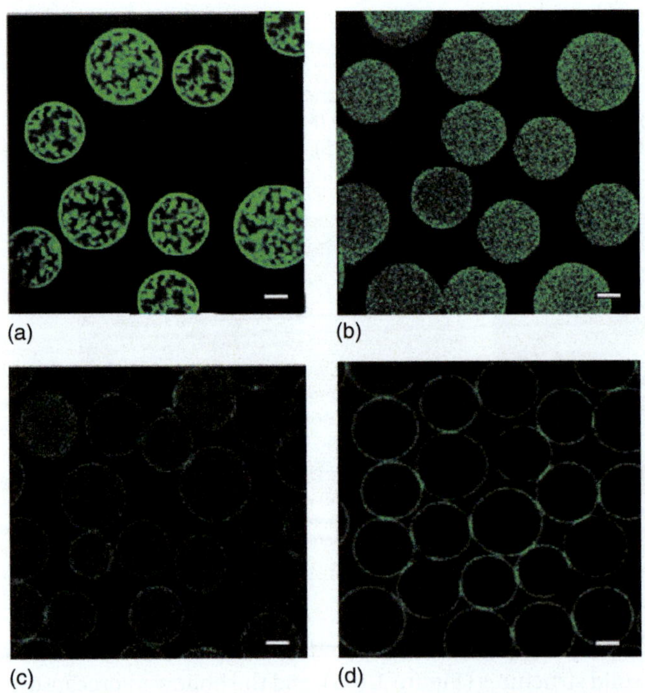

Figure 3.21 Confocal laser scanning microscopy (CLSM) images of P(NIPAM-*co*-AAc) microspheres absorbing FITC labeled trypsins. (a) AA(0 wt%); (b) AA(5 wt%), (c) AA(10 wt%); and (d) AA(15 wt%). Scale bar are 8 μm. Source: Si et al. [33]/Elsevier.

and pH-sensitivity of AAc, the microspheres possessed thermo-/pH-sensitivity. Moreover, formulas with different AAc wt% were used to prepare microspheres with different porous structures. With the increase of acrylic acid weight percentage, the pores of P(NIPAM-*co*-AAc) microspheres transformed from macropores to micropores as shown in Figures 3.20 and 3.21. The porous microspheres with different pore structures could be applied to meet various requirements in different bioapplications.

Nagashima et al. also prepared poly(acrylamide-*co*-acrylic acid) (P(AM-*co*-AAc)) microspheres from W/O emulsions by using modified SPG membranes with different membrane pore sizes [34, 35]. They firstly prepared stable monomer emulsion with SPG membrane and then polymerized the monomer droplets in the continuous phase. The average diameters of the microspheres were dependent on the pore size of the used SPG membranes and about 3.5–4 times larger than the pore size.

3.4 Preparation Microparticles from Preformed Polymer System

Microspheres and microcapsules made by natural, biodegradable, and block polymers are widely used in medical engineering and biotechnology. Hydrophilic microspheres can be prepared not only from monomer systems but also from preformed polymer systems. Conventional methods for preparing hydrophilic microspheres from preformed polymer systems include emulsification, solidification, spray drying, and coacervation technique. Due to the high viscosity of preformed polymer systems, high shear forces are required to disperse the emulsion. Therefore, the usual mechanical emulsification methods are not suitable for encapsulating biologically active substances, such as peptides and proteins, due to the harsh conditions. This problem can be solved by adopting membrane emulsification method. Various hydrophilic microspheres prepared from preformed polymer systems by direct membrane emulsification method are described below.

3.4.1 Chitosan Microspheres

Chitosan is a product of partial deacetylation of the natural polysaccharide chitin. The relative molecular mass of chitosan is about $2 \times 10^5 \sim 5 \times 10^5$ kDa. Because chitosan is biodegradable, biocompatible, nontoxic, and positively charged, it is widely investigated in many fields, such as food additives, textiles, agriculture, cosmetics, antibacterial agents, medical fibers, medical dressings, artificial tissue materials, medical absorbable materials, tissue engineering carrier materials, medical treatment, drug development, and other daily chemical industries.

The traditional preparation methods of chitosan microspheres include emulsification/solvent evaporation, spray drying, ionotropic gelation, and coacervation technique. The size of microspheres prepared by these methods is difficult to control, and the broad size distribution may bring some limitations to the application. Membrane emulsification allows for the preparation of chitosan microspheres that are uniform and size controllable. For example, our group prepared chitosan microspheres by direct membrane emulsification technique and the mechanical stirring method, respectively [6]. For membrane emulsification preparation process, the dispersed aqueous phase was pressed by the nitrogen gas through the pores of the SPG membrane into the continuous oil phase containing the emulsifier (PO-500) to form the droplets. Then the cross-linking reagent Glutaraldehyde saturated toluene, (GST) was slowly dropped into the emulsion to react with chitosan. After cross-linking, the formed chitosan microspheres were collected by centrifugation and washed several times. Figure 3.22 shows the SEM photographs of chitosan microspheres prepared by membrane emulsification (i) and mechanical stirring method (ii), respectively. Compared with stirring method, the microspheres by membrane emulsification were much more uniform.

The diameter of the chitosan microspheres prepared by different pore sizes of membranes is summarized in Table 3.5. Plotting the average diameter of the microspheres as a function of pore size of the membranes gives a linear relationship. Thus, the membrane emulsification is an ideal method to prepare uniform and size-controllable droplets or particles. The researchers can easily obtain the microspheres with required size by choosing the adequate pore size of membranes.

It is important that the cross-linked chitosan microspheres can exhibit fluorescent properties without conjugation with any fluorescent agent. This phenomenon is

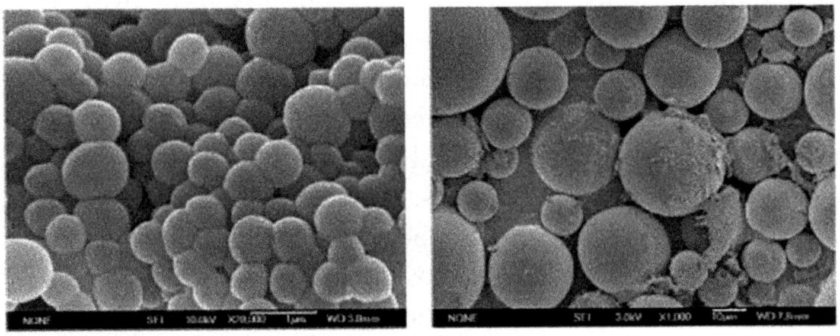

(a) (b)

Figure 3.22 SEM photographs of chitosan microspheres prepared by membrane emulsification (a) and mechanical stirring method (b). Source: Wang et al. [6]/Elsevier.

Table 3.5 Preparation results of microspheres by the membrane with different pore sizes.

Pore size of membrane (μm)	0.5	4.7	5.7	10.2	19.6	Regression equations
Critical pressure (kgf/cm^2)	2.00	0.134	0.104	0.021	0.006	
Diameter of microspheres after drying (μm)	0.39	3.92	5.30	7.26	14.38	$y = 0.6804x + 0.8764$

y stands for diameters of microspheres after drying and x stands for pore diameter of membranes.
Source: Adapted from Wang et al. [6].

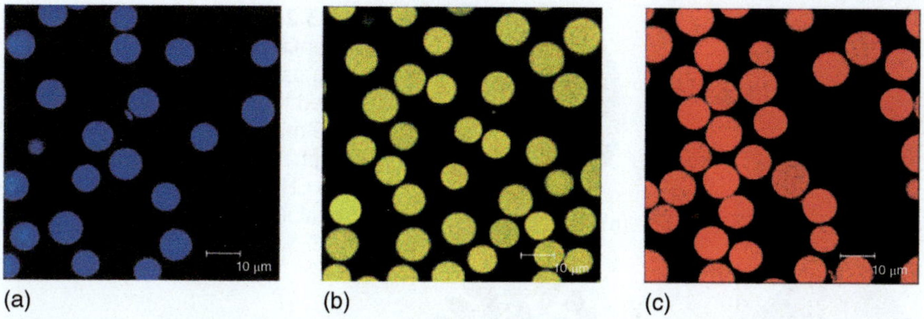

Figure 3.23 CLSM images of (a) SRCG microspheres excited at 365 nm, (b) C–G microspheres excited at 488 nm, and (c) C–F microspheres excited at 543 nm. The scale bars represent 10 μm. Source: Wei et al. [36]/John Wiley & Sons.

firstly found by our group [36]. The fluorescence color varies with different crosslinkers and can be modulated by further chemical reduction, whereas the fluorescence intensity can be controlled by tuning the particle size and degree of cross-linking. The chitosan microspheres cross-linked by glutaraldehyde (C–G microspheres) and formaldehyde (C–F microspheres) were prepared by using SPG membrane emulsification. The selectively reduced C–G microspheres (SRCG microspheres) were prepared by using $NaBH_4$ to modify the C–G microspheres and selectively reduce the C=N bonds to C—N bonds. The CLSM images of these chitosan microspheres with different fluorescence color are shown in Figure 3.23.

The differences in the chemical structures of the microspheres lead to the observed variations in autofluorescence. C–G microspheres have two kinds of unsaturated double bonds: CC bonds from glutaraldehyde, and CN bonds from the Schiff's base between the free amino groups of chitosan and the aldehyde groups of glutaraldehyde. As for the C–F microspheres, only CN bonds existed in these structures due to the cross-linking agent formaldehyde. For SRCG microspheres, most of the Schiff's bases in the C–G microspheres are converted into secondary amines and the CC bonds keep stable. The autofluorescent chitosan microspheres are uniformly sized, monodisperse, and exhibit evenly distributed strong fluorescence. Based on these unique properties, the chitosan microspheres may act as a novel detecting system for *in vivo* observing and quantifying the distribution, transport, and metabolism of these microspheres in animals [37].

Our group also utilized SPG membrane emulsification technique to prepare chitosan particles with different structures [38]. The traditional chitosan microspheres cross-linked by glutaraldehyde (C–G microspheres) are solid spherical particles without any pores on their surface and which compact structure is not convenient for loading bioactive drugs. Moreover, for different clinical therapy applications, the requirements for drug-release behavior vary significantly from case to case. By adopting different cross-linking reagents and cross-linking steps, chitosan particles with various structures were successfully obtained (Figure 3.24). The hollow C–PG microspheres were prepared by sequentially adding two kinds of cross-linking agents, first *p*-phthaldehyde and then glutaraldehyde. After cross-linking, the microspheres were washed by using the mixture of acetone and ethanol to remove

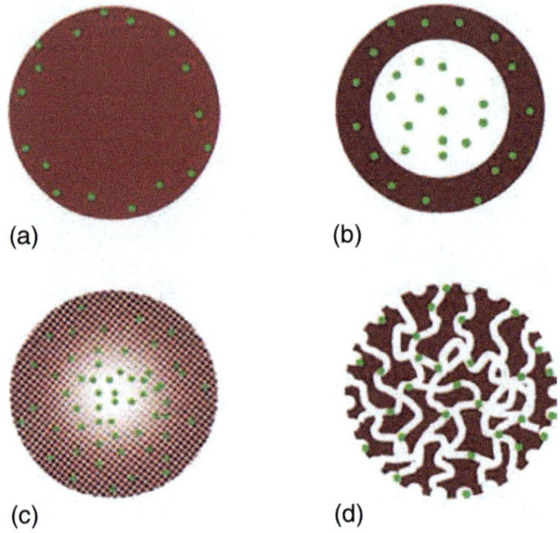

Figure 3.24 Schematic representation of BSA (green dots) loading patterns in microspheres prepared by different methods: (a) C–G microspheres, (b) C–PG microspheres, (c) CH–G microspheres, and (d) CH–PG microspheres. Source: Wei et al. [38]/with permission of John Wiley & Sons.

p-phthaldehyde. To prepare CH–G and CH–PG microspheres, the mixture polymer of chitosan and chitosan derivative (*N*-[(2-hydroxy-3-trimethyl ammonium)propyl] chitosan chloride (HTCC)) was used to replace chitosan. Glutaraldehyde as crosslinker was used to form CH–G microspheres, whereas the aforementioned two-step solidification process was used to form CH–PG microspheres. CH–G microspheres also have the hollow structure and the interior cavity is smaller than that of C–PG microspheres. The formation of this hollow structure of CH–G microspheres is attributed to the steric hindrance from quaternized groups of HTCC, which also leads to the formation of pores on the microsphere surface. By adopting the two-step cross-linking process to prepare CH–PG microspheres, the obtained microspheres possessed a macroporous structure. The aggravating phase separation occurring during the first cross-linking reaction with *p*-phthaldehyde may result in the formation of these pores.

By combining the special cross-linking strategy and SPG membrane emulsification, the size and the morphology of chitosan microspheres are tunable, which provides the feasibility to meet various specific requirements. And the hollow CH–G microspheres were investigated as the orally administered insulin delivery system in further by our group [39], which will be introduced in Chapter 4.

Akamatsu et al. also used membrane emulsification to prepare monodisperse chitosan microcapsules with hollow structures by two different methods as shown in Figure 3.25 [40]. In the first physical cross-linking method, the calcium alginate microspheres were prepared as core particles, and chitosan was coated on the microspheres and cross-linked by tripolyphosphate to form shell layer. The alginate core particles were removed at the same time because tripolyphosphate could also act as a calcium ions catcher. In the second chemical cross-linking method, the calcium alginate microspheres were also used as core particles, but the coated material was chitosan modified with azido groups and photo-cross-linked by chemical bonds under UV irradiation. Tripolyphosphate was also used to remove Ca^{2+}. In both of the two methods, SPG membrane emulsification was adopted to prepare different W/O emulsions, including the W/O emulsions containing alginic acid, calcium

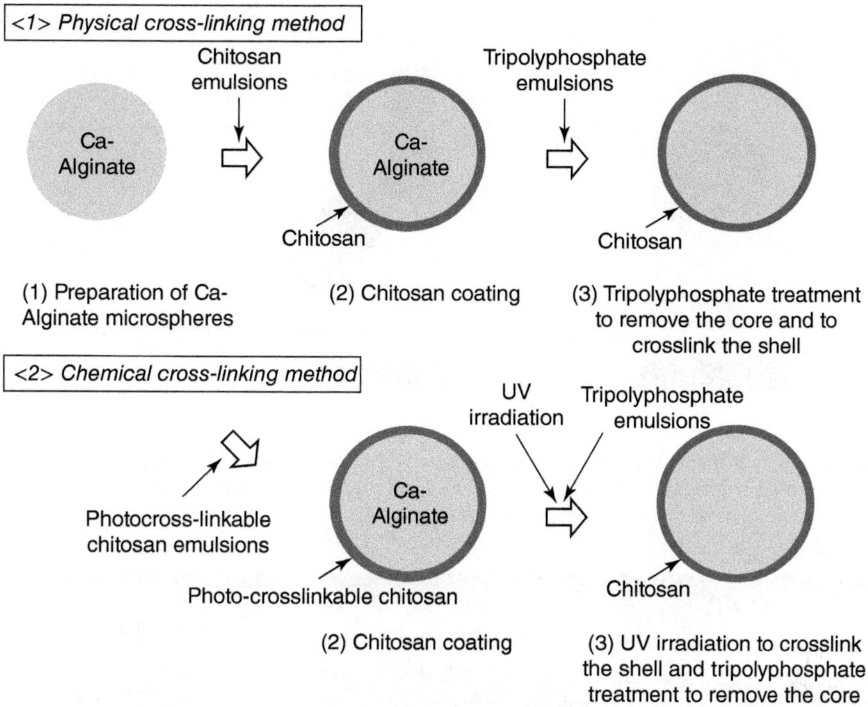

Figure 3.25 Conceptual illustration of preparation procedures for chitosan microcapsules with hollow structures. Source: Akamatsu et al. [40]/with permission of American Chemical Society.

ions, chitosan, or tripolyphosphate as the dispersed phase to form monodisperse particles.

The obtained chitosan microspheres by the two methods are shown in Figure 3.26. High contrast can be observed in the center of the microcapsules, indicating the hollow structure. However, in both methods, alginate could not be removed completely, and part of it would stay inside the microcapsules or in the layer, which might limit the subsequent application. The same group also utilized SPG membrane emulsification to prepare lysozyme-encapsulated chitosan microspheres and studied the effect of membrane pore size on chitosan microspheres [41].

Molecular weight of chitosan can influence physicochemical properties of microspheres. Niu et al. discovered that chitosan with too high or too low molecular weight was difficult to form uniform-sized microspheres [42]. As shown in Figure 3.27, chitosan microparticles exhibited a regular spherical shape and a smooth surface, but there was more extensive particle aggregation in chitosan-50 (the molecular weight of chitosan was 50 kD, the same as below), which might be explained by that the low molecular weight chitosan forms a thinner interfacial layer, and therefore the steric repulsion between the droplets is reduced, leading to more flocculation. Particles prepared with chitosan-600 demonstrated a larger size compared with those prepared with chitosan-100. This phenomenon is possibly attributed to the increased

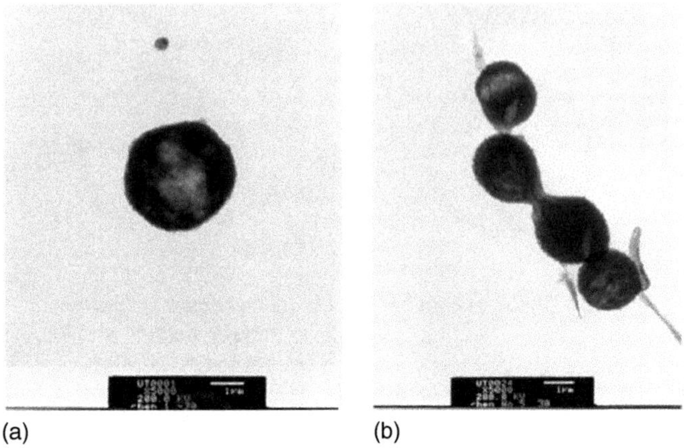

Figure 3.26 Transmission electron microscope (TEM) micrographs of chitosan microspheres prepared by physical (a) and chemical (b) cross-linking methods. Source: Akamatsu et al. [40]/American Chemical Society.

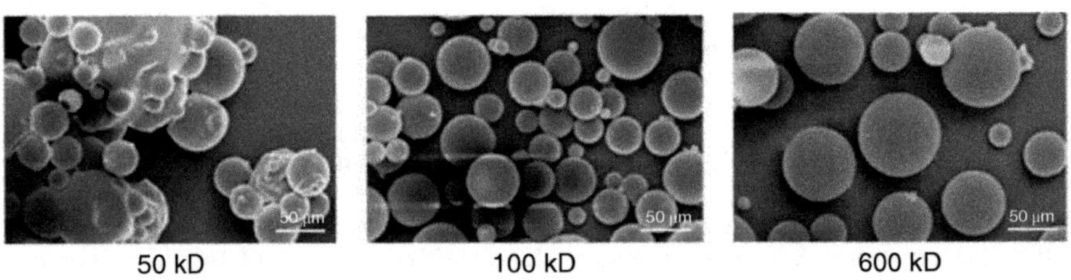

Figure 3.27 SEM of chitosan microparticles with different molecular weights. Source: Niu et al. [42]/Elsevier.

viscosity and surface tension of corresponding chitosan solution with the increase of molecular weight, and as a result of which, the larger particles were formed. Chitosan-100 was selected in this study to prepare microparticles, and the mean particle diameter of chitosan microparticles was 62.9 μm with a Span value of 1.9.

To prepare uniform-sized pH-sensitive quaternized chitosan microspheres, our group developed a novel strategy by combining thermosensitive gel and membrane emulsification method [43]. The quaternized chitosan is synthesized by the reaction of chitosan and glycidyltrimethylammonium chloride (GTMAC) and named as *N*-[(2-hydroxy-3-trimethylammonium) propyl] chitosan chloride (HTCC) [44]. HTCC can form thermosensitive hydrogel when it is mixed with glycerophosphate (GP). The phosphate group of GP can interact with quaternary amino groups on HTCC chains and form cross-linked network structure at 37 °C. The mixture is solution below or at room temperature, which allowed it to be injectable and to incorporate living cells, proteins, enzymes, or other therapeutic drugs easily. Once the surrounding temperature is up to 37 °C, the system is transformed to a nonflowing hydrogel and releases the entrapped drug as a function of pH values. In our study, HTCC was dissolved and mixed with GP solution at room temperature. Then, the mixture (water phase) was pressed by nitrogen gas through the pores of the

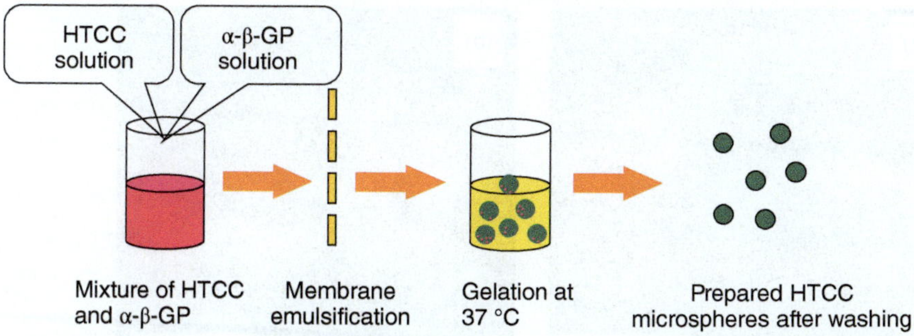

Figure 3.28 Preparation process of HTCC microspheres by combining thermogelation and membrane emulsification method. Source: Wu et al. [43]/with permission of Elsevier.

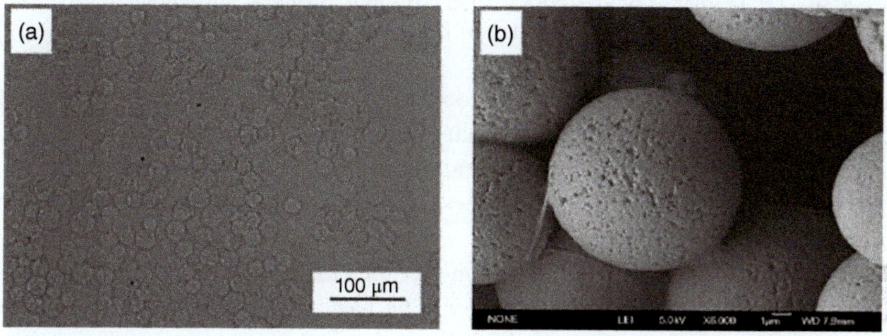

Figure 3.29 The microphotograph (a) and SEM image (b) of HTCC microspheres. Source: Wu et al. [43]/Elsevier.

membrane into oil phase to form W/O emulsion. After emulsification, the emulsion was heated to 37 °C to solidify the droplets. The preparation process is shown in Figure 3.28.

The obtained microspheres have porous structure as shown in Figure 3.29, and the pores and cavities inside microspheres can facilitate the encapsulation and diffusion of drug. Bovine serum albumin (BSA) is chosen as model drug and labeled with fluorescein isothiocyanate (FITC-BSA). Figure 3.30 shows the CLSM images of FITC-BSA loaded HTCC microspheres dispersed in PBS after 0 and 15 d, respectively. The bright area represents FITC-BSA, and the black area inside microsphere is the pore or cavity where no FITC-BSA exists. It can be observed that the BSA was distributed evenly in the microspheres. As time went on, the fluorescence weakened gradually because of the diffusion of FITC-BSA. Moreover, the microspheres behaved with obvious pH-sensitivity drug release profiles. The loaded BSA was released completely in acid solution (pH 5.0) in short time, and the release rate was lowered when the release medium was neutral buffer (pH 7.4).

3.4.2 Agarose Microspheres

Agarose is a natural polysaccharide made from agar, which is derived from seaweed. Agarose can be dissolved in water at high temperature, and the hot solution

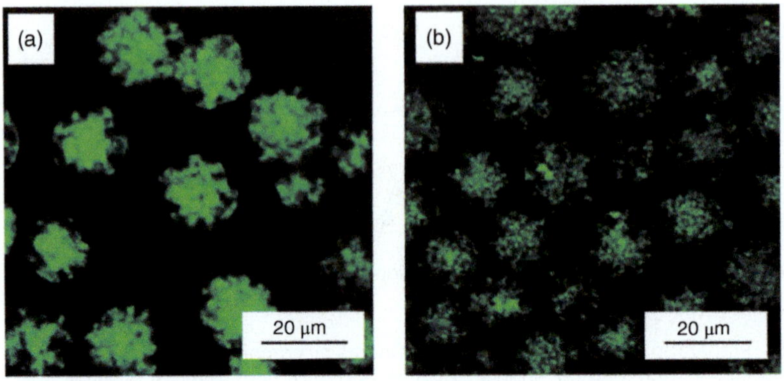

Figure 3.30 CLSM images of FITC-BSA loaded HTCC microspheres suspended in PBS (pH 7.4). (a): 0 d and (b) 15 d. Source: Wu et al. [43]/Elsevier.

of agarose can form a hydrogel when cooled. The formed gel is hydrophilic and microporous, and it does not contain charged groups or other adsorption centers. Agarose gel has been widely used in separation and purification, gel electrophoresis, immunoadsorption, immobilized enzymes, cell culture, protein renaturation, and tissue engineering.

Until now, the conventional preparation methods for preparing spherical agarose gel beads are emulsification, spraying, etc. The basic process of these methods is that the emulsion is firstly prepared by dispersing agarose aqueous solution into organic solvent containing surfactants by mechanical stirring or spraying and then the droplets are solidified by cooling. The disadvantages of these methods are that the size distribution of agarose beads is very broad and the diameters are difficult to control.

To overcome the shortcomings of traditional preparation methods, our group prepared agarose beads by membrane emulsification technique and compared the result with the agarose beads made by mechanical stirring emulsification technique [45]. The optical microscopic photographs and size distribution are shown in Figures 3.31

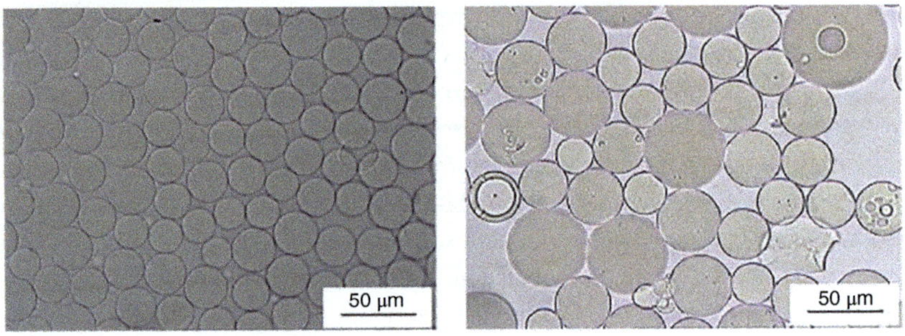

(a) (b)

Figure 3.31 The optical microscopic photographs of agarose beads: (a) prepared by membrane emulsification and (b) prepared by mechanical stirring emulsification method. Source: Zhou et al. [45]/Elsevier.

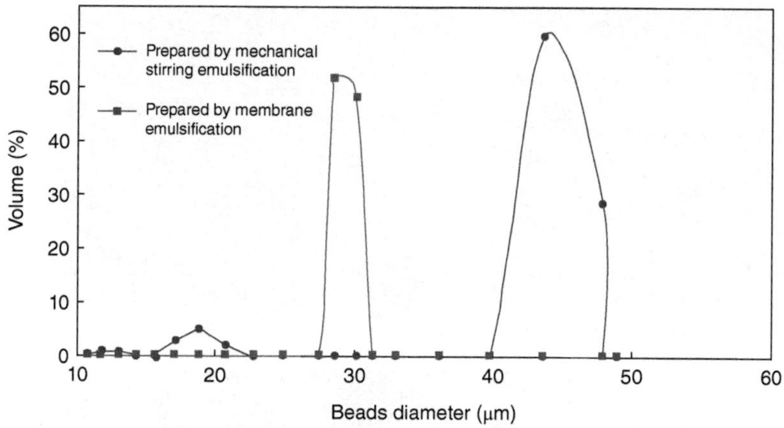

Figure 3.32 Size distribution of agarose beads prepared by membrane emulsification and by mechanical stirring emulsification method. Source: Zhou et al. [45]/with permission of Elsevier.

and 3.32. The average diameter of beads prepared by mechanical stirring emulsification method was 34 μm and CV was 71%, the size distributed from 10 to 60 μm. The average diameter of beads prepared by membrane emulsification was 30 μm and CV was 15%. The size distribution prepared by the membrane emulsification was much narrower than that by mechanical stirring emulsification method.

Our group also explored the effect of oil phase on particle size uniformity of agarose microspheres [45]. Olive oil, cottonseed oil, castor oil, and a mixture of liquid paraffin and petroleum ether (volume ratio 7 : 5) were chosen as the oil phase. Figure 3.33 is the optical micrographs of the emulsion prepared with these four groups of oil phases. It can be seen that the emulsions prepared with a mixture of liquid paraffin and petroleum ether as the oil phase have very uniform sizes than those prepared with the other four oil phases. As shown in Figure 3.34a, the interfacial tension between water phase and the mixture of liquid paraffin and petroleum ether was 37.5 mN/m, and was much higher than those between water phase and the other oil phases. As stated before, in order to prepare uniform-sized droplets, the interfacial tension between the water phase and the thin layer of the oil phase, which covered the surface of the membrane should be high enough. By this high interfacial tension, a stable droplet can be formed by higher CA with the membrane; as a result, the droplet will be taken away by a weak flow of continuous phase (oil phase). Moreover, the viscosity of oil phase is also important for emulsification. With the addition of petroleum ether, the viscosity of oil phase decreased from 10.6 to 2.0 MPa s (Figure 3.34b), which was much lower than the viscosity of olive oil, cotton seed oil, castor oil, and liquid paraffin at 55 °C. The decrease of viscosity is favorable for the water droplets to detach from the surface of the membrane timely, which is important for formation of uniform-sized emulsion droplets. After preparing uniform-sized agarose microspheres, we used these microspheres for high-throughput sorting bacteria by flow cytometry [46]. The supernatant of sludge was encapsulated in the agarose gel beads and incubated

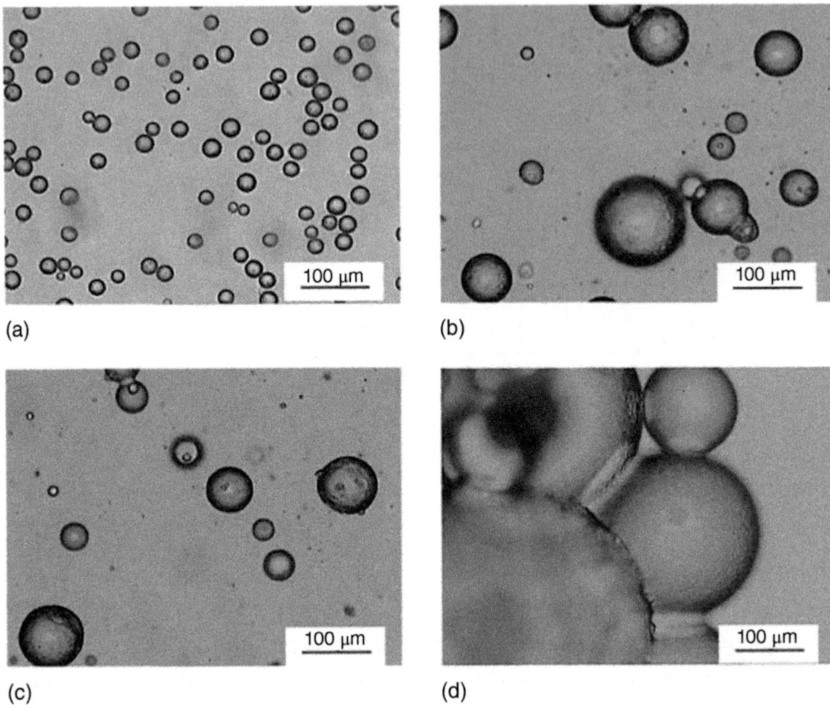

Figure 3.33 Optical micrographs of emulsion droplets prepared by using different oil phases: (a) liquid paraffin/petroleum ether (7 : 5, v/v), (b) cotton seed oil, (c) olive oil, and (d) castor oil. Source: Zhou et al. [45]/Elsevier.

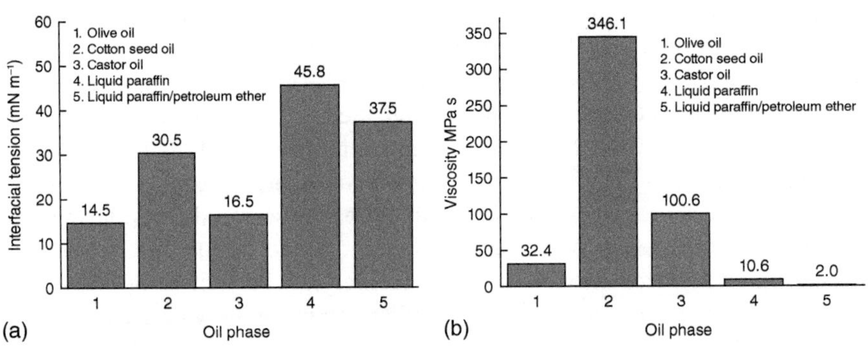

Figure 3.34 The interfacial tension between different oil phase and water phase (a) and the viscosity of different oil phase (b) at 55 °C. Source: Zhou et al. [45]/with permission of Elsevier.

for 14 days to form microcolonies. And the gel beads were detected from free-living cells and then sorted by flow cytometry. The encapsulated cells preserved activity and viability during the membrane emulsification process. More details about this study are introduced in Chapter 9.

For industrial-scale applications, the agarose microspheres with average particle diameters around 90 μm show low flow resistance at relatively high flow rates. In order to prepare large-sized agarose microspheres, the SPG membranes with large pores should be adopted. However, with increasing pore size, the optimum pressure drops to such a low level that even a subtle pressure fluctuation would significantly affect the size and size distribution of the resulting emulsion. Furthermore, the fast droplet generation induces coalescence on the surface of the SPG membrane caused by the high penetrability of the large pores. Therefore, certain obstacles have to be overcome when preparing uniform and large-sized agarose microspheres and our group developed an improved membrane emulsification technique [47]. The schematic diagram of membrane emulsification apparatus is shown in Figure 3.35. It is composed of a mechanical stirrer, a pressurized (N_2) steel tank, an emulsion container, and several tubular glass membranes, all of which are enclosed in a thermostatic tank in order to prevent gelation during the emulsification process.

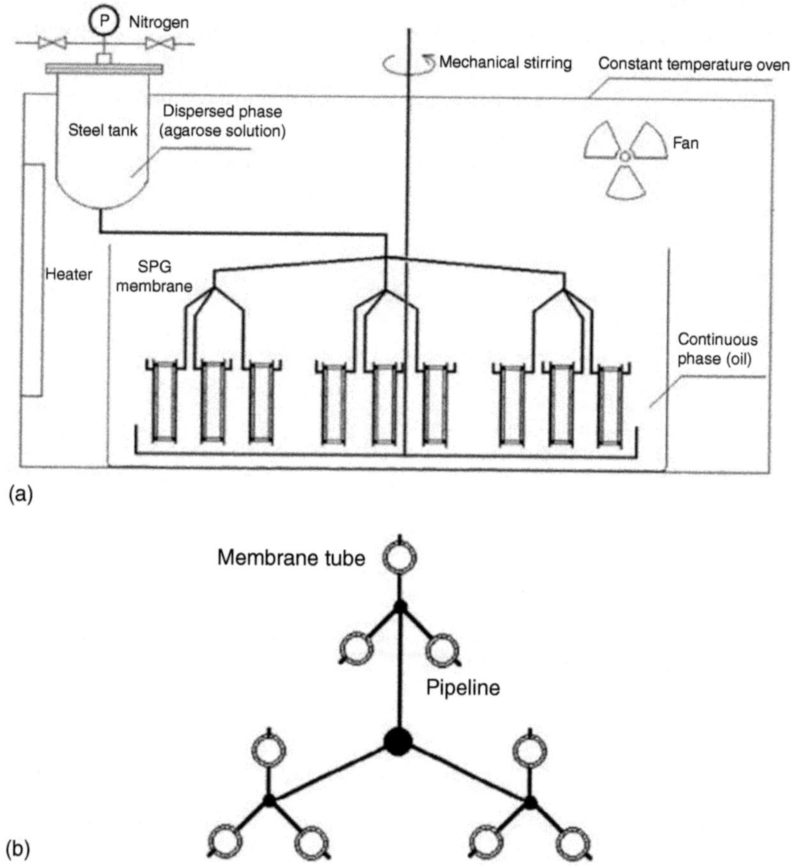

Figure 3.35 (a) Schematic diagram of the home-designed large-scale membrane emulsification apparatus. (b) Top view of the membrane module with 9 tubes attached. Source: Zhao et al. [47]/with permission of Elsevier.

The agarose solution is stored in a steel tank (500 ml), and is allowed to pass through the uniform pores of the membrane into the surrounding continuous phase. The resulting emulsion is gently agitated by a mechanical stirrer so as to avoid undesirable creaming. One to 12 membrane tubes with an individual membrane size of 5 cm (l) × 1 cm (Φ) are arranged radially inside the emulsion container, as depicted in Figure 3.35b with nine tubes. When multiple membranes are employed, the available specific surface area increases significantly, thus the throughput enhances substantially. The system is promising for the application of high-viscosity systems such as agarose solution or other polysaccharide solutions. Optical micrographs of obtained agarose microspheres and the commercial product Sepharose 4FF, the latter produced by mechanical stirring combined with a sieving process, are shown in Figure 3.36a, b, respectively. It is obvious that the uniformity of the agarose microspheres prepared by membrane emulsification without sieving process is better than that of Sepharose 4FF. The average size of uniform agarose microspheres is almost identical to that of Sepharose 4FF, however, with a significantly narrower size distribution.

We also prepared uniform polysaccharide composite microspheres with a mixture of agarose and dextran solution by membrane emulsification technique [48]. As chromatographic media, their pore structure was deliberately regulated by adjusting both the polysaccharide composition and the molecular weight of dextran. The average diameter of composite microspheres with different agarose/dextran ratios prepared by membrane emulsification was 9.4, 10.0, and 13.4 μm, and the

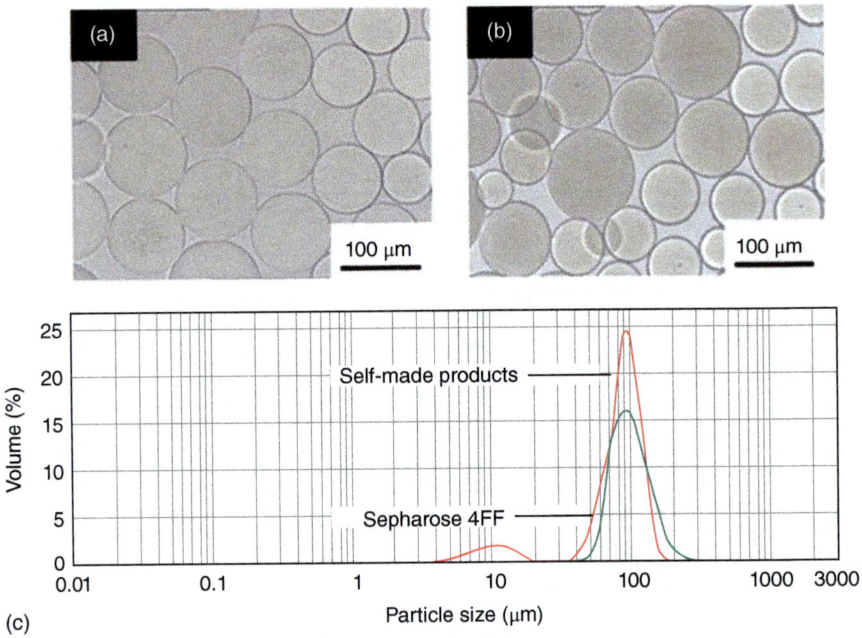

Figure 3.36 Optical micrographs of (a) homemade agarose microspheres, (b) Sepharose 4FF, and (c) particle size distribution curves. Source: Zhao et al. [47]/Elsevier.

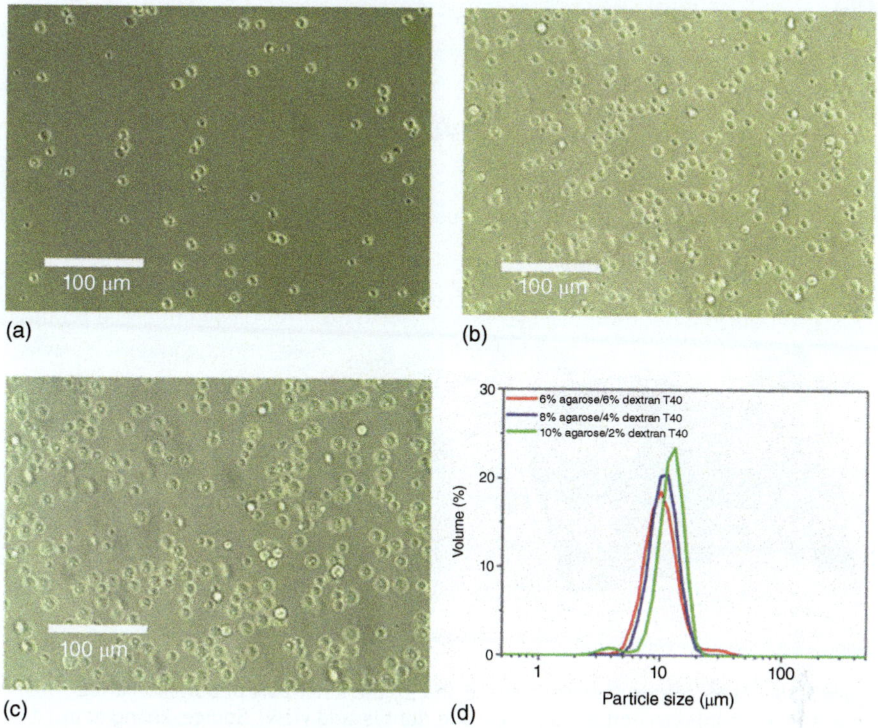

Figure 3.37 Optical micrographs (a), (b), and (c) were 6% agarose/6% dextran T40, 8% agarose/4% dextran T40, and 10% agarose/2% dextran T40, respectively, and their size distribution (d). Source: Zhang et al. [48]/Elsevier.

particle size distribution width (Span value) was 0.80, 0.67, and 0.94, respectively (Figure 3.37). The atomic force microscope (AFM) images of the surface and inner part of microspheres are shown in Figure 3.38. Peaks distributed evenly on both the surface and the inner part corresponded to the hydrogel net skeleton, while pores were distributed uniformly among these peaks, demonstrating that there were numerous pores in the composite microspheres. Polysaccharide composite microspheres prepared using membrane emulsification technique showed a great advantage in the uniformity and controllable gel network, beneficial for high-resolution chromatography. Compared with other traditional methods, such as mechanical stirring technique, membrane emulsification technique can provide uniform-sized microspheres without the need for sieving, which could reduce the cost and save the raw materials.

3.4.3 Alginate Microspheres

Alginic acid is a kind of natural polysaccharides found in brown algae. Its molecular weight ranges from 10 to 600 kDa and it exists as a mixed salt. When alginic acid meets bivalent cations or polycations, ion transfer occurs and a gel with certain strength and elasticity is formed. The chemical structure, molecular size,

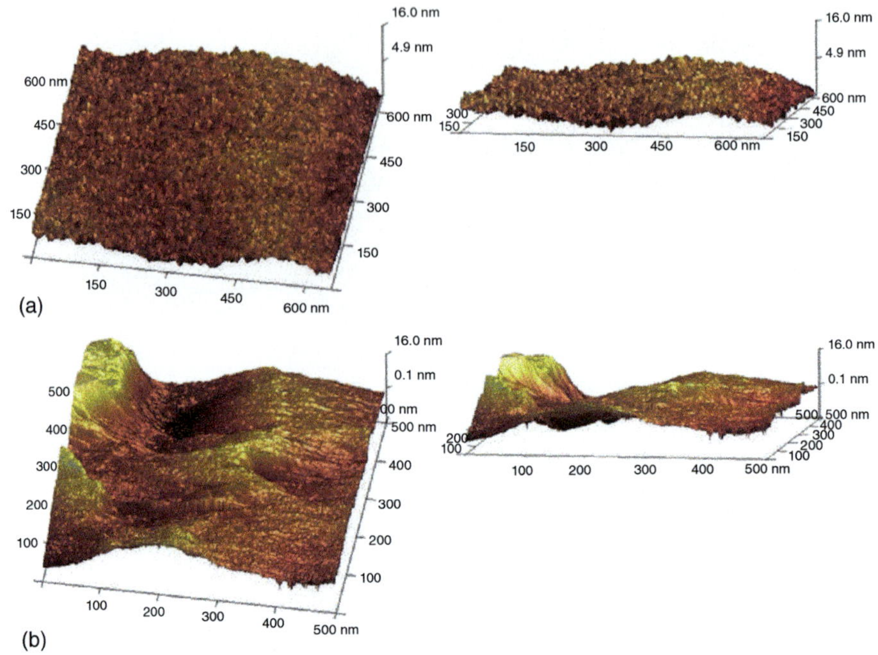

Figure 3.38 AFM images of (a) the surface and (b) the inner part of polysaccharide composite microspheres (left is front view and right is side view). Source: Zhang et al. [48]/Elsevier.

gel formation kinetics, and cations of alginate have important effects on some of its functional properties, such as porosity, swelling, stability, biodegradability, gel strength, and biocompatibility. In the pharmaceutical industry, alginate is a commonly used excipient. Its viscosity makes it a suitable binder for tablets. It is also used as a disintegrant to disperse tablets because it swells with water. It can act as a dispersant that disperses the active ingredients into a liquid to become a suspension and also as a drug-sustained release agent.

Sodium alginate is the most commonly used alginate salt. It can form gel in the presence of multivalent cations such as calcium ions in aqueous media [49, 50]. However, the formed gel has loose network, which might result in the undesired drug-leakage. Polycations, such as chitosan, can also react with alginate by the strong electrostatic interaction of amino groups of chitosan with the carboxyl groups of alginate. The complexation reduces the porosity of alginate beads and decreases the leakage of the encapsulated drugs. To prepare alginate microspheres, drop injection method, spray drying method, electrostatic granulation method, and jet shear method are the conventional ways. Membrane emulsification is also applied to obtain microspheres with more uniform particle size. Our group successfully utilized membrane emulsification technique to prepare alginate–chitosan microspheres with narrow size distribution and good sphericity by a two-step solidification process [49]. Alginate in acetic acid buffer solution was used as the water phase and pressed through the membrane into oil phase (liquid paraffin and petroleum ether in a volume ratio of 7 : 5; emulsifier: 4 wt% PO-500). Then, the miniemulsion of $CaCl_2$ prepared by homogenizer was added into the sodium alginate emulsion as the first-step solidification. After 5 hours solidification, the solidified alginate gels were collected, washed, and then dispersed in chitosan acetic acid buffer solution as the second-step solidification. Finally, the alginate–chitosan microspheres were washed and dried by lyophilization (Figure 3.39). Figure 3.40a illustrates that the microspheres size is around 7.5 μm, and the Span value is 0.731. The SEM observation (Figure 3.40b) clearly demonstrates that these microspheres are spherical, and the diameter is consistent with the results from the particle size analyzer. The surface is rough and has many micropores (Figure 3.40c). Furthermore, as shown in Figure 3.40d, the fluorescence indicates that chitosan is homogeneously distributed inside the microsphere, which reflects that chitosan could permeate into the alginate–Ca microspheres.

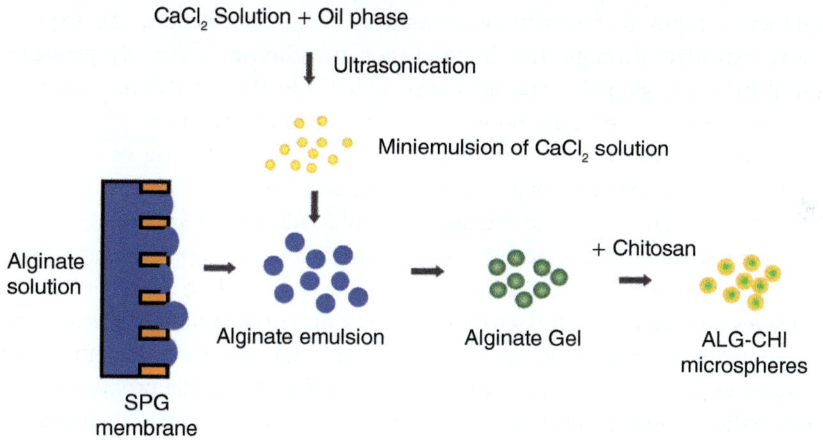

Figure 3.39 A schematic representation of the preparation process of blank alginate–chitosan microspheres. Source: Zhang et al. [49]/with permission of Elsevier.

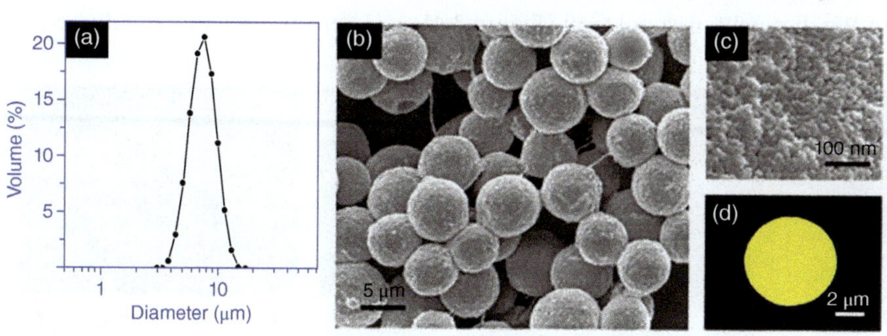

Figure 3.40 Characterization results of blank alginate–chitosan microspheres. (a) Size distribution, (b) SEM image, (c) surface morphology, and (d) CLSM image showing chitosan distribution after cross-linking with glutaraldehyde. Source: Zhang et al. [49]/Elsevier.

You et al. studied the effect of alginate concentration on the preparation of alginate microspheres by using the membrane emulsification method [50]. When the concentration of sodium alginate solution was 1 wt%, it was difficult to operate the membrane emulsification apparatus continuously, since the microporous glass membrane was easily wetted owing to the low viscosity of the dispersing phase. Also, the prepared microspheres were easily crushed due to their poor mechanical strength. On the other hand, 4 wt% sodium alginate solution was too viscous to pass through the micropores of the glass membrane, unless high pressure was applied to the disperse phase. The high pressure gave the resulting microspheres a broad size distribution. When the concentration of sodium alginate was 2 wt%, the pressure was moderate, the wetting of a microporous glass membrane was minimum, and the monodisperse microspheres with a mean size of 4 μm were obtained.

Liu et al. developed an internal gelation technology to prepare spherical Ca–alginate beads with small and uniform size [51]. Calcium carbonate (40 nm) was mixed with sodium alginate solution to form the disperse phase. The continuous phase was composed of paraffin oil containing 1.5% (v) surfactant. The disperse phase was extruded through the microporous membrane under N_2 pressure. After emulsification, glacial acetic acid was added into the emulsion, dissolving $CaCO_3$ to release Ca^{2+}. The gelation was then initiated and the microspheres were obtained. The Ca–alginate beads prepared with nickel membrane of pore size 2.9 μm possessed good sphericity and smooth surface. The beads also had narrow size distribution (Span = 0.7), and the mean size of beads was 55.3 μm.

Wu et al. investigated the effects of mixed emulsifier ratios and concentrations of cross-linking agent on the calcium alginate/sodium cellulose sulfate–chitosan microcapsules morphology [52]. Two kinds of nonionic surfactants were chosen and mixed as the emulsifier in oil phase. Span 80 (HLB = 4.3) is a lipophilic emulsifier while Tween 80 (HLB = 15.0) is a hydrophilic emulsifier. Thus, the property of the mixed emulsifier of Span 80 and Tween 80 is between lipophilicity and hydrophilicity. The mixed emulsifiers with volume ratios (Span 80 : Tween 80) of 5 : 5, 7 : 3, and 9 : 1 were chosen. As shown in Figure 3.41, the volume ratios of two emulsifiers had major influence on microsphere's morphology. With an equal volume of Span 80 and Tween 80 (5 : 5), the microspheres had a relatively spherical shape, but the dispersity was not well (Figure 3.41a). Comparatively, at the volume ratio

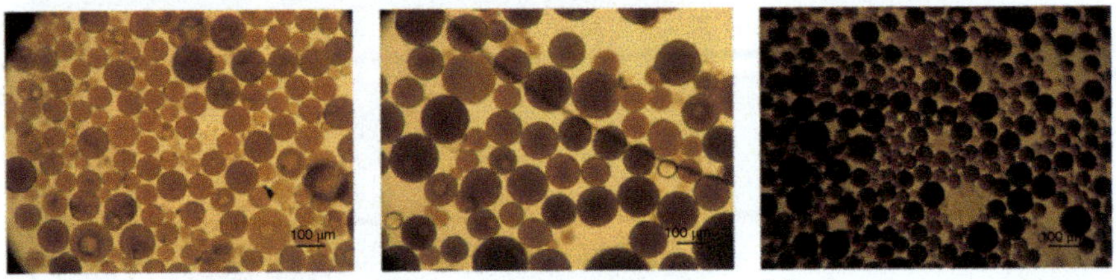

Figure 3.41 Effects of emulsifier ratios on microsphere's morphology. (a) $V_{Span\ 80} : V_{Tween\ 80} = 5 : 5$, (b) $V_{Span\ 80} : V_{Tween\ 80} = 7 : 3$, and (c) $V_{Span\ 80} : V_{Tween\ 80} = 9 : 1$. Source: Wu et al. [52]/Elsevier.

of 7 : 3 the microspheres had spherical shapes with good dispersity (Figure 3.41b). However, when the proportion of Span 80 continued to increase (9 : 1), the size of microspheres became smaller with some coagulation and breakage (Figure 3.41c). Therefore, it can be inferred that the HLB value of the mixed emulsifier with volume ratio of 7 : 3 was more favorable to stabilize the emulsion system.

3.4.4 Cellulose Microspheres

Cellulose is the most abundant and sustainable material in nature and has been used in various forms such as film and microspheres. With unique structures and properties, such as porosity, low cost, biocompatibility, and cellulose microspheres, have been widely used in food, medicine, and other fields. Obrien et al. reported on the continuous manufacturing of cellulose microspheres as a sustainable alternative to plastic microparticles [53]. As shown schematically in Figure 3.42a, the disperse phase of cellulose dissolved in an organic electrolyte solution is extruded through the pores of a tubular porous membrane into the lumen through which a continuous carrier phase of sunflower oil containing 2 wt% Span 80 flows. Recirculation of the formed emulsion (Figure 3.42b) through the interior of the tubular membrane enabled an increase in concentration of droplets, and the formed emulsion was extracted into a measuring cylinder containing ethanol, which resulted in the formation of cellulose microbeads by antisolvent precipitation or phase inversion

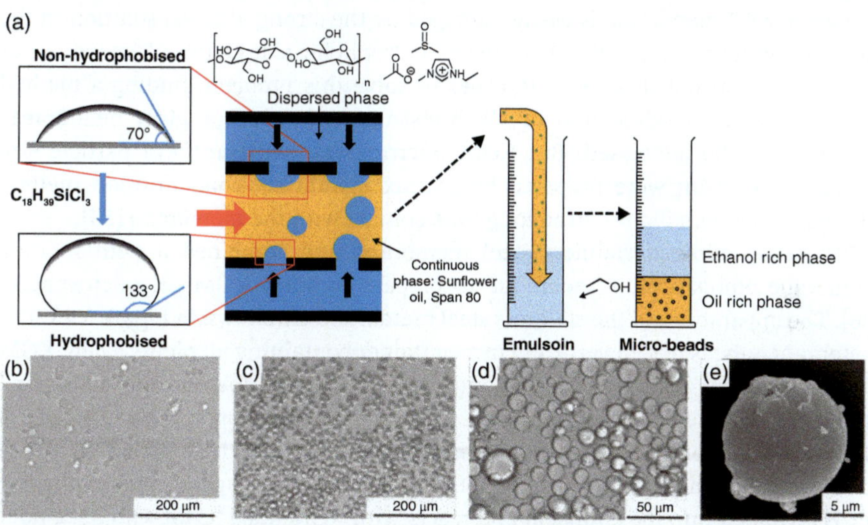

Figure 3.42 (a) Schematic representation of the continuous membrane emulsification–phase inversion process for the generation of cellulose microbeads along with the contact angles of the disperse phase before and after hydrophobization of the SPG membranes. (b) Optical micrograph of emulsion droplets of 8 wt% microcrystalline cellulose (dissolved in 30 : 70, w/w, 1-ethyl-3-methylimidazolium acetate: dimethyl sulfoxide) dispersed in sunflower oil-2 wt% Span 80 mixture. (c, d) Cellulose microbeads formed by phase inversion with ethanol. (e) SEM image of a single cellulose bead. Source: Coombs Obrien et al. [53]/American Chemical Society/CC BY.

(Figure 3.42c,d,e). Wet bead diameters in the range from 17 to 65 μm are reliably accessible using the 10 μm pore size SPG membranes applied here.

3.4.5 Glucomannan Microspheres

Glucomannan is a kind of water-soluble polysaccharide, which is generally considered to be an excellent dietary fiber. The traditional preparation method of glucomannan microspheres is mainly the stirring emulsification and solidification method. The microspheres prepared by this method have wide particle size distribution and need to be separated by sieving, resulting in low product yield and poor repeatability of preparation. In addition, it is difficult to produce small microspheres (<10 μm) by mechanical stirring, however, small microspheres are necessary basic materials in high-performance liquid chromatography.

Konjac glucomannan (KGM) is a polysaccharide extracted from the konjac plant, which is extracted by heating konjac flour in the presence of alkali. KGM is composed of β-1,4-linked D-mannosyl and D-glucosyl residues with a small degree of β-1,3-branching linked at the C-3 position of mannose. For KGM, alkali treatment is crucial for its gelation since alkali facilitates the deacetylation and decreases solvation of KGM chains [54]. However, the preparation of KGM microspheres by conventional membrane emulsification has not achieved ideal results because the hydrophobic modification of SPG membranes is not stable in alkaline solution [55]. The silicon–ether bond formed between the modifier and the Si-OH on the surface of SPG membrane is easily damaged by the strong alkaline solution in the alkaline solution, causing the shedding of hydrophobic side chains ultimately leads to the emulsion polydispersity. In order to solve this problem, finding a method for hydrophobic modification of alkali-resistant microporous glass membranes still needs to be addressed. But KGM microspheres with uniform particle size of less than 10 μm were prepared by premix membrane emulsification method successfully, which did not need long time contact with the membrane [55].

Our group adopted stainless steel membrane and combined it with rotating membrane emulsification technique to prepare uniform KGM gel microspheres [56]. The apparatus and the stainless steel membrane are shown in Figure 3.43. The membrane tube is placed vertically in a container containing oil phase. Then, KGM alkaline aqueous solution is added into the membrane tube and the membrane is rotating by the homogenizer. Due to the centrifugal force, the water phase is thrown out from the pores of the membrane tube into the external oil phase to form emulsion. After emulsification, the cross-linking agent epichlorohydrin (ECH) is added into the emulsion and reacts with KGM at 60 °C for 4 hours. Under the optimum conditions, the prepared KGM gel microspheres had the uniform diameter of about 70 μm (Figure 3.44) and size distribution index Span<1.0. And the rotating membrane emulsification technique also was used to prepare large-size agarose microspheres [57]. The mutual effects between variables and optimum values of the variables were investigated. Under the optimum conditions, the obtained agarose microspheres possessed average size of 220 μm and the Span value was 0.76.

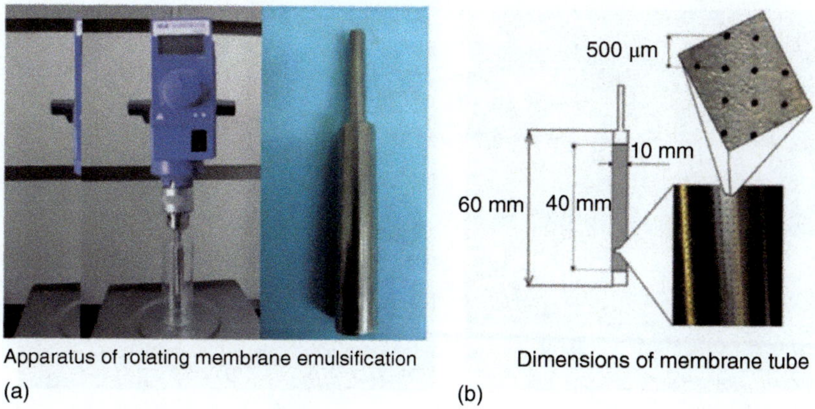

Figure 3.43 Experimental apparatus of the rotating membrane emulsification. Source: Li et al. [56]/The Chinese Journal of Process Engineering.

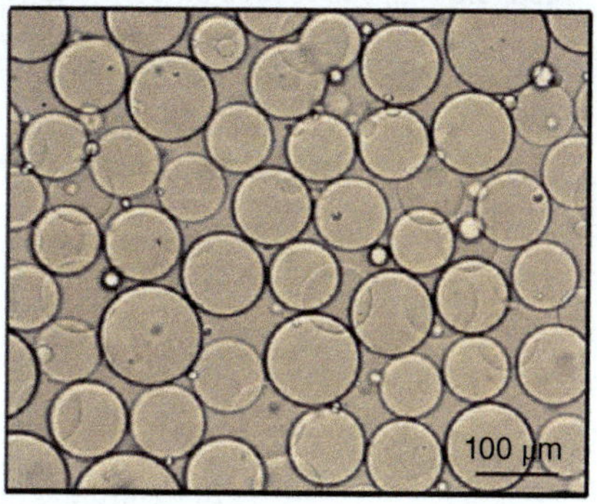

Figure 3.44 Microscopic photographs of KGM microspheres prepared under the optimum conditions (d_{50} = 67.9 μm, span = 0.457). Source: Li et al. [56]/The Chinese Journal of Process Engineering.

3.5 Other Hydrophilic Microspheres Prepared by Membrane Emulsification

3.5.1 PVA Microspheres

PVA has been recognized as promising and suitable candidate for drug delivery system and has different dosage forms depending on the route of drug administration, such as membranes, fibers, and particles. Piacentini et al. prepared PVA microspheres by membrane emulsification as enzyme carriers [58]. The dispersed phase was PVA solution with 0.1 N HCl and the continuous phase was isooctane containing Span 80. Emulsification was performed using hydrophobic nickel membrane with 10 μm circular holes. The emulsions were reacted with glutaraldehyde-saturated toluene and the microspheres formed were washed and collected. The enzyme was immobilized during particle formation (by entrapment) or after particle production (by adsorption or cross-linking). By using a PVA and glutaraldehyde ratio of 1 : 0.2, it was possible to produce uniform, spherical solidified carrier particles after two hours of the cross-linking reaction (Figure 3.45).

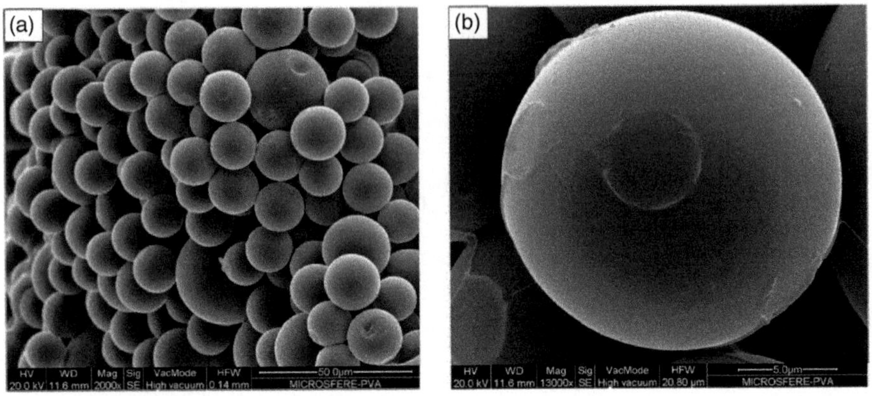

Figure 3.45 SEM's of dry cross-linked particles. (a) PVA microsphere and (b) single PVA microsphere. Source: Piacentini et al. [58]/Elsevier.

Table 3.6 Lipase-loaded amount and specific activity of lipase-loaded microspheres.

Immobilization method	Immobilization conditions	Lipase-loaded amount (mg)	Specific activity (mmol min^{-1} $mg_{lipase}{}^{-1}$)
Adsorption	*Lipase concentration* ($g\,l^{-1}$)		
	1	1.8	0.00033
	2	3.9	0.00030
	6	11.7	0.00012
Cross-linking	*Lipase concentration* ($g\,l^{-1}$)		
	GA concentration (% V/V)		
	0.4	2.1	0.00042
	0.8	2.6	0.00043
	1.2	1.9	0.00042
Entrapment	*Lipase concentration* ($g\,l^{-1}$)		
	0.1	0.16	0.0036
	1	0.75	0.0015
No immobilization (free lipase)	—	—	0.0046

Source: Piacentini et al. [58]/with permission of Elsevier.

By applying adsorption, cross-linking, or entrapment, the enzyme lipase can be successfully loaded in the microspheres (Table 3.6). The entrapment method was demonstrated to be the most suitable method, enabling an improved operational stability and promoting the intrinsic optimal interaction between enzyme and substrate. Especially the mild operating conditions of membrane

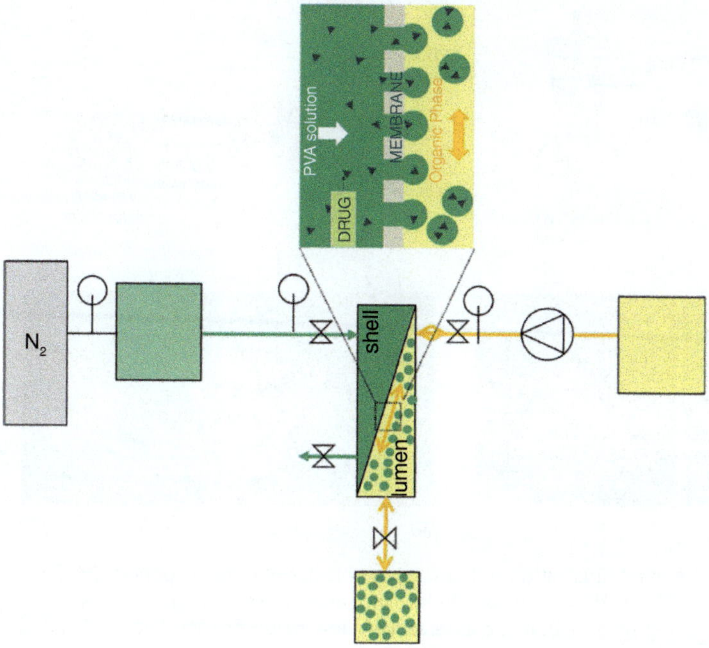

Figure 3.46 PVA liquid droplets formation by pulsed back-and-forward membrane emulsification. Source: Piacentini et al. [59]/with permission of Elsevier.

emulsification provided an efficient strategy for immobilization of biomolecules without compromising their function.

The same team also prepared PVA microparticles to encapsulate water-soluble drug via back-and-forward membrane emulsification (Figure 3.46) [59]. The dispersed phase used to produce the W/O emulsion was obtained by dissolving PVA in hot water (90 °C). The continuous phase was an isooctane solution containing Span 80. The dispersed (aqueous) phase was pressed through the hydrophobic SPG membrane under gas pressure. The continuous phase was agitated by a pulsed back-and-forward mode along the lumen side of the membrane by using a programmable peristaltic pump. The droplets were precipitated as solid microparticles by the addition of toluene saturated with glutaraldehyde.

Comparing those particles prepared by homogenization, PVA particles prepared by membrane emulsification have much more precisely controlled morphology and narrower size distribution. And membrane emulsification process also endows PVA microparticles with a higher encapsulation efficiency of diclofenac sodium.

Morelli et al. prepared chitosan and PVA complex microparticles by membrane emulsification for encapsulation and pH-controlled release [60]. The microsieve membranes (metal foils with straight through pores), which are less prone to fouling compared to the SPG and ceramic membranes, were used and operated under high transmembrane fluxes. By blending chitosan and PVA, the prepared microparticles exhibited pH-sensitive drug release profiles, which released 80% of the drug in 3

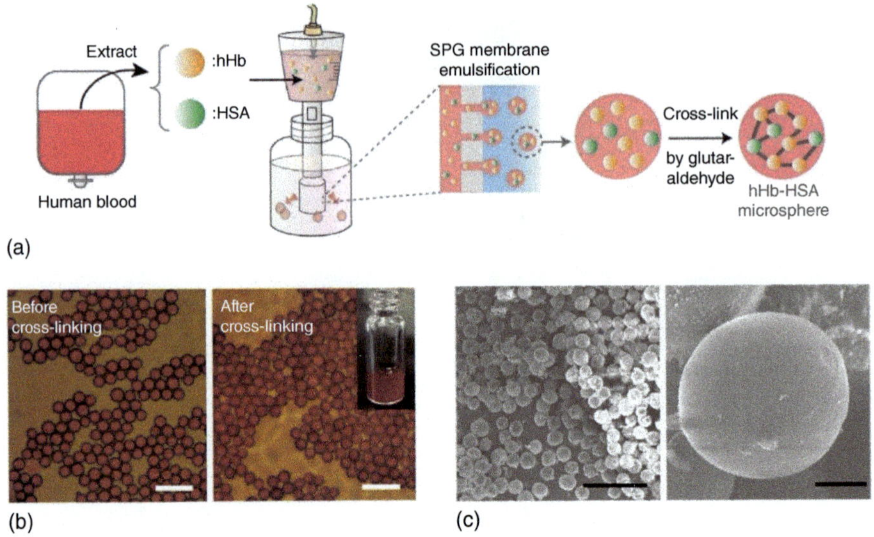

Figure 3.47 (a) Schematic illustration of fabrication of hHb–HSA microspheres. (b) Optical microscopic images of hHb–HSA emulsions before and after cross-linking. Scale bar is 50 μm. (c) SEM images of hHb–HSA microspheres. Left: low magnification image of various particles. Scale bar: 500 μm. Right: enlarged images of one representative particle. Scale bar: 3 μm. Source: Ohta et al. [61]/Elsevier.

hours under acidic conditions and less than 3% under neutral conditions. On the other hand, the drug release profile from PVA particles was not pH-sensitive with an initial "burst release."

3.5.2 Protein Microspheres

Protein is another important hydrophilic macromolecule, which has biomedical application. For example, hemoglobin (Hb) has been investigated as oxygen carrier to prevent hypoxia in regenerated tissues. Ohta et al. developed human hemoglobin (hHb) and human serum albumin (HSA)-based microspheres using SPG membrane emulsification [61]. As shown in Figure 3.47a, hHb and HSA were dissolved in PBS as the dispersed phase, while kerosene containing tetra glycerol condensed ricinoleate (TGCR) was used as the continuous phase. Glutaraldehyde was used as a curing agent and then hHb–HSA microspheres were obtained. The obtained hHb–HSA microspheres (Figure 3.47b, c) had a uniform size with an average diameter of 15.1 μm measured by optical microscope, which is similar to the diameter of human red blood cells (7–8 μm). The loading amount of hHb in the microspheres was 20 wt%, which is similar to that of red blood cells (33 wt%). The hHb–HSA microspheres showed similar oxygen dissociation behavior and methemoglobin formation resistance to native hHb.

3.6 Summary

The preparation of hydrophilic microspheres from W/O emulsions by direct membrane emulsification was mainly introduced in this chapter. Traditional emulsion preparation mostly adopts mechanical stirring method. The particle size of the emulsion droplets formed by this method is relatively large, usually in the range of micron to hundreds micron. In order to overcome the above shortcomings, microporous membrane emulsification technology was developed to prepare emulsions with uniform size and controllable particle size.

SPG membrane is a hydrophilic microporous glass membrane with high pore size uniformity. In order to prepare W/O emulsion, the hydrophilic SPG membrane should be chemically modified to obtain a hydrophobic membrane, because the dispersed phase is hydrophilic and forms a higher interfacial tension with the surface of the hydrophobic SPG membrane. The surface of SPG membrane can be hydrophobized by chemical reaction with organosilane compounds, such as chlorosilanes, or physical coating with silicone resin. Under a certain pressure, the dispersed phase passes through the membrane pores to form uniform droplets on the membrane surface and enters the continuous phase to form a uniform W/O emulsion.

However, the hydrophobic modification process is complex, and the modifiers are somewhat toxic and tend to fall off from the membrane surface. Thus, some researchers preferred to presoak the membrane into the continuous phase for a long time to increase the hydrophobicity of the membrane surface. But this hydrophobic modification cannot keep a long time and is not suitable for industrial continuous production. Other researchers promoted to prepare hydrophobic polymer membranes such as epoxy resin membranes. However, at present the alternative polymer membranes are not enough to meet various application requires, and further study and scale-up are urgent.

The hydrophilic microspheres prepared by membrane emulsification have important applications in the field of many fields, for example, as drug carrier or separation and purification medium. For these applications, the uniformity of the microspheres is especially crucial. The microspheres with broad size distribution might target to different human organs and result in severe side effects. Moreover, the processing conditions of membrane emulsification technology are mild and can meet the requirements for loading bioactive agents. In the entire emulsification process, the activity of protein and peptide drugs can be protected, and the process of microporous membrane emulsification is easy to scale-up.

The membrane emulsification technology in this chapter is suitable for preparing microspheres with larger particle size. Moreover, the production scale of commercially available equipment has been developed successfully as shown in Chapter 10. And the parameters of production process could be set using the control software. The relative standard derivation of different batches is less than 5%. We believe that with the commercialization of the membrane emulsification technique and related equipment, the clinical application of particles in biomedical field would be promoted in the near future.

References

1 Vladisavljević, G.T., Shimizu, M., and Nakashima, T. (2005). Permeability of hydrophilic and hydrophobic Shirasu-porous-glass (SPG) membranes to pure liquids and its microstructure. *Journal of Membrane Science* 250 (1–2): 69–77.

2 Omi, S., Katami, K., Taguchi, T. et al. (1995). Synthesis of uniform PMMA microspheres employing modified SPG (Shirasu porous glass) emulsification technique. *Journal of Applied Polymer Science* 57 (8): 1013–1024.

3 Cheng, C.J., Chu, L.Y., and Xie, R. (2006). Preparation of highly monodisperse W/O emulsions with hydrophobically modified SPG membranes. *Journal of Colloid and Interface Science* 300 (1): 375–382.

4 Kandori, K., Kishi, K., and Ishikawa, T. (1991). Preparation of monodispersed W/O emulsions by Shirasu-porous-glass filter emulsification technique. *Colloids and Surfaces* 55: 73–78.

5 Cheng, C.J., Chu, L.Y., Xie, R. et al. (2008). Hydrophobic modification and regeneration of Shirasu porous glass membranes on membrane emulsification performance. *Chemical Engineering & Technology* 31 (3): 377–383.

6 Wang, L.Y., Ma, G.H., and Su, Z.G. (2005). Preparation of uniform sized chitosan microspheres by membrane emulsification technique and application as a carrier of protein drug. *Journal of Controlled Release* 106 (1-2): 62–75.

7 Mi, Y.C. (2017). Preparation of Epoxy-based porous material with uniform pores and its application in membrane emulsification technology. Doctoral dissertation. University of Chinese Academy of Sceinces.

8 Wang, L.Y. (2006). Preparation and application of uniform-sized chitosan microspheres containing biodrug. Doctoral dissertation. University of Chinese Academy of Sceinces.

9 Kukuzaki, M., Fujimoto, K., Kai, S. et al. (2010). Ozone mass transfer in an ozone-water contacting process with Shirasu porous glass (SPG) membranes-A comparative study of hydrophilic and hydrophobic membranes. *Separation and Purification Technology* 72: 347–356.

10 Suzuki, K., Hayakawa, K., and Hagura, Y. (1999). Preparation of high concentration O/W and W/O emulsions by the membrane phase inversion emulsification using PTFE membranes. *Food Science and Technology Research* 5 (2): 234–238.

11 Yamazaki, N., Yuyama, H., Nagai, M. et al. (2002). A comparison of membrane emulsification obtained using SPG (Shirasu porous glass) and PTFE [poly(tetrafluoroethylene)] membranes. *Journal of Dispersion Science and Technology* 23 (1–3): 279–292.

12 Bottino, A., Capannelli, G., and Comite, A. (2005). Novel porous poly(vinylidene fluoride) membranes for membrane distillation. *Desalination* 183 (1–3): 375–382.

13 Mi, Y.C., Zhou, W.Q., Li, Q. et al. (2015). Detailed exploration of structure formation of an epoxy-based monolith with three-dimensional bicontinuous structure. *RSC Advances* 5 (68): 55419–55427.

14 Zhou, Q.Z., Ma, G.H., and Su, Z.G. (2009). Effect of membrane parameters on the size and uniformity in preparing agarose beads by premix membrane emulsification. *Journal of Membrane Science* 326 (2): 694–700.

15 Huang, Y., Huang, Q.L., Liu, H. et al. (2017). Preparation, characterization, and applications of electrospun ultrafine fibrous PTFE porous membranes. *Journal of Membrane Science* 523: 317–326.

16 Yu, S.W., Huang, Q.L., Cheng, J.X. et al. (2020). Pore structure optimization of electrospun PTFE nanofiber membrane and its application in membrane emulsification. *Separation and Purification Technology* 251: 117297.

17 Vladisavljević, G.T., Brösel, S., and Schubert, H. (2000). Preparation of water-in-oil emulsions using microporous polypropylene hollow fibres: conditions for producing small uniform droplets. *Chemical Papers* 54 (6a): 383–388.

18 Vladisavljević, G.T., Tesch, S., and Schubert, H. (2002). Preparation of water-in-oil emulsions using microporous polypropylene hollow fibers: influence of some operating parameters on droplet size distribution. *Chemical Engineering and Processing* 41: 231–238.

19 Vladisavljević, G.T., Brösel, S., and Schubert, H. (2000). Characterization of water-in-oil emulsions produced with microporous hollow polypropylene fibers. *Journal of the Serbian Chemical Society* 65 (11): 829–837.

20 Mi, Y.C., Liu, S., Zhang, Y.F. et al. (2020). The roles played by DMF in the structure formation of epoxy-based porous monolith. *Polymer Science* 62 (5): 465–472.

21 Mi, Y.C., Zhou, W.Q., Li, Q. et al. (2015). Preparation of water-in-oil emulsions using a hydrophobic polymer membrane with 3D bicontinuous skeleton structure. *Journal of Membrane Science* 490: 113–119.

22 Yanagishita, T., Tomabechi, Y., Nishio, K. et al. (2004). Preparation of monodisperse SiO_2 nanoparticles by membrane emulsification using ideally ordered anodic porous alumina. *Langmuir* 20 (3): 554–555.

23 Yanagishita, T., Maejima, Y., Nishio, K. et al. (2014). Monodisperse nanoparticles of metal oxides prepared by membrane emulsification using ordered anodic porous alumina. *RSC Advances* 4: 1538–1542.

24 Silva, P.S., Morelli, S., Dragosavac, M.M. et al. (2017). Water in oil emulsions from hydrophobized metal membranes and characterization of dynamic interfacial tension in membrane emulsification. *Colloids and Surfaces, A: Physiochemical and Engineering Aspects* 532: 77–86.

25 Dragosavac, M.M., Vladisavljevic, G.T., Holdich, R.G. et al. (2012). Production of porous silica microparticles by membrane emulsification. *Langmuir* 28 (1): 134–143.

26 Kubota, K., Fujishige, S., and Ando, I. (1990). Single-chain transition of poly(*N*-isopropylacrylamide) in water. *Journal of Physical Chemistry* 94 (12): 5154–5158.

27 Abbott, L.J., Tucker, A.K., and Stevens, M.J. (2015). Single chain structure of a poly(*N*-isopropylacrylamide) surfactant in water. *The Journal of Physical Chemistry B* 119 (9): 3837–3845.

28 Stile, R.A., Burghardt, W.R., and Healy, K.E. (1999). Synthesis and characterization of injectable poly(*N*-isopropylacrylamide)-based hydrogels that support tissue formation in vitro. *Macromolecules* 32 (22): 7370–7379.

29 Osada, Y., Okuzaki, H., and Hori, H. (1992). A polymer gel with electrically driven motility. *Nature* 355 (6357): 242–244.

30 Miyata, T., Asami, N., and Uragami, T. (1999). A reversibly antigen-responsive hydrogel. *Nature* 399 (6738): 766–769.

31 Cheng, C.J., Chu, L.Y., Zhang, J. et al. (2008). Preparation of monodisperse poly (*N*-isopropylacrylamide) microspheres and microcapsules via Shirasu-porous-glass membrane emulsification. *Desalination* 234 (1–3): 184–194.

32 Cheng, C.J., Chu, L.Y., Ren, P.W. et al. (2007). Preparation of monodisperse thermo-sensitive poly(*N*-isopropylacrylamide) hollow microcapsules. *Journal of Colloid and Interface Science* 313: 383–388.

33 Si, T.B., Wang, Y.X., Wei, W. et al. (2011). Effect of acrylic acid weight percentage on the pore size in poly(*N*-Isopropyl acrylamide-*co*-acrylic acid) microspheres. *Reactive & Functional Polymers* 71: 728–735.

34 Nagashima, S., Ando, S., and Makino, K. (1998). Size dependence of polymer composition in the surface layer of poly(acrylamide-*co*-acrylic acid) hydrogel microspheres. *Journal of Colloid and Interface Science* 197: 377–382.

35 Nagashima, S., Koide, M., Ando, S. et al. (1999). Surface properties of monodisperse poly(acrylamide-*co*-acrylic acid) hydrogel microspheres prepared by a membrane emulsification technique. *Colloids and Surfaces, A: Physiochemical and Engineering Aspects* 153: 221–227.

36 Wei, W., Wang, L.Y., Yuan, L. et al. (2007). Preparation and application of novel microspheres possessing autofluorescent properties. *Advanced Functional Materials* 17: 3153–3158.

37 Wei, W., Wang, L.Y., Yuan, L. et al. (2008). Bioprocess of uniform-sized crosslinked chitosan microspheres in rats following oral administration. *European Journal of Pharmaceutics and Biopharmaceutics* 69: 878–886.

38 Wei, W., Yuan, L., Hu, G. et al. (2008). Monodisperse chitosan microspheres with interesting structures for protein drug delivery. *Advanced Materials* 20: 2292–2296.

39 Wei, W., Ma, G.H., Wang, L.Y. et al. (2010). Hollow quaternized chitosan microspheres increase the therapeutic effect of orally administered insulin. *Acta Biomaterialia* 6: 205–209.

40 Akamatsu, K., Chen, W., Suzuki, Y. et al. (2010). Preparation of monodisperse chitosan microcapsules with hollow structures using the SPG membrane emulsification technique. *Langmuir* 26 (18): 14854–14860.

41 Akamatsu, K., Ikeuchi, Y., Nakao, A. et al. (2012). Size-controlled and monodisperse enzyme-encapsulated chitosan microspheres developed by the SPG membrane emulsification technique. *Journal of Colloid and Interface Science* 371 (1): 46–51.

42 Niu, R., Yang, Y.M., Wang, S.W. et al. (2019). Chitosan microparticle-based immunoaffinity chromatography supports prepared by membrane emulsification technique: characterization and application. *International Journal of Biological Macromolecules* 131: 1147–1154.

43 Wu, J., Wei, W., Wang, L.Y. et al. (2008). Preparation of uniform-sized pH-sensitive quaternized chitosan microsphere by combining membrane emulsification technique and thermal-gelation method. *Colloids and Surfaces B: Biointerfaces* 63 (2): 164–175.

44 Wu, J., Su, Z.G., and Ma, G.H. (2006). A thermo- and pH-sensitive hydrogel composed of quaternized chitosan/glycerophosphate. *International Journal of Pharmaceutics* 315: 1–11.

45 Zhou, Q.Z., Wang, L.Y., Ma, G.H. et al. (2007). Preparation of uniform-sized agarose beads by microporous membrane emulsification technique. *Journal of Colloid and Interface Science* 311 (1): 118–127.

46 Zhou, Q.Z., Liu, X.Y., Liu, S.J. et al. (2008). Preparation of uniformly sized agarose microcapsules by membrane emulsification for application in sorting bacteria. *Industrial & Engineering Chemistry Research* 47: 6386–6390.

47 Zhao, X., Wu, J., Gong, F.L. et al. (2014). Preparation of uniform and large sized agarose microspheres by an improved membrane emulsification technique. *Powder Technology* 253: 444–452.

48 Zhang, H.W., Zhao, L., Huang, Y.D. et al. (2018). Uniform polysaccharide composite microspheres with controllable network by microporous membrane emulsification technique. *Analytical and Bioanalytical Chemistry* 410 (18): 4331–4341.

49 Zhang, Y.L., Wei, W., Lv, P.P. et al. (2011). Preparation and evaluation of alginate–chitosan microspheres for oral delivery of insulin. *European Journal of Pharmaceutics and Biopharmaceutics* 77 (1): 11–19.

50 You, J.O., Park, S.B., Park, H.Y. et al. (2001). Preparation of regular sized Ca-alginate microspheres using membrane emulsification method. *Journal of Microencapsulation* 18 (4): 521–532.

51 Liu, X.D., Bao, D.C., Xue, W.M. et al. (2003). Preparation of uniform calcium alginate gel beads by membrane emulsification coupled with internal gelation. *Journal of Applied Polymer Science* 87 (5): 848–852.

52 Wu, Q.X., Lin, D.Q., and Yao, S.J. (2016). Fabrication and formation studies on single-walled CA/NaCS-WSC microcapsules. *Materials Science and Engineering: C* 59: 909–915.

53 Coombs OBrien, J., Torrente-Murciano, L., Mattia, D. et al. (2017). Continuous production of cellulose microbeads via membrane emulsification. *ACS Sustainable Chemistry & Engineering* 5 (7): 5931–5939.

54 Zhou, Y., Jiang, R.S., and Perkins, W.S. (2018). Morphology evolution and gelation mechanism of alkali induced konjac glucomannan hydrogel. *Food Chemistry* 269: 80–88.

55 Mi, Y.C., Li, J., Zhou, W.Q. et al. (2016). Improved stability of emulsions in preparation of uniform small-sized konjac glucomanna (KGM) microspheres with epoxy-based polymer membrane by premix membrane emulsification. *Polymers* 8 (3): 53.

56 Li, X.Q., Gong, F.L., and Ma, G.H. (2014). Preparation of uniform konjac glucomannan gel microspheres by rotating membrane emulsification. *The Chinese Journal of Process Engineering* 14 (2): 296–303.

57 Li, X.Q., Li, Q., Gong, F.L. et al. (2015). Preparation of large-sized highly uniform agarose beads by novel rotating membrane emulsification. *Journal of Membrane Science* 476: 30–39.

58 Piacentini, E., Yan, M., and Giorno, L. (2017). Development of enzyme-loaded PVA microspheres by membrane emulsification. *Journal of Membrane Science* 524: 79–86.

59 Piacentini, E., Bazzarelli, F., Poerio, T. et al. (2020). Encapsulation of water-soluble drugs in Poly(vinyl alcohol) (PVA)-microparticles via membrane emulsification: influence of process and formulation parameters on structural and functional properties. *Materials Today Communications* 24: 100967.

60 Morelli, S., Holdich, R.G., and Dragosavac, M.M. (2016). Chitosan and poly(vinyl alcohol) microparticles produced by membrane emulsification for encapsulation and pH controlled release. *Chemical Engineering Journal* 288: 451–460.

61 Ohta, S., Hashimoto, K., and Fu, X. (2018). Development of human-derived hemoglobin–albumin microspheres as oxygen carriers using Shirasu porous glass membrane emulsification. *Journal of Bioscience and Bioengineering* 126 (4): 533–539.

4

Preparation of Uniform Microcapsules and Microspheres from W/O/W Double Emulsion

4.1 Introduction

Double emulsion is complex system, also called "emulsion in emulsion," in which the droplets of the dispersed phase contain one or more types of smaller dispersed droplets, especially water-in-oil-in-water (W/O/W) double emulsion shows great potential as compartmentalized system to encapsulate multiple components for protecting their activities. The main issue when using double emulsion is their production in a well-controlled manner, with uniform droplet size.

For membrane emulsification, two kinds of methods are developed, direct membrane emulsification and premix membrane emulsification (rapid membrane emulsification). In the latter method, the coarse double-emulsion droplets pass through the membrane to be broken up into finer droplets. The resulting droplet size distribution is slightly wider than that obtained with direct membrane emulsification, but higher fluxes can be obtained. In this chapter, we will focus on the direct membrane emulsification (cross-flow membrane emulsification), and will discuss premix membrane emulsification in Chapter 5.

In the direct membrane emulsification, the to-be-dispersed phase is pressed through microporous membrane while the continuous phase flows along the membrane surface. Droplets grow at pores and detach at a certain size, which is determined by the balance between the adhesion torque and detachment torque, as described in Chapter 1. Double emulsion with uniform size can be produced by pressing a single emulsion (dispersed phase) through a microporous membrane while the continuous phase flows along the membrane surface (Figure 4.1). Hydrophilic membrane is needed to prepare W/O/W double emulsion and hydrophobic membrane for O/W/O double emulsion [2]. The primary emulsion may be produced by means of a conventional method (sonication and high-shear mixing) or also by membrane emulsification [3].

Our group prepared lipiodol W/O/W double emulsion as drug carrier by direct membrane emulsification [4]. Mitomycin C and Adriamycin were used as model drugs, respectively. The double-emulsion droplet size distribution prepared by direct membrane emulsification was significantly improved compared with mechanical stirring method (Figure 4.2 and Figure 4.3a). The size distribution of droplets by

Novel Membrane Emulsification: Principles, Preparation, Processes, and Bioapplications,
First Edition. Guanghui Ma.

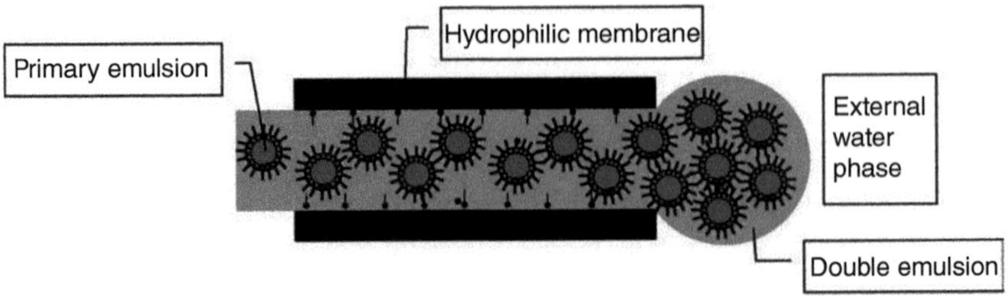

Figure 4.1 Formation of double emulsion using direct membrane emulsification system. Source: Liu et al. [1]/with permission of Elsevier.

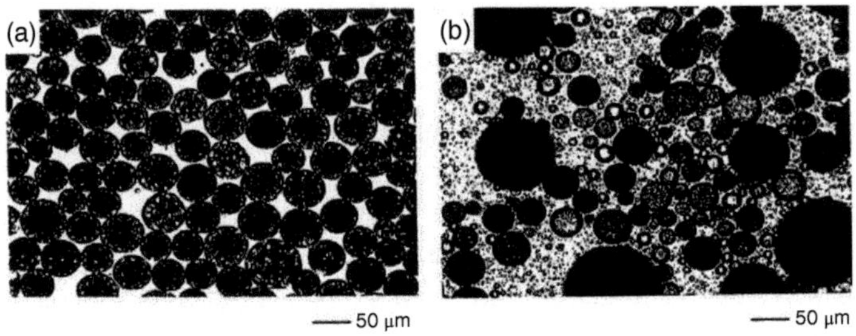

Figure 4.2 Optical micrographs of lipiodol W/O/W double emulsion prepared by (a) direct membrane emulsification and (b) mechanical stirring method. Source: Shao [4]/with permission from Chinese Academy of Sciences.

direct membrane emulsification was lower than 15%, and after 30 days of storage at 4 °C, there was no obvious coalescence between the double-emulsion droplets and no significant change in droplet size. It can be seen from Figure 4.3b, the size distribution of the double emulsion before and after storage is relatively narrow. Encapsulation efficiency of lipiodol W/O/W double emulsion by direct membrane emulsification was high up to 90%, and shelf stability at 4 °C attained to 30 days with lower than 10% of drug released, and the drug can be released steadily at 37 °C.

It was also found that the addition of BSA macromolecule in inner aqueous phase enhanced the mechanical strength of the film between lipiodol and inner aqueous phase, which improved the encapsulation efficiency and the stability of lipiodol double emulsion. By adding cholesterol and phosphatidylcholine together into the oil phase, a structure like liposome was probably formed, which also improved the encapsulation efficiency and the stability of lipiodol double emulsion.

J.D. Ilić et al. compared the W/O/W double emulsion prepared by direct membrane emulsification and homogenization, and found that the span value was decreased to 0.6 ± 0.02 for W/O/W double emulsion by direct membrane emulsification, while the corresponding double emulsion prepared by homogenization had span of about 1.15 [5]. Visual inspection of these micrographs can confirm the improvement of uniformity with the use of direct membrane emulsification (Figure 4.4).

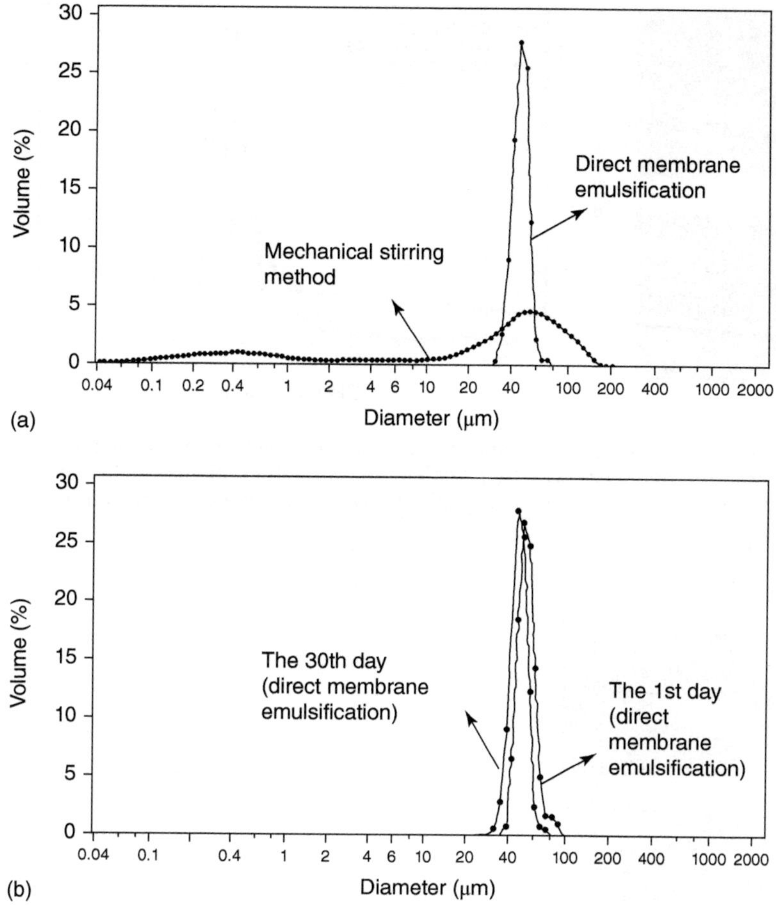

Figure 4.3 The size distributions of lipiodol double emulsion prepared by direct membrane emulsification and mechanical stirring method. (a) direct membrane emulsification vs. mechanical stirring method; (b) the 1st day vs. the 30th day (direct membrane emulsification). Source: Shao [4]/Chinese Medical Association.

4.2 Preparation of Uniform Microcapsules

The most often used double-emulsion technique for preparation of microcapsules is double emulsion-solvent evaporation technique. Formation of double emulsion is followed by evaporation of the organic solvent (O) from the dispersed phase, leading to a point of insolubility and consequently, hardening the polymer. The solvent may be evaporated under reduced pressure via rotary evaporator or by simple stirring at ambient temperature depending upon the boiling point of organic solvent.

Our group prepared uniform-sized biodegradable poly(lactide acid) (PLA) and poly(dl-lactic-*co*-glycolic acid) (PLGA) microcapsules by combining direct membrane emulsification technique and double emulsion-solvent evaporation method [1, 6, 7]. Compared with the stirring method, it was found that the size was more

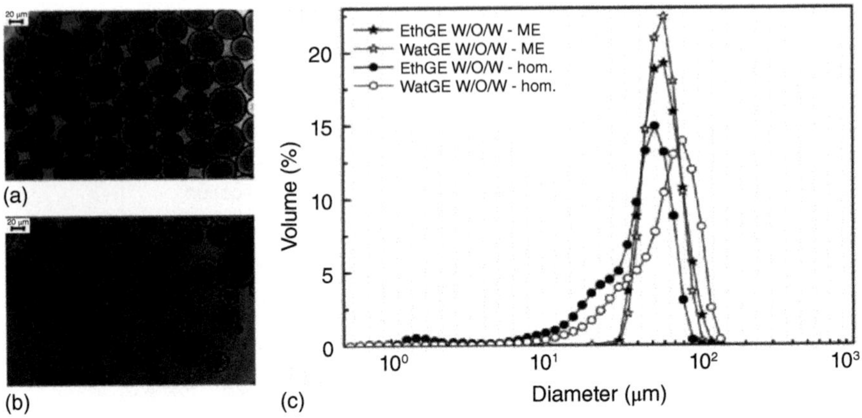

Figure 4.4 Micrographs of W/O/W double emulsion obtained by (a) direct membrane emulsification and (b) homogenization; (c) droplet size distributions of W/O/W double emulsion prepared by direct membrane emulsification and homogenization. Source: Ilić et al. [5]/Elsevier.

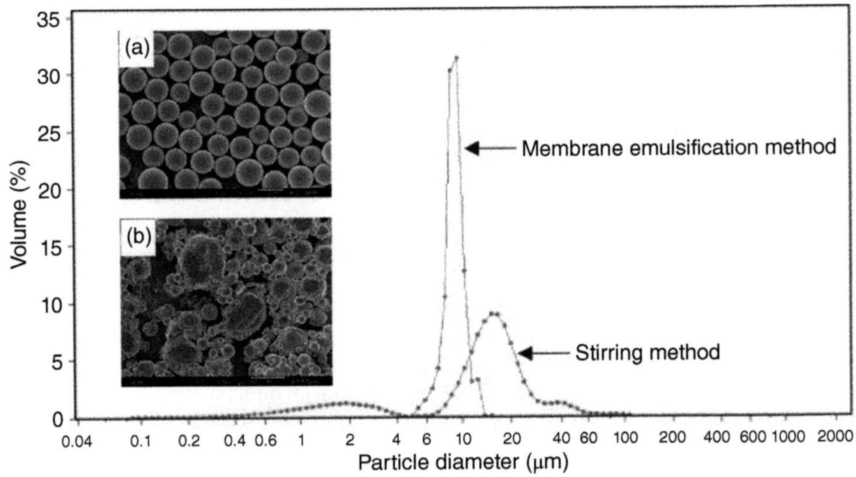

Figure 4.5 Size distributions of PLA microcapsules prepared by direct membrane emulsification and stirring method, respectively. SEM photographs of PLA microcapsules prepared by direct membrane emulsification (a) and stirring method (b). Source: Liu et al. [6]/Elsevier.

uniform when the microcapsules were prepared by direct membrane emulsification technique (Figure 4.5).

It is a big challenge to prepare uniform-sized PLA or PLGA microcapsules by combining the direct membrane emulsification and W/O/W double-emulsion method. Compared with single emulsion (oil/water (O/W) or water/oil (W/O) emulsion), it is more difficult to prepare uniform W/O/W double emulsion for following reasons: (i) the double-emulsion system is more complex, there are more factors affecting size distribution of the double emulsion and (ii) the double emulsion was prepared by pressing W/O primary emulsion through the pores of the membrane into the outer phase, the primary emulsion should be stable during the membrane emulsification process, which had never been encountered in the preparation of single emulsion. Although the addition of oil-soluble emulsifiers can improve the stability of the primary W/O emulsion, it was concerned that it would wet the membrane and leading to polydispersed droplets. So, various factors influencing the microcapsules preparation were investigated [1].

4.2.1 Oil-Soluble Emulsifier on the Size Distribution of Microcapsules

The essential conditions for preparing W/O/W double emulsion with narrow size distribution by membrane emulsification are that the W/O primary emulsion should be stable during the emulsification process, and the interfacial tension between the W/O primary emulsion and the pores of the membrane should be high enough. Although adding a large amount of emulsifier to the oil phase could improve the stability of W/O primary emulsion, that is, to retard the phase separation between the inner aqueous phase and oil phase, it will reduce the interfacial tension between the W/O primary emulsion and the microporous membrane, and broaden the size distribution of W/O/W double-emulsion droplets.

We investigated the effect of the oil-soluble emulsifier concentration on the particle size distribution for the preparation of W/O/W double emulsion-templated microcapsules. The results are shown in Figure 4.6, the prepared microcapsules

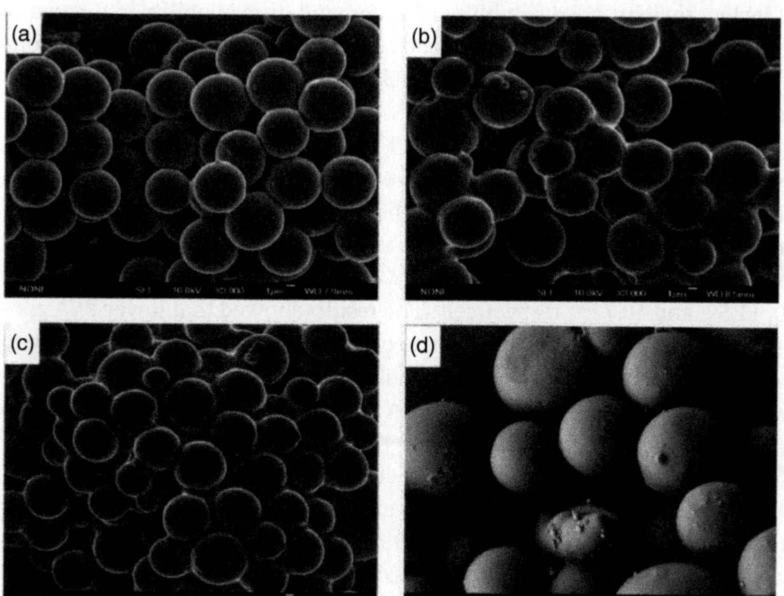

Figure 4.6 (a–c) SEM photographs of PLA microcapsules prepared by combining direct membrane emulsification technique and W/O/W double-emulsion evaporation method. The concentration of Arlacel 83 in the oil phase: (a) 0.25 wt%; (b) 1.0 wt%; (c) 2.0 wt%; and (d) SEM photographs of PLA microspheres prepared by combining direct membrane emulsification and O/W single emulsion evaporation method, Arlacel 83 in the oil phase was 0.25 wt%. Source: Liu et al. [1]/Elsevier.

were all spherical in shape with smooth surfaces. With increasing the amount of Arlacel 83, from 0.25 to 2.0 wt%, the particle size distribution became broader, and the CV value increased from 13.90% to 23.41%. This result confirmed that lower concentration of oil-soluble emulsifier was beneficial for the formation of more uniform-sized microcapsules prepared by direct membrane emulsification. However, a unique phenomenon was observed by comparing the cases of O/W and W/O/W double emulsion. At the same concentration of Arlacel 83, the CV value of the microcapsules was higher in O/W system than in W/O/W system. For example, the CV values were 13.90% and 28.67% for these two cases when 0.25 wt% of Arlacel 83 was added (Figure 4.6d). This can be explained that in O/W system, a large fraction of oil-soluble emulsifier was adsorbed on and wetted the pore surface of the membrane, whereas, in W/O/W system, much less of oil-soluble emulsifier was adsorbed on the membrane because most of the oil emulsifier was used to keep the W/O primary emulsion stable. This unique phenomenon enabled the successful preparation of uniform-sized double emulsion.

4.2.2 PVA Concentration in the Outer Aqueous Phase

The presence of a stabilizer in the outer aqueous phase is essential for the successful formation of spherical microcapsules during the emulsification process [8]. The role of the stabilizer is to prevent the coagulation of microcapsules during emulsification and solvent removal process. PVA is the most commonly used stabilizer, mainly because of its low toxicity, good solubility in water, and its availability in a range of molecular weights [9]. Here, PVA was chosen as the stabilizer in the outer aqueous phase and the effect of its concentration on the size distribution of microcapsules was investigated. As shown in Figure 4.7a, when PVA concentration was 0.1 wt%, the CV value of microcapsules was rather high (39.50%), but with the increase of PVA concentration, the CV value of microcapsules decreased gradually from 39.20 to 10.72%. After PVA concentration exceeded 1.0 wt%, the particle size distribution obtained was rather narrow, and the CV values were all below 14.0% and the decreasing trend became less apparent. For 1.0, 1.5, and 2.0 wt% PVA concentrations, the CV values were 13.90, 11.59, and 10.72%, respectively. The results showed that lower PVA concentration (0.1 or 0.5 wt%) was not sufficient to prevent the coagulation of the formed emulsion droplets, eventually leading to the polydispersity of microcapsules.

As shown in Figure 4.7b, the lower the PVA concentration, the larger the mean diameters of microcapsules, which indicated that the polydispersity of microcapsules was mainly caused by the coalescence of emulsion droplets when PVA concentration was lower than 1.0 wt%. Although the above results suggested that a higher concentration of PVA was beneficial for the formation of more uniform microcapsules, it was better to use PVA as lower as possible because PVA tends to adhere to microcapsules and is difficult to be washed away [9].

4.2.3 Transmembrane Pressure on the Size Distribution of Microcapsules

Transmembrane pressure (P_{tm}) plays an important role in the membrane emulsification process. Firstly, the P_{tm} must be kept constant during the whole emulsification

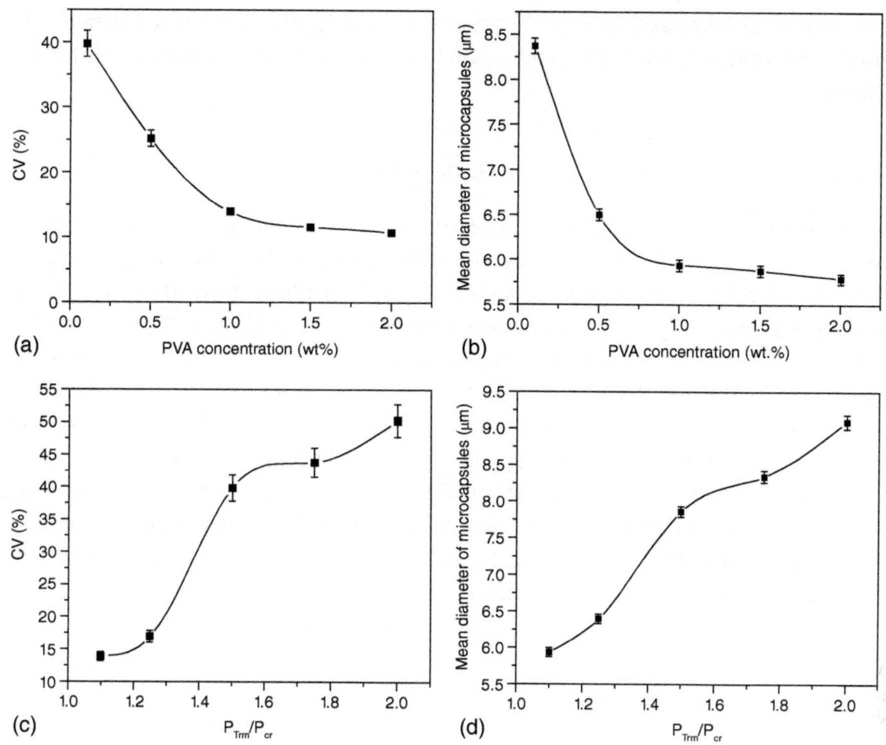

Figure 4.7 (a, b) Effect of PVA concentration in the outer aqueous phase on the size distribution of PLA microcapsules and mean diameters of microcapsules prepared by direct membrane emulsification technique; (c) effect of P_{tm}/P_{cr} on the size distribution of PLA microcapsules prepared by direct membrane emulsification technique; and (d) relationship between the mean size of PLA microcapsules and the pore size of membrane used. Source: Liu et al. [1]/with permission of Elsevier.

process, otherwise, polydispersion of the double-emulsion droplets will occur. Secondly, a suitable P_{tm} must be chosen, and too high P_{tm} leads to jet-like steam of dispersed phase and very large droplets. On the other hand, if the P_{tm} is too low, the emulsification rate will be very slow, resulting in low productivity. In order to select a suitable P_{tm}, in our study, the W/O primary was pressed into the outer aqueous phase with different P_{tm}, and the relationship between the P_{tm} and the size distribution of microcapsules for the membrane with pore size of 2.8 μm was investigated. The PVA concentration in the outer aqueous phase was 1.0 wt%. As shown in Figure 4.7c, the particle size distribution of microcapsules prepared was rather narrow (CV value was 13.90%) when P_{tm} was slightly higher than the critical pressure (P_{cr}, P_{tm}/P_{cr} = 1.1). However, when the P_{tm}/P_{cr} exceeded 1.25, the CV values increased dramatically from 16.90 to 50.20%. For the transmembrane pressure, there exist different results with different emulsification systems. Williams et al. [10] have suggested that between 2 and 10 times the minimum membrane pressure is probably a usable range. The broader size distribution of microcapsules was mainly due to the formation of larger emulsion droplets during the membrane emulsification process, which could be confirmed from Figure 4.7d, the mean diameter of microcapsules increased apparently with increasing P_{tm}/P_{cr}. The mechanism for P_{tm} effect on droplets distribution can be explained by Torque balance model in 1.2.3 of Chapter 1.

4.2.4 The Membrane with Different Pore Size

Next, the PLA microcapsules with different sizes were prepared under the optimized conditions by using the membrane with different pore sizes. PVA and Arlacel 83 concentrations in the outer aqueous phase and oil phase were fixed at 1.0 and 0.25 wt%, respectively. The transmembrane pressure was slightly higher than the critical pressure and remained constant during the emulsification process. As shown in Figure 4.8, the PLA microcapsules were rather uniform, and their CV values were all under 14.0%. The relationship between the pore size of membrane and the particle size of the microcapsules is shown in Figure 4.9. A linear relationship was observed, similar to the result for the O/W single emulsion system [11]; that is, the particle size of the PLA microcapsules mainly depended on the pore size of the membrane. PLA microcapsules with desired sizes could be prepared easily by choosing a membrane with appropriate pore size. Controlling the diameter of PLA microcapsules is very important for the application of drug delivery systems, because the bioavailability and release rate of the drug largely depend on the diameter of the microcapsules.

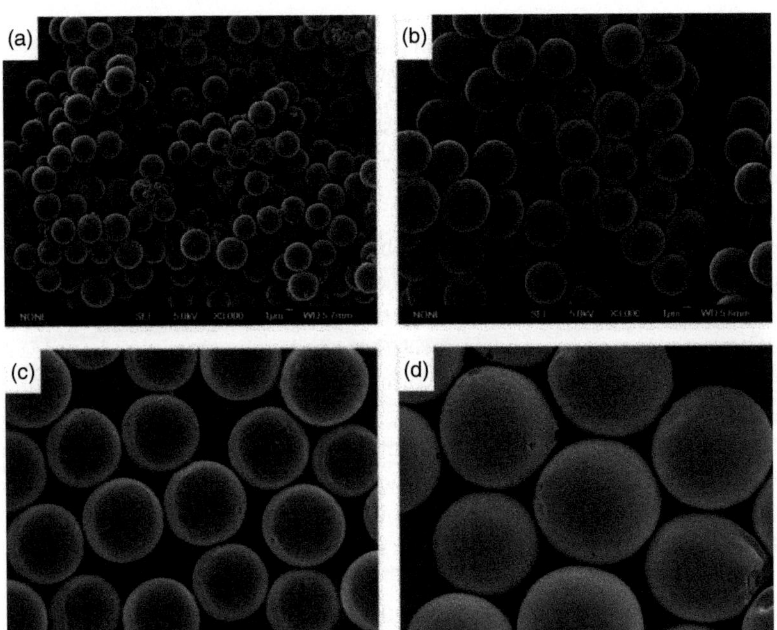

Figure 4.8 SEM photographs of PLA microcapsules prepared by membrane with different pore sizes: (a) 1.4 μm, (b) 2.8 μm, (c) 5.2 μm, and (d) 10.2 μm. Source: Liu et al. [1]/Elsevier.

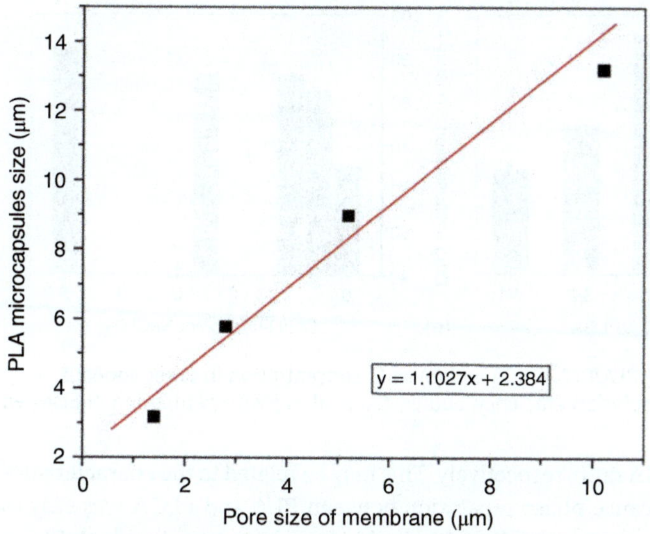

Figure 4.9 Relationship between the mean size of PLA microcapsules and the pore size of membrane used. Source: Liu et al. [1]/with permission of Elsevier.

4.2.5 Microcapsules for Drug Encapsulation

Double emulsion-templated microcapsules for drug encapsulation is performed in two steps. In the first step, water soluble drugs are incorporated in the inner aqueous phase (W) and polymer/lipophilic drugs are added into oil phase (O), then both phases are homogenized by proper agitation (mechanical stirring, sonication, or membrane emulsification) to form the primary emulsion (W/O). Then, the primary emulsion is emulsified with the outer aqueous phase containing appropriate stabilizer to form double emulsion (W/O/W) by membrane emulsification. Then, the solid polymer microcapsules were obtained by simply evaporating the solvent or cooling the melted oil phase.

Our group prepared uniform-sized biodegradable PLA/PLGA microcapsules for drug encapsulation by combining membrane emulsification technique and double emulsion-solvent evaporation method [7]. Various factors of the preparation process influencing the drug encapsulation efficiency and the drug cumulative release were investigated.

4.2.5.1 Composition of Polymers

PLA and PLGA are the most investigated polymers in terms of toxicological and clinical data. Here, due to the slow degradation rate of PLA, PLGA with lower molecular weight was added to increase the cumulative release rate of drugs encapsulated, and the effect of composition of polymers (PLA/PLGA ratio) on both drug encapsulation efficiency was investigated. The total amount of PLA and PLGA was 0.25 g. As shown in Figure 4.10a, the highest drug encapsulation efficiency was obtained when the PLA/PLGA ratio was equal to 1/1, and subsequently decreased with the

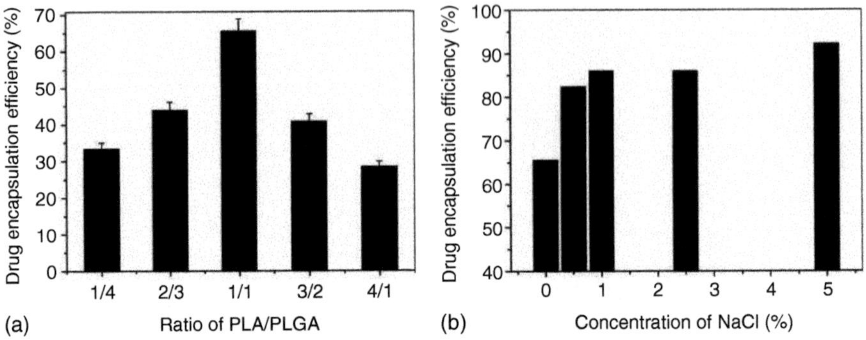

Figure 4.10 Effect of (a) PLA/PLGA ratio and (b) NaCl concentration in outer aqueous phase on the drug encapsulation efficiency. Source: Liu et al. [7]/with permission of Elsevier.

increase of PLA or PLGA ratio, respectively. This may be related to the characteristics of PLA and PLGA. Because phase separation between PLA and PLGA was easy to occur, it tended to form microcapsules with double-layer structure, in which PLGA formed the inner-layer wall that faced inner aqueous phase and PLA formed the outer-layer wall [12]. As PLGA is more hydrophilic than PLA, hydrophilic drugs tended to locate in the inner-layer PLGA phase, while the outer-layer PLA phase formed a hydrophobic wall to retard the leakage of drugs, thus the drug encapsulation efficiency was enhanced. When the PLA/PLGA ratio was lower than 1/1, the PLGA inner layer was thicker, and the distance between the drug located in inner-layer phase and the surface of microcapsules was shortened, leading to the leakage of the drugs into the outer aqueous phase and led to the decrease of drug encapsulation efficiency. Interestingly, when the PLA/PLGA ratio was higher than 1/1, the drug encapsulation efficiency was also decreased. It is generally believed that there are three main factors that are considered to affect the drug encapsulation efficiency in the process of emulsification and solidification of double emulsion: (i) coalescence between inner aqueous phase and outer aqueous phase, leading to the leakage of the drug into the outer aqueous phase. Therefore, the more stable the inner aqueous phase, the higher the drug encapsulation efficiency; (ii) coalescence and breakup of double emulsion, also leading to the leakage of drug; and (iii) diffusion of drug from inner to outer aqueous phase through oil phase. During the membrane emulsification process, the probability for coalescence and breakup of the double emulsion was little because of their uniform size and stability. Therefore, the decrease in drug encapsulation efficiency with increase in PLA/PLGA ratio was mainly due to the coalescence between inner aqueous phase and outer aqueous phase.

4.2.5.2 The Inner Aqueous Phase Volume

To investigate its influence on the drug encapsulation efficiency and cumulative release profile, the volume of oil phase was maintained at 2.5 ml, and the inner aqueous phase volume was changed. When the inner aqueous phase volume increased from 250 to 625 μl, the drug encapsulation efficiency decreased sharply

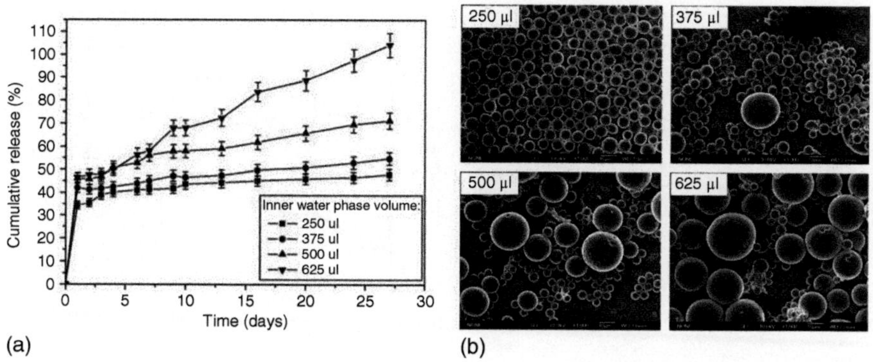

Figure 4.11 (a) Effect of the inner aqueous phase volume on the drug cumulative release profile; (b) SEM photographs of the PLA/PLGA microcapsules with different inner aqueous phase volume. Source: Liu et al. [7]/Elsevier.

from 65.61% to 14.19%, and initial release increased from 34.43% to 46.54%, respectively. In addition, the cumulative release rate of recombinant human insulin (rhI) after initial release was accelerated as the volume of the internal aqueous phase increased (Figure 4.11a). As mentioned above, the loss of drug entrapment during the preparation of drug-loaded microcapsules is mainly due to the coalescence between inner aqueous phase and outer aqueous phase during the double-emulsion formation and the solvent removal process. Increasing the inner aqueous phase in the primary W/O emulsion results in an increased coalescence probability between inner and outer aqueous phase and thus inducing the decrease in drug encapsulation efficiency. In addition, increasing the volume of inner aqueous phase also increased the internal cave volume of microcapsules, shortened the distance from drug to the surface of the microcapsules, and accelerated the cumulative release of drugs. In addition, from Figure 4.11b, it was obvious that increasing the volume of inner aqueous phase led to the formation of a few larger microcapsules. This implied that the coalescence and breakup between the droplets occurred more frequently during the membrane emulsification and solidification process when a large volume of inner aqueous phase was used. Therefore, the drugs were prone to escape from the droplets and the drug encapsulation efficiency was decreased.

4.2.5.3 NaCl Concentration in Outer Aqueous Phase

In the process of preparing microcapsules by double emulsion-solvent evaporation method, NaCl was usually added into the outer aqueous phase to adjust the osmotic pressure difference between the inner and outer aqueous phases of W/O/W double emulsion. Different osmotic pressure affects microcapsules formation, the drug encapsulation efficiency, and the cumulative release profile. The rhI encapsulation efficiency increased from 65.61% to 92.21% when NaCl concentration in outer aqueous phase increased from 0 to 5.0 wt% (Figure 4.10b). When the osmotic pressure of the outer aqueous phase is greater than that of the inner aqueous phase, there would be an influx of water from inner aqueous phase, as a result, the microcapsules

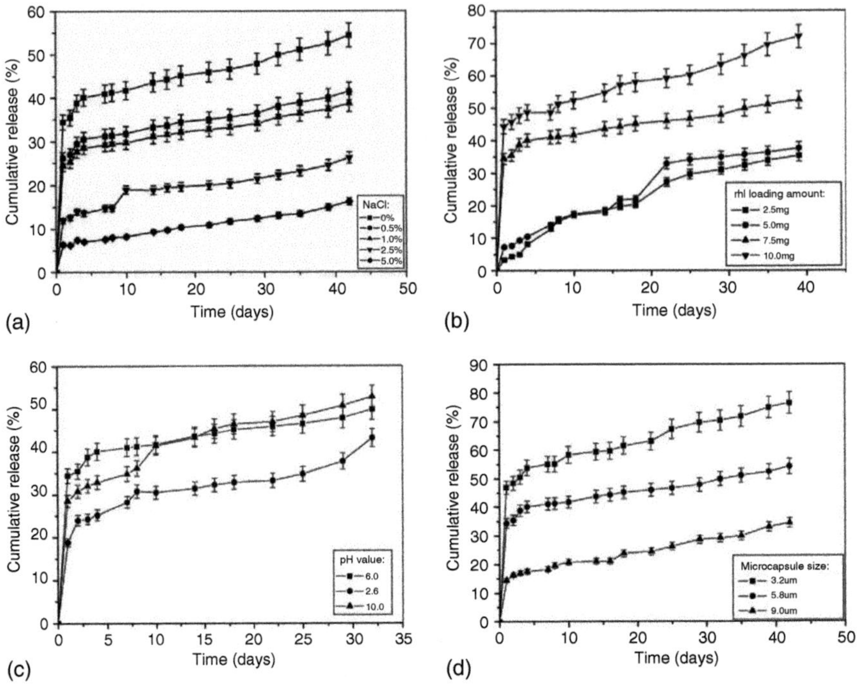

Figure 4.12 Effect of (a) NaCl concentration in outer aqueous phase, (b) rhI-loading amount, (c) pH value in outer aqueous phase, and (d) the microcapsules size on the drug cumulative release profile. Source: Liu et al. [7]/with permission of Elsevier.

are denser and have smaller particle sizes. This, in turn, inhibits drug leakage and is considered to be favorable for the enhancement of the drug encapsulation efficiency.

Due to the influence of the osmotic pressure, the concentration of NaCl in outer aqueous phase also affects the rhI cumulative release profile (Figure 4.12a). The initial release on the first day decreased from 34.43% to 6.33% when NaCl concentration increased from 0 to 5.0 wt%. It was obvious that the initial release was decreased when NaCl concentration in outer aqueous phase was higher. When NaCl concentration in outer aqueous phase was lower, it was easier for the drugs to diffuse to the surface of microcapsules, resulting in the faster "burst effect." In addition, similar release rates after the "burst effect" were found. This suggested that the phenomenon was mainly due to the release of drugs located near the surface of microcapsules. Therefore, it is necessary to adjust osmotic pressure of outer aqueous phase according to the amount of drug in the inner aqueous phase in order to prepare microcapsules with high drug encapsulation efficiency and lower initial release by membrane emulsification technique.

4.2.5.4 Drug-Loading Amount

Drug-loading amount is a key factor not only influencing the drug encapsulation efficiency but also affecting the drug cumulative release profiles. Figure 4.12b shows that the cumulative release profiles were also related to the rhI-loading amount. At higher loading amount, rhI followed a slow and pseudo-linear release after a larger initial burst. However, at lower loading amount, rhI released little by little and increased as a function of time following a smaller initial burst. The difference in drug release rate is due to the fact that there is a larger rhI concentration gradient between the microcapsules and the outer aqueous phase when the rhI-loading amount is higher. Since the gradient is the driving force for rhI diffusion, 4.0 wt% loading resulted in a higher initial burst and faster release rate. Moreover, the higher the actual loading level, the more rhI is distributed near the surface area of the microcapsules. This led to a greater initial release. As the rhI was released, it left more pores and interconnected channels for the release of the remaining rhI. However, rhI has to travel a longer path under a lower concentration gradient. As a result, the release rate of rhI from the microcapsules decreased at the later stage (Figure 4.12b). These results suggested that drug-loading amount should be taken into account when trying to achieve a desired release profile.

4.2.5.5 pH Value in Outer Aqueous Phase

The pH value of outer aqueous phase has been considered to be an important factor affecting the drug encapsulation efficiency and cumulative release. The drug encapsulation efficiency was significantly improved when the pH value of the outer aqueous phase was 6.0, which was close to the isoelectric point of rhI encapsulated. At this isoelectric point, the solubility of the drug in the outer aqueous phase decreased, and the drug is more easily adsorbed to the surface of the microcapsule, which is favorable to the improvement of the encapsulation efficiency of the drug. However, if the improvement of the drug encapsulation efficiency was just due to the adsorption of the drug onto the surface of the microcapsules, it would inevitably produce the high "burst effect." Figure 4.12c shows that at pH 6.0, the initial release of rhI exceeded 34.0%, which is larger than those prepared in the acidic (pH 2.6) or basic (pH 10.0) outer aqueous phase. This may indicate that at pH 6.0, a large amount of the drug was on the surface of the microcapsules.

4.2.5.6 Microcapsules Size

The size of drug-loaded microcapsules plays an important role in the cumulative release. With the membrane emulsification technique, the size of microcapsules can be controlled accurately by selecting the appropriate pore size of the membrane, because there exists a linear relationship between the microcapsules size and the pore size of membrane [11]. As shown in Figure 4.12d, the smaller the size of the microcapsule is, the faster the drug is released from the microcapsule. This was mainly due to the larger specific surface between the microcapsules and the outer aqueous phase. As the size of microcapsules prepared can be controlled accurately and the repeatability of the experiment was rather good, it is convenient to control the drug cumulative release accurately.

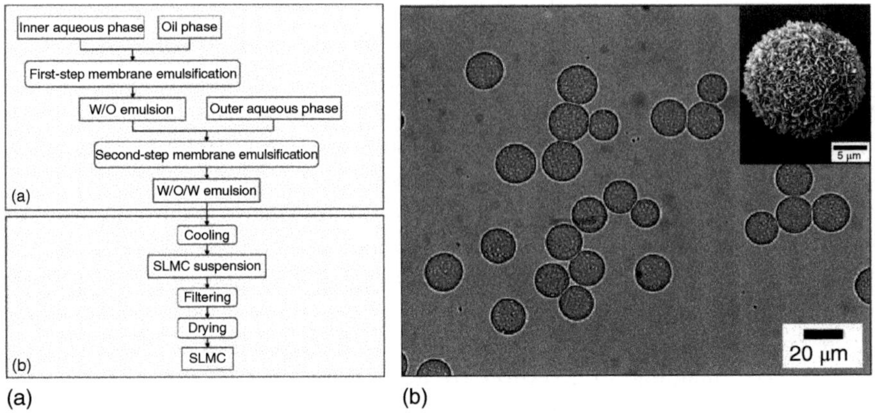

Figure 4.13 (a) Preparation diagram of SLMCs using the direct membrane emulsification technique; (b) Optical micrograph of a W/O/W double emulsion prepared by direct membrane emulsification technique, and SEM of the resultant solid lipid microcapsules. Source: Kukizaki and Goto [13]/Elsevier.

Microcapsules for drug encapsulation can also be prepared by cooling the high-melting-point oil phase of the W/O/W double emulsion. M. Kukizaki et al. prepared solid lipid microcapsules (SLMCs) with uniform size by temperature-controlled direct membrane emulsification using a high-melting-point triglyceride (tripalmitin) as the encapsulating material (Figure 4.13) [13]. W/O/W double emulsions with a narrow diameter distribution and encapsulating vitamin B12 as a model drug were prepared by a two-step direct membrane emulsification technique using Shirasu porous glass (SPG) membranes. In the first emulsification step, the inner aqueous phase containing vitamin B12 was permeated into the hydrophobic porous membrane (mean pore diameter 0.3 μm) into the oil phase using a magnetic stirrer at a transmembrane pressure of 85 kPa, and uniform-sized W/O emulsion droplets were prepared. In the second emulsification step, the resultant W/O emulsion with narrow droplet diameter distribution was dispersed into the hydrophilic porous membrane into the outer aqueous phase, and W/O/W double emulsion with a narrow droplet size distribution was prepared. During both the first and second emulsification steps, the temperature was kept at 343 K, and the oil phase containing β crystalline form tripalmitin was entirely melted at this temperature. The stirring speed of the magnetic stirrer was kept at 150 rpm. The resultant W/O/W double emulsions were immediately cooled to solidify the oil phase. The resultant SLMC suspensions were filtered through a membrane filter. The SLMCs were obtained after being rinsed with pure water and dried at 293 K.

The encapsulation yield of VB12 in SLMCs with various particle diameters is listed in Table 4.1. The SLMCs had a high encapsulation yield of 93.2–97.7% (weight ratio [w/w]), which was almost independent of the mean SLMC particle diameter, and almost no VB12 leakage was observed during the preparation process. Compared with conventional emulsification techniques, such as homogenization

Table 4.1 Relationship between percentage weight of inner aqueous phase to oil phase of W/O/W double emulsion and water content of SLMCs.

SPG membrane	W/O/W emulsions	SLMCs			
Mean pore diameter (μm)	Mean diameter of W/O droplets [a] (μm)	Calculated percentage weight of inner aqueous phase to oil phase (% w/w)	Mean particle diameter (μm)	Water content (% w/w)	Encapsulation yield of VB12 (% w/w)
1.0	3.1	40	3.2	39.7	93.5
3.2	11.3	40	11.3	40.1	93.2
5.2	17.8	40	17.7	39.8	94.6
7.3	25.2	40	25.3	39.9	95.2
9.9	32.8	40	33.1	40.0	97.7

a) W/O droplets: oil droplets containing inner aqueous phase of W/O/W double emulsion.
Source: Kukizaki and Goto [13]/with permission of Elsevier.

and the rotating stirrer method, the mechanical stress of double emulsion prepared by direct membrane emulsification method is low [11]. Therefore, a high encapsulation yield of the SLMCs can be explained by the mild action of droplet formation during membrane emulsification.

4.2.6 Microcapsules with Controllable Structure

Double emulsion-templated microcapsules tend to have easily controllable inner and surface structure, such as macropores and multicore microcapsules. A. Imbrogno et al. produced highly uniform multicore PCL microcapsules between 40 and 120 μm by direct membrane emulsification and solvent evaporation method. Various factors influencing the microcapsule structure were investigated [14].

4.2.6.1 Polymer Concentrations in Oil Phase

Figure 4.14 showed the W/O/W double emulsion produced with different PCL concentrations (16–30%). Coalescence of the inner aqueous droplets occurred when PCL concentration was 16% and 20% as indicated by the size and the low concentration of the inner aqueous droplets. However, stable concentrated droplets of inner aqueous phase were observed when 30% of PCL was used. The increase of PCL concentration increased the viscosity of the liquid film where the aqueous droplets are dispersed. As a result, their coalescence was reduced, and more inner compartments in microcapsules reveal after droplet solidification.

4.2.6.2 Solidification Rate of the Droplets

Microcapsule structure can be controlled by adjusting solidification rate of the droplets, which is controlled by changing the volume of solidification water used

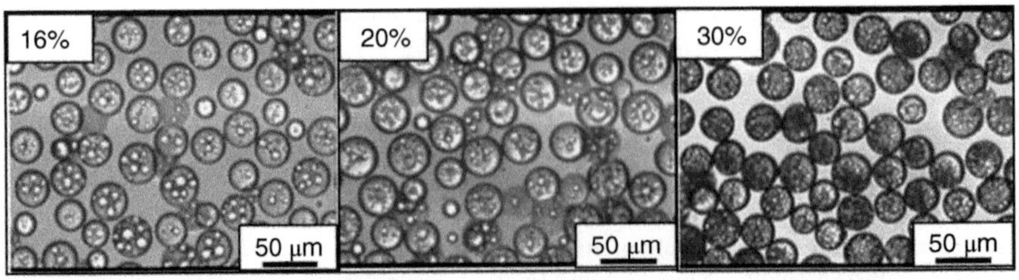

Figure 4.14 Images of W/O/W double emulsion with different concentrations of PCL in the organic phase. Source: Imbrogno et al. [14]/Elsevier.

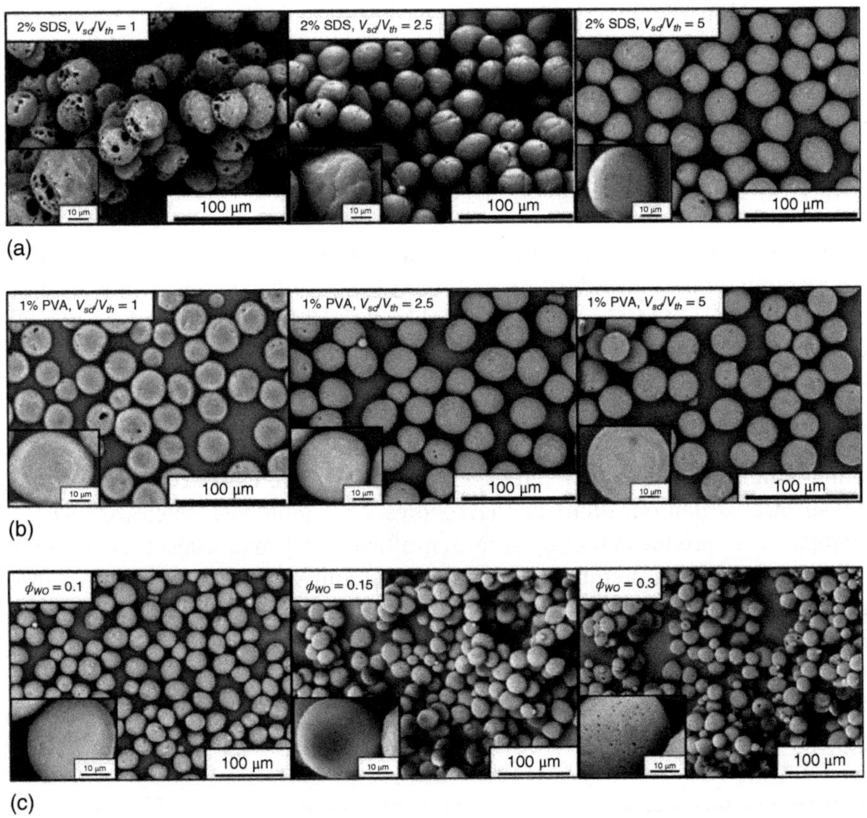

Figure 4.15 SEM photographs of multicore microcapsules produced with (a, b) different ratios between the used (V_{sd}) and the theoretical volume (V_{th}) and different stabilizer (PVA and SDS), and (c) different volume fractions of inner aqueous phase (φ_{WO}). Source: Imbrogno et al. [14]/Elsevier.

for solvent diffusion (V_{sd}) relative to the theoretical volume (V_{th}) (Figure 4.15a). The rapid formation of the polymeric shell is of great importance to reduce the loss of the inner aqueous phase during the microcapsule solidification. When the lowest volume of water ($V_{sd}/V_{th} = 1$) was used for the solidification, porous microspheres were produced. This structure was a result of the diffusion of water between the outer and inner aqueous phases when the organic phase was in a semisolid state for a long time. The solidification time of PCL can be shortened by increasing the solidification water ($V_{sd}/V_{th} = 2.5$), and the formation of pores and holes was avoided. However, since the formation of the polymeric shell at the water droplet interface is not instantaneous, some cracks and slight roughness are observed on the surface of the microcapsules. By further increasing the volume of solidified water ($V_{sd}/V_{th} \geq 5$), the surface of the microcapsule is dense and smooth due to the instant solidification of PCL at the interface, which preserved the integrity of the microcapsules and avoiding the loss of the internal aqueous phase through the droplets.

4.2.6.3 Stabilizer Type

The effect of the stabilizer type to microcapsules morphology was investigated using different stabilizers. Sodium dodecyl sulphate (SDS) (2%) or PVA (1%) was used due to their different chemical properties. Compared with viscosity-enhancing stabilizer PVA, SDS is a low molecular weight anionic emulsifier. In Figure 4.15b, morphologies of microcapsules after solidification with PVA as stabilizing agent at different volumes of water were reported. The particles solidified using PVA as stabilizer are characterized by a dense surface without pores no matter what volume of aqueous phase is used for the solidification of the particles, while SDS tend to form porous structures under same solidification water. PVA is a surface-active polymer that forms a protective film at the water/oil droplet interface [15]. Therefore, when the solidification of the PCL matrix was carried out in low volume of water ($V_{sd}/V_{th} = 1$), the adsorbed PVA polymer layer can slow down the diffusion of water between the outer and inner aqueous phase through the organic phase. This is very important for the scaling-up of the process as less volume of water can be used to maintain the integrity of particle morphology.

4.2.6.4 Volume Fraction of the Inner Aqueous Phase

When the volume fraction of the inner aqueous phase (φ_{WO}) was increased from 0.10 to 0.15, variation of the multicore microcapsules morphology was observed (Figure 4.15c). Under these conditions, the loss of the inner aqueous phase did not occur, and microcapsules are characterized by a dense polymer surface. However, a further increase of (φ_{WO}) to 0.3 resulted in the formation of multicore microcapsules with porous surface produced as a result of the water diffusion through the PCL film during the solidification step.

4.3 Preparation of Composite Microspheres

Functional composite polymer microspheres have attracted extensive attention, and have a wide variety of applications, including in medical diagnostics [16, 17], chemical separation [18, 19], photonics [20], and biological probes [21]. Composite polymer microspheres not only be fabricated via the phase separation of polymers based

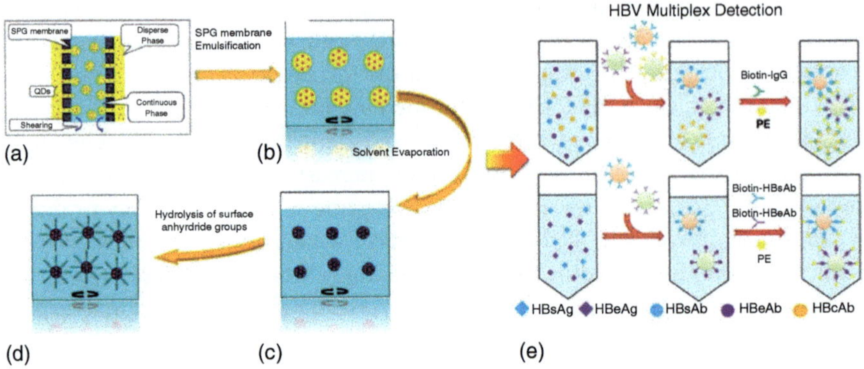

Figure 4.16 Schematic illustration showing the preparation of QD-encoded composite microspheres with surface carboxyl groups by the membrane emulsification-solvent evaporation approach. These steps include (a, b) formation of QD-encapsulating S/O/W double-emulsion droplets; (c) solvent evaporation of S/O/W double-emulsion droplets to form QD microspheres; and (d) hydrolysis treatment to yield QD microspheres bearing carboxyl groups on their surfaces. (e) Flowchart showing the sequential steps involved in HBV multiplexed detection (PE: dye phycoerythrin). Source: Wang et al. [22]/with permission of American Chemical Society.

on O/W or W/O single emulsion droplets but also prepared by double-emulsion droplets (including W/O/W, O/W/O, and S/O/W).

Wang et al. have developed direct membrane emulsification combined solvent evaporation approach to fabricate quantum dots (QDs) encoded composite microsphere with controllable sizes and accurate encoding (Figure 4.16) [22]. Firstly, hydrophobic QDs were dispersed in mixture of poly(styrene-*co*-maleic anhydride) (PSMA) and toluene to form S/O emulsions, and were further emulsified into an aqueous surfactant solution through microporous membrane to form S/O/W double emulsion. The emulsification step was followed by solvent evaporation and surface functionalization of the microspheres in order to generate carboxyl groups, to which antibodies can covalently attach via an EDC/NHS-mediated amidation reaction. Barcode library consisted of 12 barcodes using only one type of QDs, combined through four strength levels and three size levels. Different intensity levels were obtained using various concentrations of QDs in the oil phase (S/O emulsion) prior to direct membrane emulsification, while different size QD-PSMA microspheres were achieved using microporous membranes with different pore sizes. QD barcodes generated by the membrane emulsification approach can provide feasible platforms for both singleplex and multiplex HBV detection.

Our group prepared uniform-sized polystyrene-polyacrylamide (PST-PAAm) hydrophobic–hydrophilic composite microspheres from W/O/W double emulsion by using direct membrane emulsification technique [23]. An AAm aqueous solution was emulsified into an oil phase containing ST, initiator, Span 85, and PST by ultrasonification to obtain W/O primary emulsion. Then, the W/O emulsion was pressed through the uniform pores of microporous membrane into an outer aqueous solution to form uniform-sized W/O/W double emulsion (Figure 4.19a). Various factors influencing preparation of composite microspheres were investigated.

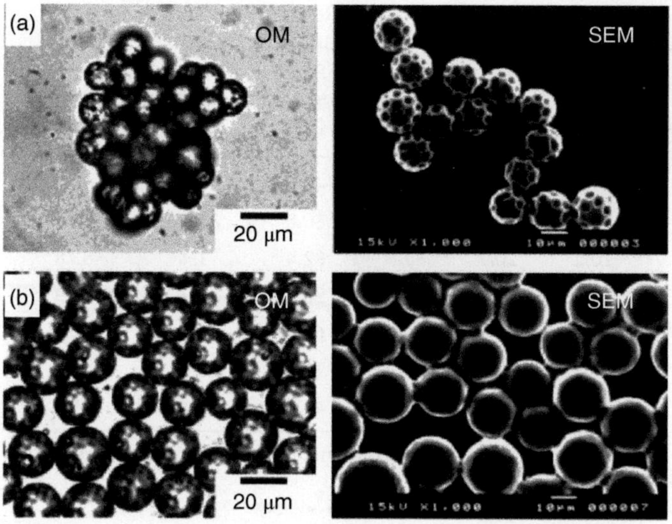

Figure 4.17 OM and SEM photographs of composite microspheres showing effect of water-soluble inhibitor. Inhibitor: (a) HQ and (b) DAP. Source: Ma et al. [23]/American Chemical Society.

4.3.1 Water-Soluble Inhibitor

In this research, we investigated the effect of water-soluble inhibitor on the preparation of composite microspheres. It was found that the monomer would not only polymerize in the droplet but also diffuse into the aqueous phase to produce secondary nucleation. Therefore, it is necessary to add the water-soluble inhibitor into the aqueous phase to inhibit the secondary nucleation. Water-soluble inhibitors hydroquinone (HQ) and diaminophenylene (DAP) were tested separately. The SEM photographs are shown in Figure 4.17. It was found that the diameter decreased largely, from 21.4 μm of the droplet to 14.9 μm of the microspheres when HQ was used. This result suggested that a large amount of AAm and ST diffused into the aqueous phase to form the secondary microspheres, and it is interesting to note that microspheres with dimpled surfaces can be obtained when HQ was used. This was because the inner PAAm droplets escaped to the surface of microsphere and adhered there during polymerization. After washing, the PAAm droplets were removed, leaving dimples on the surface. On the other hand, when the DAP was used, the diameter of the microspheres did not change much compared to the droplet before polymerization. The droplet size before polymerization was 22.3 μm with a CV value of 10.5%, and the diameter of microspheres was 20.6 μm with a CV value of 11.5%. In addition, the surface of the composite microspheres was smooth compared with the case of HQ.

4.3.2 Stabilizer in Outer Aqueous Phase

It was found that the additional amount of stabilizer polyvinyl pyrrolidone (PVP) showed an apparent effect on the stability of composite particles after

Table 4.2 Effect of PVP Amount on the Polymerization Results of Composite Particles.

Run no.	PVP amount (g)	de (μm)	CV of de (%)	dp (μm)	CV of dp (%)	Particle yield (%)	Coagulum (g)	Total conversion (%)
19	2.0	22.1	7.0	21.6	8.0	48.8	42.4	91.2
18	3.0	21.1	10.4	20.9	9.1	80.8	17.2	98.0
13	4.0	21.7	10.2	20.7	8.6	82.4	1.4	83.8
21	5.0	21.4	10.5	21.4	10.7	90.2	8.7	98.9
16	6.0	20.8	9.2	20.8	9.2	68.2	24.0	92.1
17	7.0	22.0	8.2	20.9	14.9	18.9	58.8	77.8
19	2.0	22.1	7.0	21.6	8.0	48.8	42.4	91.2

de; number-average diameter of droplet before polymerization. dp; number-average diameter of polymer particle after polymerization.
Source: Ma et al. [23]/with permission of American Chemical Society.

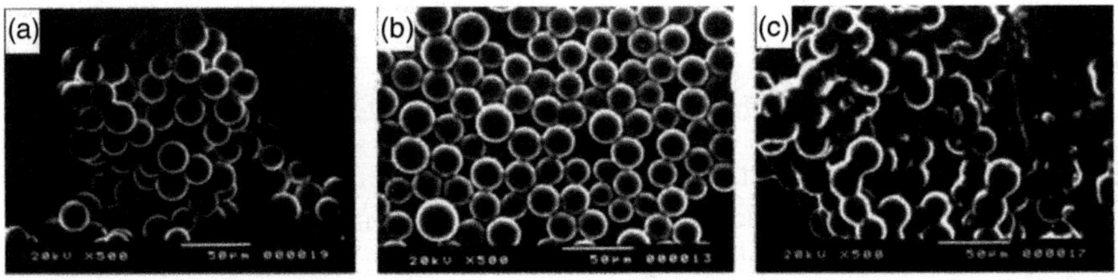

Figure 4.18 SEM photographs of composite particles showing effect of PVP amounts. PVP amount in 225 g of water (g): (a) 2; (b) 4; and (c) 7. Source: Ma et al. [23]/American Chemical Society.

polymerization. PVP was varied from 2 to 7 g in 225 g of water. The polymerization results are summarized in Table 4.2. The SEM photographs of obtained particles are shown in Figure 4.18. It was evident that the formation of coagulum was prevented effectively when PVP increased to 4 g. However, when PVP was increased further, the amount of coagulum increased again, probably due to the bridging effect of too much free PVP.

4.3.3 Cross-Linking Agent

We also investigated how the cross-linking agent affects the morphology of hydrophilic-hydrophobic composite microspheres. The cross-linking agent will increase the crosslinking rate so that the increased viscosity and the network inside the microspheres will affect the morphology (Figure 4.19b). The OM photograph shows that some tiny spherical domains of PAAm were formed. The SEM photograph depicts that no PAAm tiny domain exists, and the surface of the microspheres is golf ball-like, being covered with tiny dents (Figure 4.19c). This surface feature indicated that some of the PAAm domains have localized on

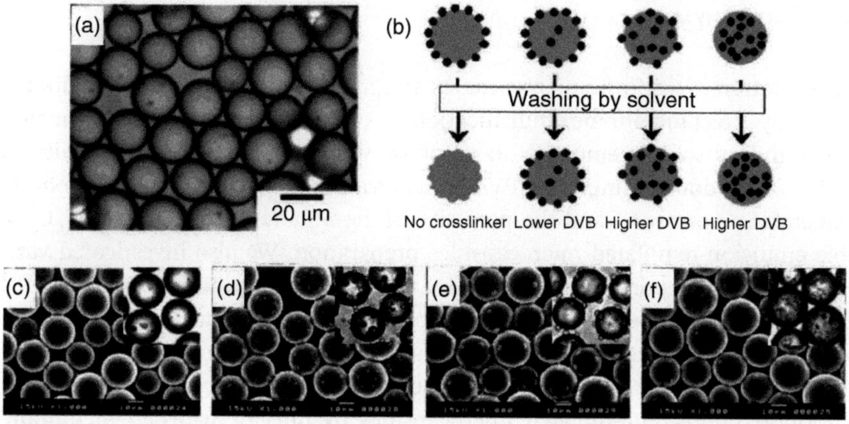

Figure 4.19 (a) Typical OM photograph of W/O/W double emulsion prepared by the direct emulsification technique, (b) Scheme showing effect of crosslinking agent on particle morphology, (c–f) SEM photographs of composite microspheres showing effects of crosslinker DVB amounts. DVB amount (wt% in oil phase): (c) 0; (d) 2; (e) 6; and (f) 10. Source: Ma et al. [23]/American Chemical Society.

the surface of composite microspheres. When they were washed away from the microspheres, tiny dents were formed. Since PAAm is hydrophilic, it tends to move more toward the surface during polymerization. When the PST matrix was not cross-linked, PAAm microdomains move to the surface more easily.

To restrain the movement of PAAm to the surface and eventually into the aqueous phase, cross-linking of PST phase was attempted. Figure 4.19d–f shows the OM and SEM photographs of cross-linked PST-AAm composite microspheres with an increasing amount of DVB in the oil phase. By comparing the OM and SEM photographs, it was known that some of tiny PAAm microdomains were localized on the surface of the composite microspheres. With the increase of DVB amount, the number of PAAm microdomains on the surface of composite microspheres decreases, which means that the PAAm domains entrapped inside the microspheres increased. Furthermore, it can be known that the PAAm microdomains decreased and the microspheres became spherical with a smooth surface when the DVB amount increased to 10 wt%, while an OM photograph showed clear phase separation inside the microspheres. This phenomenon confirmed again that the PAAm was retained inside the microspheres by the cross-linking network of PST matrix. Therefore, due to rapid formation of the PST network, the movement of PAAm and AAm through the PST matrix was gradually suppressed, and more PAAm domains remained inside.

According to the above results, it is clear that the number of the PAAm domains can be controlled by the amount of cross-linking agent added in the oil phase. Since the amides can be easily hydrolyzed to carboxylic acids, the PST-AAm cross-linked composite microspheres have potential applications for adsorption or immobilization media of active species, such as proteins and cells.

4.4 Summary

The preparation of uniform microcapsules and microspheres from W/O/W double emulsion by direct membrane emulsification was mainly introduced in this chapter. It was found oil-soluble emulsion in oil phase was crucial for obtaining uniform size of W/O/W double emulsion. PVA concentration in the outer aqueous phase, transmembrane pressure, and the pore size of membrane influenced the W/O/W double emulsion-templated microcapsules preparation. We also investigated various factors of the preparation process influencing the drug encapsulation efficiency and the drug cumulative release of microcapsules systemically, such as composition of polymers, the inner aqueous phase volume, NaCl concentration in outer aqueous phase, drug-loading amount, pH value in outer aqueous phase, and microcapsules size. Double emulsion-templated microcapsules by one- or two-step membrane emulsification had a higher encapsulation compared with conventional emulsification techniques, such as homogenization and mechanical stirrer method. It can be explained by the mild action of droplet formation during membrane emulsification. The unique advantage of preparing drug-loaded microcapsules by membrane emulsification technique is that the size of microcapsules can be controlled accurately, and thus the drug cumulative release profile can be adjusted flexibly.

Double emulsion-templated microcapsules also tend to have easily controllable inner and surface structures by direct membrane emulsification, such as macropores and multicore microcapsules. Various factors influencing the microcapsules structure were investigated, such as polymer concentrations in oil phase, solidification rate of the droplets, stabilizer type, and volume fraction of the inner aqueous phase. Uniformed microcapsules with smooth or porous surface structures and controllable inner compartments can be obtained.

Composite polymer microspheres can also be prepared by double-emulsion droplets with direct membrane emulsification technology. Various factors influencing preparation of composite microspheres were investigated, such as water-soluble inhibitor, stabilizer in outer aqueous phase, and cross-linking agent.

The advantages of membrane emulsification routes for double emulsion-templated microcapsules/microspheres synthesis are in uniform particle size that can be controlled over a wide range, low shear environment during droplet generation. However, the throughput of production is limited. It is only suitable for the production of high-value products such as drug delivery systems. Therefore, increasing the dispersed phase flux of direct membrane emulsification is still a challenge. Nevertheless, with the commercialization of the membrane emulsification technique and related equipment, the double emulsion and double emulsion-templated microcapsules/microspheres products in biomedical, food, and cosmetic products field will be developed in the near future.

References

1 Liu, R., Ma, G.H., Wan, Y.H., and Su, Z.G. (2005). Influence of process parameters on the size distribution of PLA microcapsules prepared by combining membrane emulsification technique and double emulsion-solvent evaporation method. *Colloids and Surfaces B-Biointerfaces* 45 (3-4): 144–153.

2 Kai, T., Suma, Y., Ono, S. et al. (2006). Effect of the pore surface modification of an inorganic substrate on the plasma-grafting behavior of pore-filling-type organic/inorganic composite membranes. *J. Polym. Sci. Pol. Chem.* 44 (2): 846–856.

3 Vladisavljevic, G.T. (2015). Structured microparticles with tailored properties produced by membrane emulsification. *Adv. Colloid. Interfac.* 225: 53–87.

4 Shao, H. (2004). *Preparation of Lipiodol Double Emulsion as Drug Carrier by Microporous Membrane Emulsification Technique*. Beijing: Institute of Process Engineering, Chinese Academy of Sciences.

5 Ilić, J.D., Nikolovski, B.G., Petrović, L.B. et al. (2017). The garlic (*A. sativum* L.) extracts food grade W1/O/W2 emulsions prepared by homogenization and stirred cell membrane emulsification. *Journal of Food Engineering* 205: 1–11.

6 Liu, R., Ma, G.H., Meng, F.T., and Su, Z.G. (2005). Preparation of uniform-sized PLA microcapsules by combining Shirasu porous glass membrane emulsification technique and multiple emulsion-solvent evaporation method. *Journal of Controlled Release* 103 (1): 31–43.

7 Liu, R., Huang, S.-S., Wan, Y.-H. et al. (2006). Preparation of insulin-loaded PLA/PLGA microcapsules by a novel membrane emulsification method and its release in vitro. *Colloids and Surfaces B-Biointerfaces* 51 (1): 30–38.

8 Parikh, B.V., Upadrashta, S.M., Neau, S.H., and Nuessle, N.O. (1993). Estrone loaded poly(l-lactic acid) microspheres-preparation, evaluation and invitro release kinetics. *Journal of Microencapsulation* 10 (2): 141–153.

9 Singh, M. and O'Hagan, D. (1998). The preparation and characterization of polymeric antigen delivery systems for oral administration. *Advanced Drug Delivery Reviews* 34 (2-3): 285–304.

10 Williams, R.A., Peng, S.J., Wheeler, D.A. et al. (1998). Controlled production of emulsions using a crossflow membrane part II: Industrial scale manufacture. *Chemical Engineering Research & Design* 76 (A8): 902–910.

11 Nakashima, T., Shimizu, M., and Kukizaki, M. (2000). Particle control of emulsion by membrane emulsification and its applications. *Advanced Drug Delivery Reviews* 45 (1): 47–56.

12 Matsumoto, A., Matsukawa, Y., Suzuki, T. et al. (1997). The polymer-alloys method as a new preparation method of biodegradable microspheres: principle and application to cisplatin-loaded microspheres. *Journal of Controlled Release* 48 (1): 19–27.

13 Kukizaki, M. and Goto, M. (2007). Preparation and evaluation of uniformly sized solid lipid microcapsules using membrane emulsification. *Colloids and Surfaces A: Physicochemical and Engineering Aspects* 293 (1): 87–94.

14 Imbrogno, A., Dragosavac, M.M., Piacentini, E. et al. (2015). Polycaprolactone multicore-matrix particle for the simultaneous encapsulation of hydrophilic and hydrophobic compounds produced by membrane emulsification and solvent diffusion processes. *Colloids and Surfaces B: Biointerfaces* 135: 116–125.

15 Boury, F., Ivanova, T., Panaïotov, I. et al. (1995). Dynamic properties of poly(DL-lactide) and polyvinyl alcohol monolayers at the air/water and dichloromethane/water interfaces. *J. Colloid. Interf. Sci.* 169 (2): 380–392.

16 Han, S.W., Jang, E., and Koh, W.-G. (2015). Microfluidic-based multiplex immunoassay system integrated with an array of QD-encoded microbeads. *Sensors and Actuators B: Chemical* 209: 242–251.

17 Xiang, Y., Yan, H., Zheng, B. et al. (2021). Light-regulated natural fluorescence of the PCC 6803@ZIF-8 composite as an encoded microsphere for the detection of multiple biomarkers. *ACS Sensors* 6 (7): 2574–2583.

18 Wang, Y., Zou, B., Gao, T. et al. (2012). Synthesis of orange-like Fe_3O_4/PPy composite microspheres and their excellent Cr(VI) ion removal properties. *Journal of Materials Chemistry* 22 (18): 9034–9040.

19 Elsaidi, S.K., Sinnwell, M.A., Devaraj, A. et al. (2018). Extraction of rare earth elements using magnetite@MOF composites. *Journal of Materials Chemistry A* 6 (38): 18438–18443.

20 Liu, X.-M. and Li, Y.-S. (2009). One-step facile fabrication of Ag/gamma-Fe_2O_3 composite microspheres. *Materials Science & Engineering C-Biomimetic and Supramolecular Systems* 29 (4): 1128–1132.

21 Yang, Y., Zhu, X., Bao, J. et al. (2020). Preparation of SiO_2/YPO_4:Nd/SiO_2 composite microspheres with near-infrared luminescence and surface functionalization. *Journal of Nanoparticle Research* 22 (11): 320–330.

22 Wang, G., Leng, Y., Dou, H. et al. (2013). Highly efficient preparation of multiscaled quantum dot barcodes for multiplexed hepatitis B detection. *ACS Nano* 7 (1): 471–481.

23 Ma, G.H., Sone, H., and Omi, S. (2004). Preparation of uniform-sized polystyrene-polyacrylamide composite microspheres from a W/O/W emulsion by membrane emulsification technique and subsequent suspension polymerization. *Macromolecules* 37: 2954–2964.

5

Rapid Membrane Emulsification Process for Preparation of Small Microspheres

5.1 Introduction

The optimum controllable particle size prepared by the aforementioned crossflow or called direct membrane emulsification technique (MET) is between 5 and 100 μm, and smaller particle is difficult to be prepared especially for dispersed phase with high viscosity. Our group further developed rapid MET, also called premix MET to prepare uniform small particles, as it is a two-step emulsification process. As shown in Figure 5.1, the coarse droplets (premixed emulsion) with larger diameters were prepared at the first step by membrane emulsification or conventional stirring/homogenization methods, then they were pressed through the membrane quickly. At the optimum condition, the larger droplets could be broken up into smaller uniform droplets. Here, the membrane played the role of homogeneous shearing field to the coarse emulsion. The diameter of the coarse emulsion was not necessarily uniform but should be larger than the membrane pore size. Otherwise, the small droplets passed through the membrane freely and broadened the size distribution. In addition, the interfacial tension of the continuous phase and the membrane pore should be lower, and the uniformity is related to the transmembrane pressure, pore length, viscosity, and so forth. By using rapid MET, particle size can be controlled between 0.1–30 μm with a coefficient of variation (CV) around 15% by membrane pore size [1].

Our group has developed this process from oil in water (O/W) to water in oil (W/O) and double emulsion (O/W/O or W/O/W), and successfully prepared uniform microspheres and microcapsules as small as 100 nm either for monomer or for preformed polymer systems (polylactide material, chitosan, agarose, alginate, glucomannan, etc.) [2–4]. Using this technique, divergent attributes (e.g. porosity property, drug loading capacity, adjuvanticity, and stimuli-responsive or biomimetic merit) can also be endowed with these particles, which are potential in separation media, controlled drug release, antitumor therapy, or vaccine delivery.

Novel Membrane Emulsification: Principles, Preparation, Processes, and Bioapplications, First Edition. Guanghui Ma.

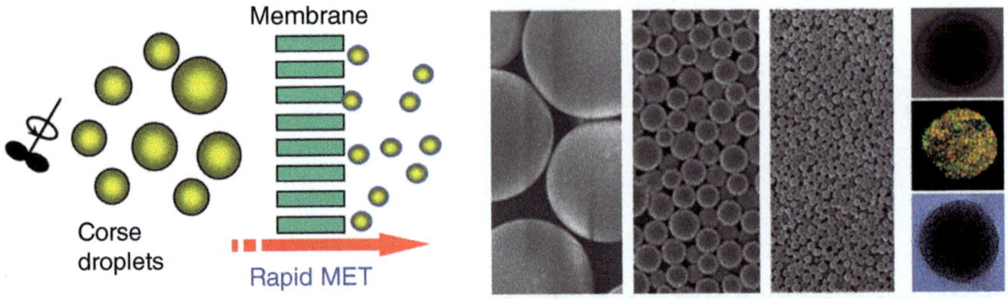

Figure 5.1 Rapid membrane emulsification process (left) and the as-prepared particles (right). Source: Ma and Yue [1]/John Wiley & Sons.

5.2 Preparation of Hydrophobic Microspheres from O/W Emulsion

The O/W emulsion process has been developed along with rapid MET to prepare hydrophobic microspheres that are sourced from polymer or monomer system. The biodegradable polymers, such as poly(lactic acid) (PLA), poly(lactic-*co*-glycolic acid) (PLGA), and poly(caprolactone) (PCL), are ideal materials for preparing biomedical particles. Since polyester is usually insoluble in water and soluble in organic solvents like dichloromethane (DCM), chloroform, and ethyl acetate (EA), O/W type emulsions are usually used to prepare particles for encapsulating oil-soluble drugs or adsorbing water-soluble drug/antigen. For the monomer system like styrene, rapid MET could also be employed to prepare smaller polystyrene (PST) spheres (especially <5 μm) via O/W type emulsions. The formulation is similar to that for the particles prepared by direct MET, and a mixture of ST monomer, initiator, crosslinker, and porogen is usually used as O phase (dispersed phase). The as-prepared micro/nano PST spheres have good rigidity and are resistance to organic solvents, which could be exploited in various areas, like enzyme immobilization, diagnosis carrier, and separation medium.

5.2.1 Preparation of Polylactide-Based Particles from O/W Emulsion

To prepare PLA nanoparticle (NP), PLA/EA mixture was usually used as the oil phase, and poly(vinyl alcohol) (PVA) aqueous solution was used as the aqueous phase. The coarse O/W emulsions were prepared by low-speed stator homogenization and then poured into the premix reservoir. The nanodroplets were achieved by extruding the coarse emulsions through the microporous membrane with a pore size of 1.4 μm. The nanodroplets were stirred overnight to evaporate the organic solvent, and finally, the particles were washed and collected [5]. It was found that PLA particles with a particle size of approximately 320 nm (Figure 5.2) and a homogeneous particle size CV of 16.9% were obtained at an oil/water volume ratio of 1 : 5 and a permeation pressure of 1000 kPa.

In a similar manner, the rapid MET has been developed to prepare PLGA particles. PLGA was dissolved in DCM solvent to obtain a homogeneous oil phase

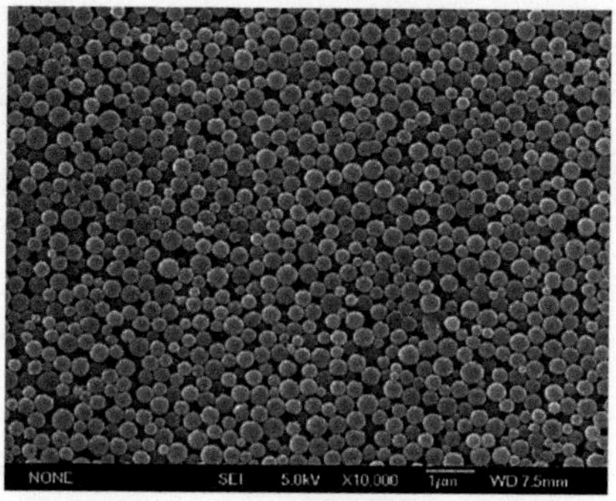

Figure 5.2 SEM photographs of PLA nanoparticles. Source: Wei et al. [5]/Elsevier.

and then mixed with an aqueous phase containing PVA to obtain a coarse emulsion. A membrane pore size of 35 μm was used for membrane emulsification to obtain homogeneous droplets. Once the DCM was removed, the homogeneous PLGA microspheres of about 18 μm with a Span value of 0.469 could be obtained. Besides PLGA, PELA [poly(ethylene glycol-*co*-lactide acid)] is another polylactide material that can be employed to prepare particles with uniform size via O/W type emulsion. During the fabrication process, insoluble drugs can be dissolved in the oil phase with the particle material simultaneously, and finally embedded in the particles after solidification via solvent evaporation.

Except for different molecular weight and the copolymer ratio, the type of PLGA could also be characterized as hydroxy terminal group (PLGA–OH), carboxyl-terminal group (PLGA–COOH), and ester terminal group (PLGA–COOR) based on the available free terminal groups. We employed rapid MET to fabricate PLGA microspheres around 17 μm with –OH, –COOH, and –COOR terminal groups [6]. The coarse O/W emulsions were prepared as the aforementioned and were repeatedly pressed through the pore (35 μm) of membrane under proper pressure. Then, the emulsion was solidified under vacuum pressure by a rotary evaporator for six minutes. Although the smooth surface and regular sphericity were observed for all microspheres (Figure 5.3), the influence on encapsulation efficiency and drug release behavior of active pharmaceutical ingredients were different. It was worth mentioning that the narrow size distribution of the particles guaranteed the good reproducibility and reliability of the subsequent biological/pharmaceutical research.

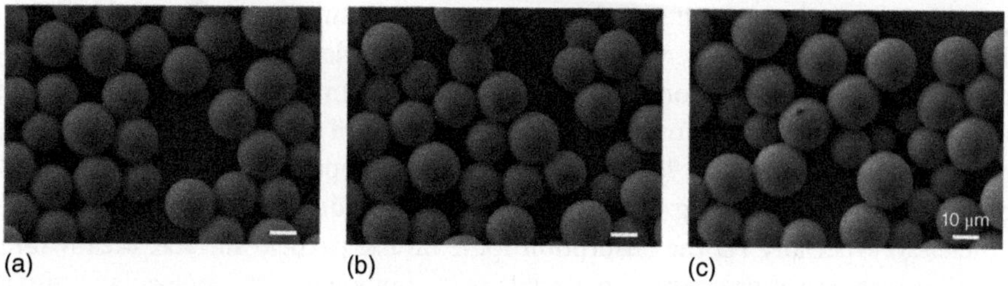

Figure 5.3 SEM images of PLGA microspheres with –OH (a); –COOH (b); and –COOR (c) terminal groups. Source: Li et al. [6]/Elsevier.

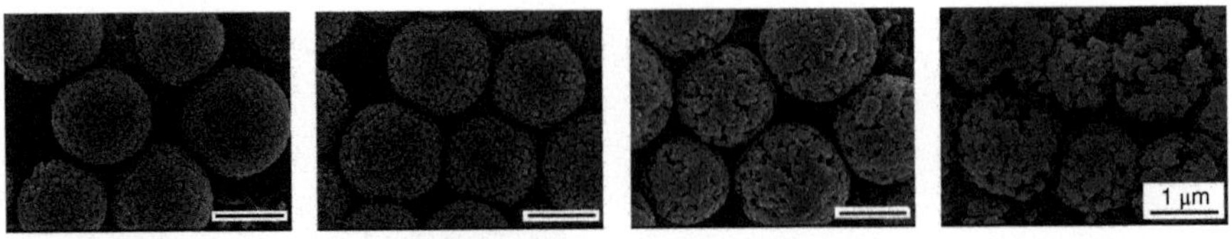

Figure 5.4 SEM images of PST particles with monomer/n-hexadecane ratio of 1 : 0.2; 1 : 0.4; 1 : 0.6; and 1 : 0.8 (from left to right).

5.2.2 Preparation of PST Particles from O/W Emulsion

Uniform PST particles with controlled size and heterogeneous structure were synthesized by rapid MET via O/W emulsion [7, 8]. In a typical preparation, the monomers styrene and divinylbenzene, the initiator benzoyl peroxide (4.3 wt% of total monomers), and different porogen (e.g. heptane, liquid paraffin, and hexadecane) were mixed together and used as O phase. Deionized water containing PVA and sodium dodecyl sulfate was used as W phase. After mixing O and W phases, this O/W coarse emulsion was extruded through the microporous membrane under a certain N_2 pressure for repeated 3–7 times. The polymerization was then carried out for 20 hours under N_2 atmosphere at 70 °C.

By using rapid MET, PST micro/nanoparticles with diameters from 700 nm to 7 μm and with pore sizes from 7 to 200 nm were synthesized [8]. The uniformity of the PST particle was affected by three factors: transmembrane pressure, number of emulsification cycles, and membrane pore size, while the structures of particles were found to depend on both porogen system and particle size. For example, the membrane pore size (1.0, 1.4, 2.8, 5.2, and 9.2 μm) was linearly correlated with the average diameter of PST particles and in inverse proportion to the optimum emulsification pressure. Moreover, the pore size increased with the porogen concentration (e.g. *n*-hexadecane), especially for preparing smaller size particles (around 1 μm) (Figure 5.4). Such an important role was indispensable for the phase separation process between the porogen and PST polymer.

5.2.3 Preparation of Drug-Loaded Particles from O/W Emulsion

The polylactide-based particles exhibit the ability to adsorb water-soluble drugs (protein/antigen/etc.) on the particle surface or encapsulate water-insoluble drugs. The drug loading behavior was affected by the particle-drug interaction environment like pH value, salt concentration, particle amount, drug concentration, and interaction temperature. For example, the potential of unmodified PLA particles is usually negative, which leads to an electrostatic repulsion with the negatively charged DNA/protein drug. Such a property may hinder a more satisfied loading efficiency, especially via the adsorption route on the particle surface. Meanwhile, the physiochemical property of materials (e.g. component and molecular weight) and fabrication conditions also affect the drug encapsulation efficiency, which can be optimized for a specific application.

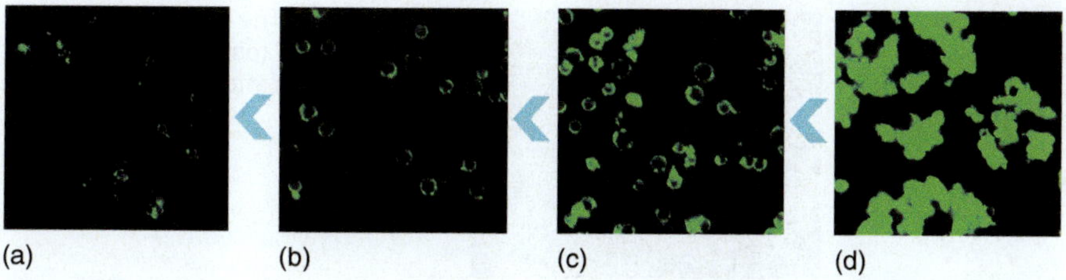

Figure 5.5 CLSM images of FITC-HBsAg adsorbed on different PLA microspheres: blank PLA (a); chitosan modified PLA (b); chitosan chloride modified PLA (c); and PEI modified PLA (d).

5.2.3.1 Preparation of Drug-Loaded Particles via Adsorption

To gain an improved drug adsorption capacity, surface modification strategies have been developed in combination with the preparation of uniform particles by rapid MET. For example, the surface of PLA particles was modified by adsorption of oppositely charged polyelectrolytes (chitosan, chitosan chloride, and polyetherimide) to investigate the effect of particle charge on the activation of cellular immune response by adsorbing HBsAg (Hepatitis B surface antigen) on particle surface. The adsorption efficiency of unmodified PLA particles could only reach 60% under optimized conditions of pH 6.0 phosphate buffer, $20\,g\,l^{-1}$ NaCl concentration, $20\,mg\,ml^{-1}$ particles, and $10\,\mu g\,ml^{-1}$ HBsAg concentration at 37 °C. In comparison, an adsorption efficiency as high as 100% was easily achieved for HBsAg in the presence of polycations shells around the particles, with the most brilliant fluorescence compared with other treatments (Figure 5.5). Albeit efficient, the poly(ethylene imine) (PEI) modification on particles was also related to a compromise of cell viability. In this sense, an appropriate modification that could improve the antigen loading efficiency as well as ensure the biocompatibility is highly appreciated.

In an alternative study, we prepared uniform MPs with insertion of cationic lipid didodecyldimethylammonium bromide (DDAB), which endowed a positive physiochemical property and efficient antigen loading merit for 1 μm particles [9]. In particular, briefly, PLA and DDAB were dissolved into DCM as the oil phase. The oil phase was then quickly added into 1.75% (m/v) PVA and stirred at 400 rpm for three minutes to form pre-emulsion. Then the pre-emulsion was quickly poured into the rapid MET reservoir and pressed through a 2.9 μm porous membrane at 0.6–0.7 MPa nitrogen pressure for three times. After solidification, a size of 1 μm particles (Figure 5.6) with a zeta potential of particles (+34.6 mv) was achieved, which exhibited an HBsAg adsorption efficiency of around 90%. In addition, DDAB causes no change to cell viability and possessed better biocompatibility than other positive agents (like PEI), which was more applicable for surface modification. Thus, the incorporation of cationic properties in this DDAB–PLA particle (DP) group was a promising and efficient manner for antigen/drug adsorption.

5.2.3.2 Preparation of Drug-Loaded Particles via Encapsulation

To gain an improved drug encapsulation, the physiochemical property of materials (e.g. component and molecular weight) and fabrication conditions are under

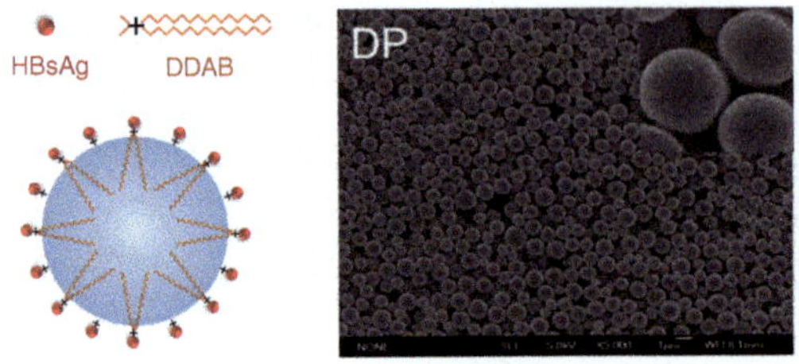

Figure 5.6 The scheme and SEM images of DP (particles with cationic DDAB modification). Source: Lu et al. [9]/Elsevier.

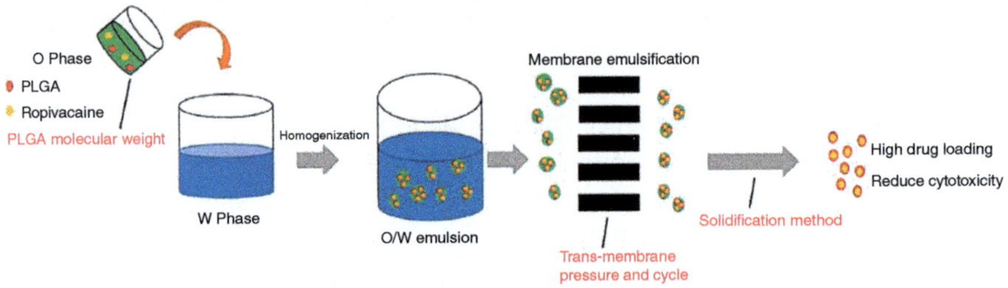

Figure 5.7 A scheme of preparing the ropivacaine-loaded PLGA microspheres via rapid MET. Source: Li et al. [11]/with permission of Elsevier.

optimized. For example, different materials (PLA, PLGA, and PELA) were selected to prepare paclitaxel (PTX) loaded microspheres by rapid MET, and their encapsulation efficiency and cumulative release profiles were studied [10]. The mean diameters of PTX-loaded PLA, PLGA, and PELA microspheres were 0.906, 0.987, and 1.015 μm with a satisfied polydispersity (PDI) of 0.005. The corresponding drug loading efficiencies were 3.89%, 4.93%, and 3.18%, while the encapsulation efficiencies were 63.2%, 71.6%, and 51.3%, respectively. The cumulative release rates of PLA, PLGA, and PELA microspheres *in vitro* for 60 days were 83.87%, 50.25%, and 41.27%.

For the insoluble drugs, corresponding encapsulation strategy has also been developed along with the particle fabrication. Our group prepared ropivacaine (RVC, a small molecule local anesthetic in clinical) loaded microspheres with narrow size distribution (Span 0.469) and a drug loading efficiency (49%) after fabrication optimization [11]. Similarly, ropivacaine and PLGA could be codissolved into the oil phase for subsequent emulsification (Figure 5.7). It was worth noting that, the codissolved drug affected the character of the dispersion phase during the rapid MET process. In particular, the solidification process affected the internal structure of microspheres and the physical state of encapsulated ropivacaine. The ambient evaporation method resulted in the honeycomb structure with macroscopic crystals embedded in the internal channels, while the vacuum negative pressure evaporation resulted in a compact inner structure without pores and crystals. Furthermore, it was interesting that lower polymer molecular weight led to higher encapsulation efficiency, which was mainly ascribed to the interaction between the -COOH terminal group of PLGA and ropivacaine.

The aforementioned terminal group of PLGA was also found to exercise a significant effect on ropivacaine incorporation and content in the microspheres. An encapsulation efficiency of 94.6% was obtained by using PLGA with the –COOH

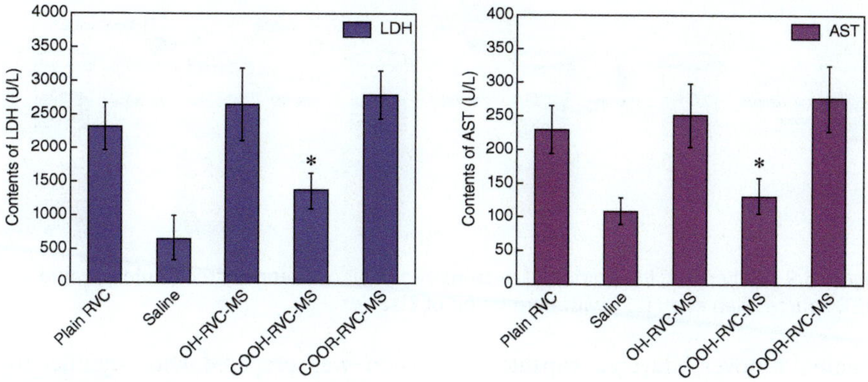

Figure 5.8 The serum biochemical analysis on the cardiotoxicity of different RVC formulations. (LDH: lactate dehydrogenase; AST: glutamic oxalacetic transaminase). Source: Li et al. [6]/with permission of Elsevier.

terminal group, as compared to 83.8% with the –OH terminal group and 80.5% with the –COOR terminal group. The COOH–PLGA exhibited a stronger interaction with ropivacaine than that of OH–PLGA and COOR–PLGA, while the OH–RVC–MS group showed the best long-acting anesthetic effect, in terms of both intensity and duration. By encapsulating ropivacaine into COOH–PLGA, the cardiotoxicity (indicated by biochemical markers LDH and AST) was significantly reduced because of steady and slow release behavior *in vivo* (Figure 5.8).

5.2.3.3 Preparation of Polydopamine Microcapsules

Constructing and functionalizing the polydopamine (PDA) layer onto the nano-sized soft templates as microcapsules have been reported in past decade for food, medicine, and cosmetic. However, very few studies focus on the macroemulsion samples, which are less kinetically stable due to Ostwald ripening but are more applicable for industrial applications like cosmetics and personal care products. Now, the rapid membrane emulsification is demonstrated as an ideal solution to prepare such stable O/W macroemulsion with narrow size distributions and formulate the micronsized polydopamine microcapsule [12].

As shown in Figure 5.9, the benzyl acetate, glyceryl tributyrate, and surfactants as essential oils (EOs) phase were firstly mixed into coarse emulsions in Tris–HCl buffer, which was subsequently refined by extruding through the membrane (pore size, 12 μm) under 0.03 mPa for five times to produce narrow sized emulsion droplets. The obtained emulsion was further dropped with dopamine (DA) solution in Tris–HCl for preliminary coating. After reaction for four hours, the branched polyethyleneimine (PEI) was added as a crosslinker for another eight hours. Subsequently, PDA capsules were prepared by depositing PDA onto the monodispersed oil droplets through oxidative self-polymerization in alkaline conditions. Then, the PDA layer constructed on oil-based soft templates in macroemulsion was further functionalized as an interlayer to anchor gelatin (GE) and gum arabic (GA) via Michael addition/Schiff reaction and complex coacervation, respectively.

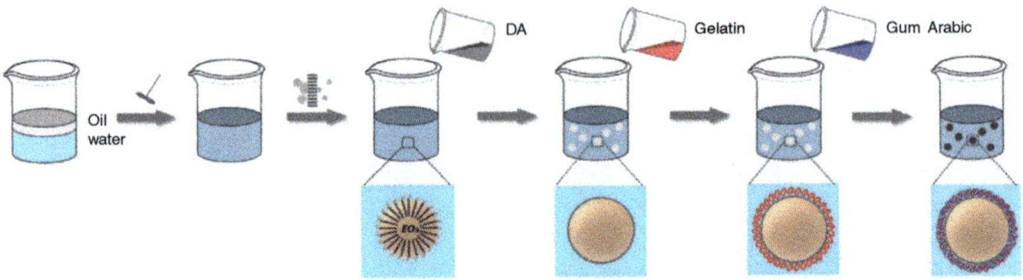

Figure 5.9 Schematic illustration of micronsized polydopamine microcapsule by rapid MET. Source: Tian et al. [12]/with permission of Elsevier.

Finally, a novel bilayered capsule suspension was prepared with significantly improved structural stability, thermal stability, and better protection of EOs with sustained release properties. Since the droplet diameters produced by PME are quite similar, the Laplace pressure difference of system and the driving force of Ostwald ripening in emulsion are relatively small according to Young–Laplace equation. The emulsion is thus more stable and the surfactant concentration can be further reduced by 3–10 times.

5.3 Preparation of Hydrophilic Microspheres from W/O Emulsion

5.3.1 Preparation of Chitosan Particles

Chitosan is a kind of biocompatible hydrophilic polysaccharides, which shows excellent mucoadhesive and permeation enhancing effect across biological surfaces. Although microspheres have been successfully prepared by direct MET, it is difficult to prepare small uniform W/O droplets due to the ultrahigh transmembrane pressure and dramatically decreased emulsification rate. In this sense, rapid MET was developed to prepare uniform and size-controllable chitosan particles with high production efficiency. Moreover, by using different chitosan derivates and specific cross-link strategies, we have fabricated solid/gel chitosan particles for divergent biomedical applications.

5.3.1.1 Preparation of Chitosan Solid Particles

By using rapid MET, the chitosan NPs from 200 nm to 1.85 μm were successfully prepared by changing the pore size of the membrane, and the polydispersity index could be as low as 0.027 [2]. A typical process under the optimum conditions via rapid MET is listed as follows. Primarily, the chitosan aqueous solution (0.5 wt%) was used as a dispersed phase, and the mixture of liquid paraffin and petroleum either [1 : 2 (v/v)] containing 4 wt% PO-500 emulsifier was used as a continuous phase. The volume ratio of water to the oil phase was 1 : 30. Nanodroplets were achieved by extruding the coarse emulsions through the microporous membrane (with a pore size of 1.4 μm) at a transmembrane pressure of 0.95 MPa. The nanodroplets were further cross-linked (the molar ratio between amino groups and aldehyde groups was 1 : 1) for 10 hours to obtain 200–300 nm chitosan particles. Chitosan particles prepared by a single W/O emulsion have been successfully utilized to encapsulate the soluble proteins (e.g. insulin) or antigens (e.g. HBsAg) via the aqueous phase.

In our previous work, we have discovered that the chitosan microsphere that is cross-linked by glutaraldehyde possesses the property of autofluorescent, which can serve as bright, inert, durable, and extremely photostable tracers [13]. In combination with the controllable particle size and autofluorescent merit, we investigated how the physical properties of particles regulate specific biological responses [14]. Cellular uptake study demonstrated that nanosized particles (430 nm) accumulated in the cells at a faster rate, in terms of particle surface area. Once the data are converted into the expression of particle volume, the higher values were found with 1.9 or 4.8 μm particles instead of NPs (Figure 5.10). Furthermore, nanosized particles are preferred to promote the secretion of Th1-specific molecule signals (e.g. IFN and IL-12) rather than immune-suppressive cytokines. This knowledge was expected, in principle, to establish the basis of design concepts for particle-based biomedical applications, which also highlighted the importance of the uniformity and controllable size of the particles.

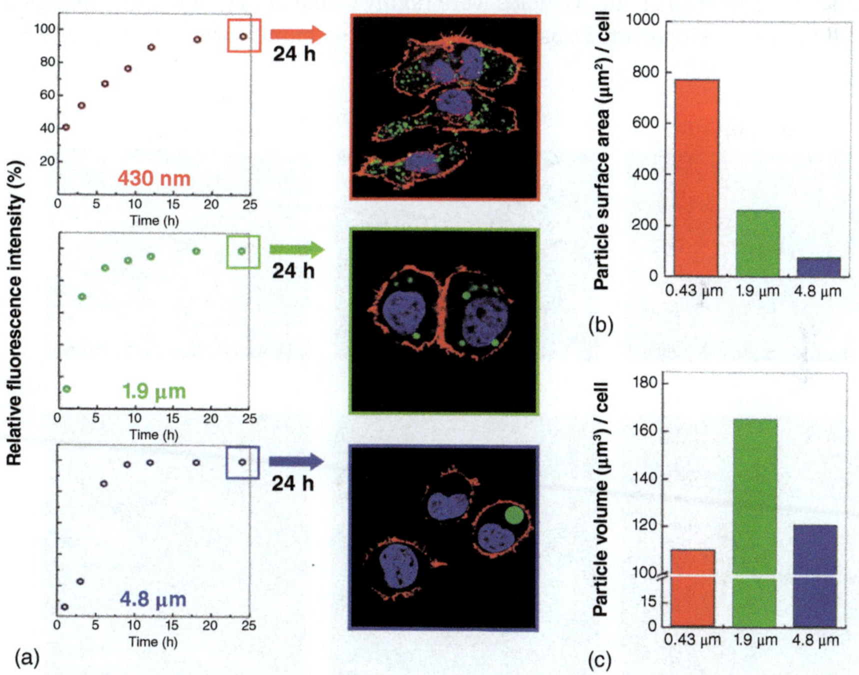

Figure 5.10 The uptake profile of chitosan particles by antigen presenting cells (APCs). (a) The internalization kinetic curve. The calculated surface area (b) and volume (c) for the internalized particles with different sizes. Source: Yue et al. [14]/Elsevier.

5.3.1.2 Preparation of Chitosan gel Particles

Apart from the solid particles, chitosan gel NPs with a pH-sensitive property have also been developed. Based on a technique for constructing a thermosensitive chitosan/glycerophosphate gel [15], uniform chitosan sol droplets can be prepared by rapid MET, and then the droplets can be solidified to gel particles via sol–gel transition at 37 °C [16, 17]. During the formation of uniform W/O emulsion, the mixture of quaternized chitosan (HTCC) solution and α-β-glycerophosphate (α-β-GP) was used as aqueous phase, and the mixture of liquid paraffin and petroleum either at 7 : 5 (v/v) containing PO-500 was used as the oil phase. In particular, the subsequent gelation method for droplets solidification was without using any chemical cross-link agents. After choosing different membrane sizes, the prepared gel microparticles (MGels) were around a diameter of 4.30 μm, 2.51 μm, and 807 nm with the pore sizes of the porous membrane, 9.2, 5.2, and 2.8 μm, respectively [18] (Figure 5.11). The CV of obtained particle diameters was well controlled below 15%. Antigen (H5N1 split vaccine) could be loaded on MGels by adsorption method, and the antigen loading capacity of MGels increased when particle diameter decreased, which was probably due to the increased surface area of MGels with smaller particle size.

The aforementioned chitosan particle with pH sensitivity was believed to have great advantage when used as a drug or vaccine carrier. On one hand, a controllable release was achieved as the particles were highly stable at pH 7.4 but disintegrated rapidly at pH 5. Moreover, a sustained release of ~1 month was detected for BSA

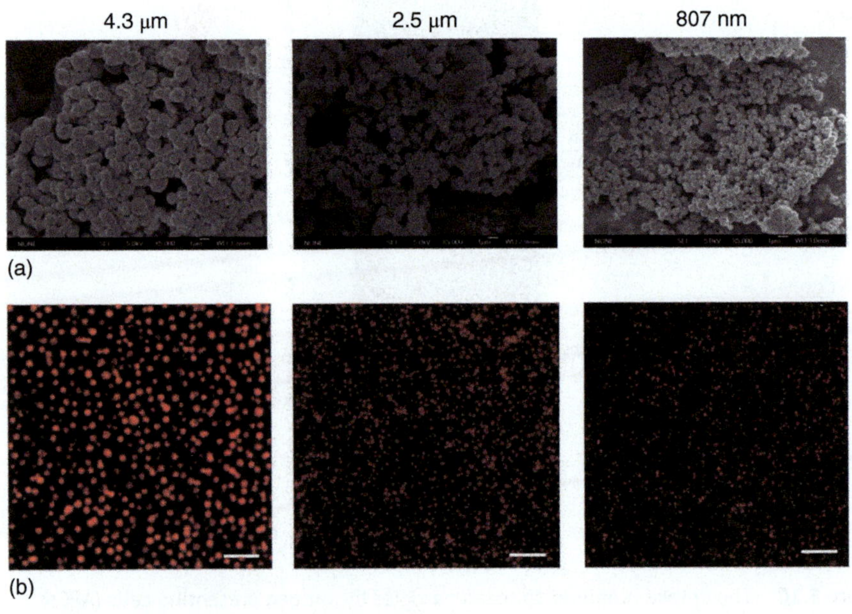

Figure 5.11 CLSM (a) and SEM (b) images of quaternized chitosan MGels with different sizes. Scar bars in CLSM: 20 μm. Red color is sourced from Cy5 labeled MGel via electrostatic interaction. Source: Wang et al. [18]/Elsevier.

(a model protein drug), as high activity of protein drug could be acquired in the absence of chemical crosslinkers.

5.3.2 Preparation of Stimuli-Responsive PNIPAM Particles

Stimuli-responsive polymer microspheres have attracted great attention due to their applications in drug delivery system that can offer considerable versatility for encapsulation and triggered release. Our group prepared thermo-sensitive poly(*N*-isopropyl acrylamide-*co*-acrylic acid) [P(NIPAM-*co*-AA)] microspheres with narrow size distribution using rapid MET combined with low-temperature polymerization [19]. In brief, the coarse W/O emulsion with broad size distribution was repeatedly pressed through a microporous membrane to form the fine emulsion with narrow size distribution. To avoid phase transition-triggered aggregation of thermo-sensitive PNIPAM microspheres during polymerization, tetramethylethylenediamine was introduced to polymerization system by adding it into W/O emulsion to initiate polymerization at low temperature (25 °C). The optimized preparation conditions were determined as follows: transmembrane pressure was 250 kPa and transmembrane passes were three times when using a membrane of 5.2 μm pore size. The obtained microspheres were around 5.11 μm with PDI at 0.0312 (Figure 5.12).

5.3.3 Preparation of Agarose Particles

A big disadvantage of agarose microspheres is that the microspheres are not pressure resistant. Although the cross-linking strategy can considerably improve the strength of agarose microparticles, it is still unsatisfactory. In industry, 90 μm agarose microspheres are generally employed to ensure a certain flow rate, however, the relatively large size of these microparticles limits the column efficiency and resolution. To improve the resolution, improved strength or smaller size of the particles is expected. Nevertheless, particle strength requires an increased agarose concentration, which may lead to high viscosity and put hurdles for particle preparation.

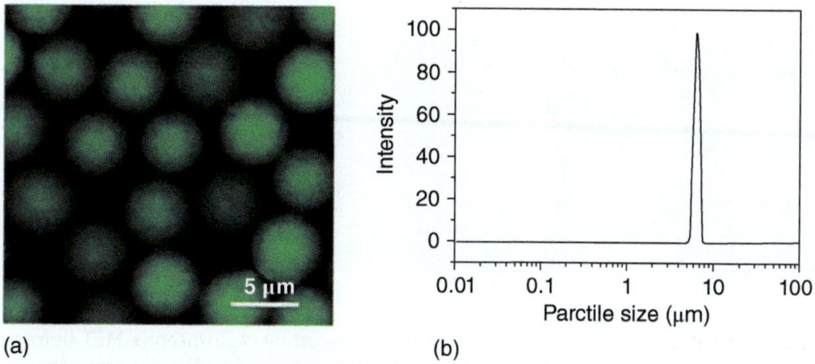

Figure 5.12 CLSM images (a) and the size distributions (b) of microspheres prepared under optimized conditions. Source: Wang et al. [19]/Elsevier.

To resolve this problem, our group fabricated agarose microspheres with smaller particle sizes and higher strength using the rapid/premix membrane emulsification method [3]. In brief, agarose aqueous solution was used as the dispersed phase, while a mixture of liquid paraffin and petroleum ether containing PO-500 was used as the continuous phase. The water phase was mixed with the oil phase at 60 °C, and a coarse W/O emulsion was produced in a homogenizer. Then, the coarse emulsion was extruded through a hydrophobic membrane under high pressure to form an emulsion, which was slowly cooled under gentle agitation to form gel beads. The formulation preparation is similar to the direct membrane emulsification method, but the agarose concentration increased up to 15%, and the particle size was controlled between several micrometers to 50 μm, with an optimum particle size CV value of 9.8%. Uniform agarose beads of diameter below 10 μm were also prepared via tuning the MET-related parameters (Figure 5.13). Since a high concentration of agarose substantially improved the particle strength, the speed of bioseparation media was correspondingly improved (~twofolds) at the same resolution.

In a systematic view, the results of rapid MET can be described in terms of droplet rupture in the membrane pore, which is related to the wall shear stress in the membrane pore as shown in the following equation:

$\sigma_{w,p} = 8\eta_e J\xi/\varepsilon\ d_m \sigma_{w,p}$ is the shear stress in the membrane pore, η_e is the average viscosity of the emulsion in the membrane pore, ξ is the curvature of the membrane pore, ε is the average porosity, d_m is the average membrane pore diameter,

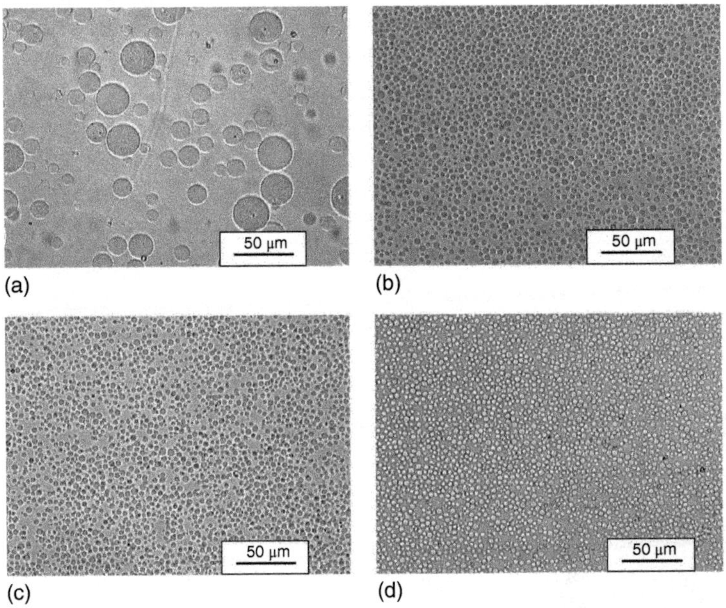

Figure 5.13 Optical micrographs of agarose beads prepared by rapid/premix MET before and after different passes through the membrane. (a) Before MET, (b) after the first pass, (c) after the second pass, and (d) after the third pass. Source: Zhou et al. [3]/Elsevier.

and J is the transmembrane flux. In this sense, the relationship between the component of the oil phase and the uniformity of droplets could be deduced from this equation. For example, with an increased dosage of petroleum ether in liquid paraffin/petroleumether, the viscosity η_e of the emulsion decreases, while the transmembrane flux J increases at the same transmembrane pressure. Therefore, the parameters η_e and J together dictate the shear stress $\sigma_{w,p}$. In the corresponding experiment, the results show that the particle size decreases and the uniformity increases with the increased dosage of petroleum ether, which means that the droplet fragmentation in the membrane pores (indicated by $\sigma_{w,p}$) increases. In addition, either η_e or J increases, $\sigma_{w,p}$ increases, leading to an increased degree of droplet fragmentation (decreased size and increased uniformity). With the arm of the aforementioned knowledge, microbeads with both uniform sizes and high agarose contents are obtained by rapid MET rather than conventional emulsion methods.

5.3.4 Preparation of Alginate Particles

In contrast to chitosan, alginate bears carboxyl groups and is an anionic polymer, which consists of linear copolymers composed of β-D-Mannuronate (M) and α-L-Guluronate (G). It possesses a unique property of gel-formation in the presence of multivalent cations (e.g. calcium ions), as the high G content yields tight junctions ("egg-box junctions") because of strong coulombic interactions between the carboxylate anions of alginate and the metal cations. The alginate particles are usually prepared via the W/W method, in which aqueous sodium alginate is added dropwise or sprayed into an aqueous $CaCl_2$ solution to solidify the gel microspheres in solution. In this sense, conventional method cannot be directly applied in the MET that requires two immiscible phases. Meanwhile, the pristine alginate microcapsules are prone to drug leakage due to their large pores, which should also be addressed prior to their further application.

To resolve the fabrication problem, we proposed a miniemulsion strategy for preparing alginate microspheres by the membrane emulsification method [20]. In detail, uniform W/O emulsions are prepared by membrane emulsification with aqueous sodium alginate solution (W phase) and liquid paraffin/petroleum ether-containing emulsifier PO-500 (O phase). Afterward, $CaCl_2$ miniemulsions were prepared by ultrasonication with the same oil phase. Once the two phases were mixed, the $CaCl_2$ droplets were small enough to be engulfed by large alginate droplets, and the solidification process started to form the alginate microspheres. To avoid the drug leakage from the pristine alginate microspheres, a cationic polyelectrolyte (e.g. chitosan) was usually introduced to form a compact polyelectrolyte complex with the anionic alginate.

In combination with the aforementioned solidification method and the rapid MET technique, alginate particles can be prepared and modified further for particular applications [21]. In brief, microdroplets were obtained by extruding the coarse emulsions through the membrane pores under high pressure. Afterward,

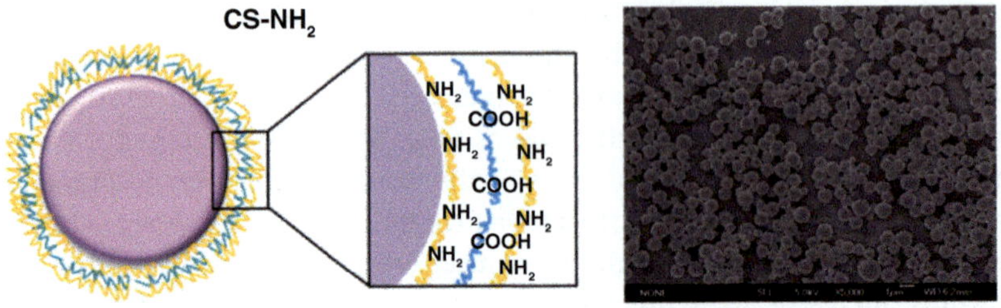

(a) (b)

Figure 5.14 The scheme of surface chemical structure (a) and SEM images (b) of the alginate particles CS–NH_2 particles. Source: Liu et al. [21]/American Chemical Society.

a miniemulsion of $CaCl_2$ was added to the sodium alginate emulsion as the first step of solidification (for five hours). The solidified alginate particles were collected, washed, and then dispersed in a 1.4 wt% chitosan acetic acid buffer solution. After adsorption of the first chitosan layer, the particles were centrifuged, washed, and then incubated in 1.0 wt% alginate solution, followed by centrifugation and another washing step. This LbL procedure was repeated once more to deposit a total of two chitosan bilayers on 1 μm CS–NH_2 particles (Figure 5.14). The formation of the chitosan layer was ascribed to the strong electrostatic interaction of amino groups of chitosan with the carboxyl groups of alginate, leading to the formation of a chitosan–alginate complex. These particles with a high surface density of amino groups were favorable for complement activation, which showed great potential for vaccine delivery.

In another study, calcium alginate microspheres by using the membrane with the pore sizes of 1.4, 2.8, and 5.2 μm were prepared for the particle diameters at 370, 700, and 1200 nm, respectively, and the optimal permeation pressures were 1.0 MPa, 250 kPa, and 150 kPa, respectively [22]. Once combined with the chitosan shell, the final particle size would be reduced by cross-linking, showing a size of 230, 550, and 1200 nm, respectively. These uniform coated particles could effectively stabilize the O/W Pickering emulsion. Based on this emulsion that was stabilized by chitosan-coated alginate particles, monodispersed colloidosomes with a high yield and low permeability were prepared by combining the rapid MET and polymer deposition method (Figure 5.15) [23]. In detail, chitosan-coated alginate particles were dispersed in water (pH 6.5) and stirred (800 rpm, 1 minute) with EA containing 5 wt% PLGA. The resultant oil-in-water emulsion had an oil phase consisting of PLGA and was stabilized by an interfacial monolayer of the deposited particles. After extruding through the rapid MET, uniform emulsions were formed and were added into water with ten times volume for solidification. During this process, the assembled particles were locked into a robust polymeric shell through the precipitation of the PLGA polymer at the interface [24], which was named as colloidosome. These colloidosomes showed high drug encapsulation efficiency (up to 96.7%) and an obvious pH-sensitive release profile. Present work provided an effective strategy to prepare uniform Pickering emulsion or colloidosomes, which can be applied to design potent carriers in the biomedical area.

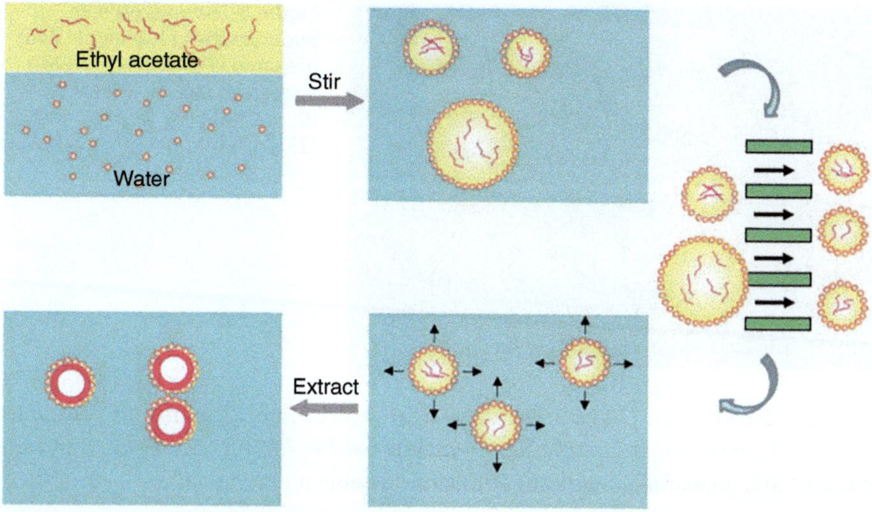

Figure 5.15 Schematic diagram of the preparation of colloidosomes by combining the premix membrane emulsification technique and the polymer deposition method. Source: Nan et al. [23]/with permission of Royal Society of Chemistry.

5.3.5 Preparation of Konjac Glucomannan Particles

Since KGM should be dissolved in alkaline solvent, it requires the membrane to have a hydrophobic surface resistant to alkali. Therefore, our group [25] developed a method to fabricate polymeric hydrophobic membranes (epoxy-based polymer membrane) with homogeneous membrane pores as introduced in Chapter 3, with which the preparation of uniform W/O emulsions can be performed directly. It was found that emulsion coalescence and flocculation occurred frequently due to the high interface energy and sedimentation velocity of KGM emulsions. Emulsion stability had a significant influence on the uniformity and dispersity of the final KGM microspheres. To improve the stability of the emulsions, the effects of the concentration of the emulsifier, the viscosity of the KGM solution, the oil phase composition, and the feeding method of epoxy chloropropane (EC) on the preparation results were systematically investigated.

For example, four different viscosities (KGM I, II, III, and IV with viscosities of 66.0, 88.4, 145.6, and 180.3 mPa s, respectively) and 8% KGM were used as the dispersed phase with different proportions of liquid paraffin petroleum ether (containing different concentrations of emulsifier PO-500). The W/O type coarse emulsion was prepared by homogenization, and then a polymer hydrophobic membrane (membrane pore size 14.8 μm) was used for rapid MET to obtain uniform droplets. Subsequently, the cross-linking agent epichlorohydrin was added dropwise at 60 °C for eight hours. The results showed that the viscosity had an important effect on the particle size distribution. The particle size distribution of the KGM I emulsion was wider (CV value of 22.55%), while the particle size distribution of the other three preparations was more uniform with CV values of 17.47%, 15.35%, and 15.17%, respectively (Figure 5.16). Under optimal preparation conditions (emulsifier

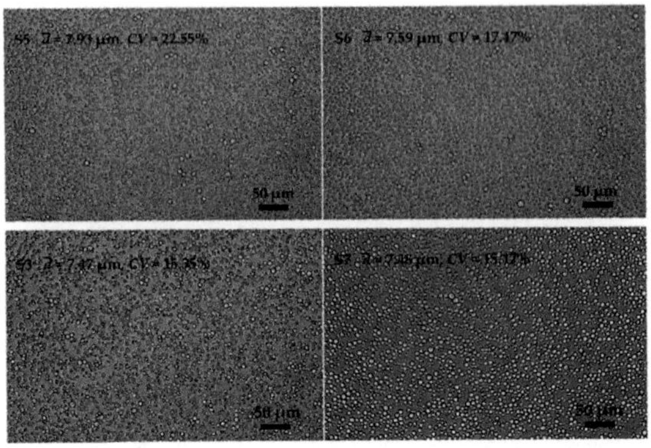

Figure 5.16 Microscopic photographs of KGM emulsions prepared with different KGM viscosities (KGM I, KGM II, KGM III, and KGM IV). Source: Mi et al. [25]/MDPI/CC BY 4.0.

5 wt% PO-5s, KGM III [145.6 mPa s], a weight ratio of liquid paraffin to petroleum ether 11 : 1), uniform and stable KGM emulsions (*d* = 7.47 μm, CV = 15.35%) were obtained and cross-linked without emulsion-instable phenomena.

5.4 Preparation of Microcapsule from Double Emulsion

For single emulsion-based particles, the encapsulation of the drug/antigen can only be fulfilled when the solubility property is the same as the inner phase of the emulsion. To extend the applications and improve the drug loading efficiency of particles sourced from different materials, double emulsions like W/O/W or O/W/O together with rapid MET have been developed. By using these methods, either the hydrophobic drugs or the hydrophilic drugs can be loaded into the particles. For example, the W/O/W type emulsions are usually applied for encapsulating water-soluble drugs, while O/W/O type emulsions are capable of loading water-insoluble drugs. Moreover, the microcapsule with a unique structure (especially for the hollow or porous structure) could be designed for various applications or biological responses (e.g. pH-responsive).

5.4.1 Preparation of Drug/Antigen-Loaded Microcapsules via W/O/W Emulsions

5.4.1.1 Preparation of Particles for Encapsulating Water-Soluble Antigen

To encapsulate hydrophilic protein drugs or vaccines, our group constructed a series of particles based on W/O/W double emulsion. Uniform-sized polylactide-based microcapsules can be directly prepared using the W/O/W emulsion–solvent extraction method combined with rapid MET. For example, we prepared PLA and PELA polymers microcapsules that were used for loading soluble antigens [26]. Since recombinant hepatitis B surface antigen (HBsAg) was a multimeric complex that is easily inactivated, the amphiphilic PELA material was employed to provide a mild preparation environment for maintaining the activity of the antigens. Briefly, the aqueous HBsAg solution was mixed with PELA/ EA, and the W/O primary emulsion was prepared by ultrasonication (15 seconds) or homogenization (6000 rpm,

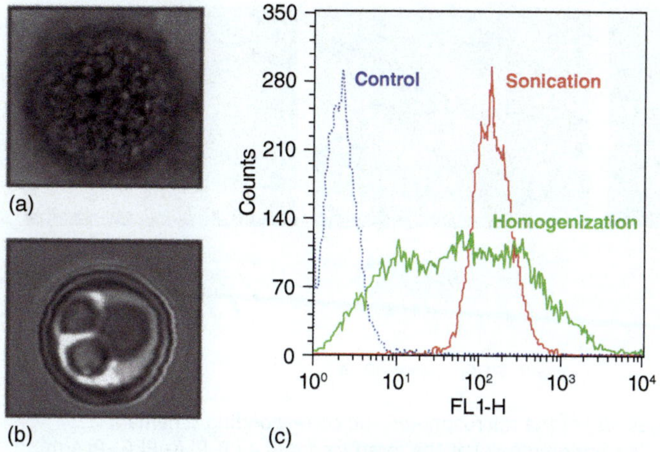

Figure 5.17 Confocal laser scanning microscope (a–b) and flow cytometer analysis (c) showed the comparison of the preparation method of primary emulsion: (a) sonication technique, (b) homogenization technique, and (c) corresponding flow cytometer analysis. Source: Wei et al. [26]/Elsevier.

15 seconds), respectively. Subsequently, the W/O emulsion was dispersed further in the external aqueous phase containing 1% PVA and 1% NaCl to obtain the coarse W/O/W double emulsion, which was then pressed through the membrane pores to obtain a uniform emulsion. This emulsion was finally poured into a large amount of water to remove EA, and uniform-sized microcapsules could be solidified and prepared.

Under an optimum condition, the size of the prepared particles was about 1 μm, and the CV value was 18.9%. Compared with the emulsion prepared by sonication method, the emulsion prepared by homogenization was unstable, and the inner water droplet was easily fused with the external water phase, resulting in a low antigen encapsulation of microcapsules. The inner droplets were small and evenly dispersed, and the antigen was loaded uniformly in each microcapsule when the sonication technique was used to prepare the primary emulsion (Figure 5.17). Notably, the distribution pattern of the PEG segment played an important role in the properties of microspheres. Compared with triblock copolymer PLA–PEG–PLA, the diblock copolymer PLA-mPEG yielded a more stable interfacial layer at the interface of the oil and water phase and thus was more suitable to stabilize the primary emulsion and protect coalescence of inner droplets and external water phase, resulting in high encapsulation efficiency (90.4%) (Figure 5.18).

5.4.1.2 Preparation of Particles for Encapsulating Water-Soluble Drugs

To encapsulate soluble drug exenatide, our group prepared large-sized PLGA microspheres by using rapid MET combined with double emulsion ($W_1/O/W_2$)-solvent evaporation method (Figure 5.19a) [27, 28]. The membrane with a large pore size (50.2 μm) was used, and the formulation optimized with transmembrane pressure and PVA concentration in the external aqueous phase (W_2) led to PLGA

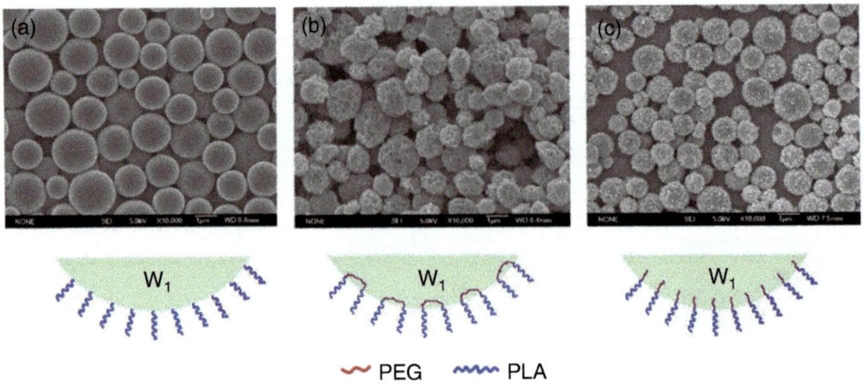

Figure 5.18 SEM images (up) of the microspheres and corresponding schematic illustration of polymer distribution (down) at the interface for PLA (a), PLA–PEG–PLA (b), and PLA–mPEG (c). Source: Wei et al. [26]/Elsevier.

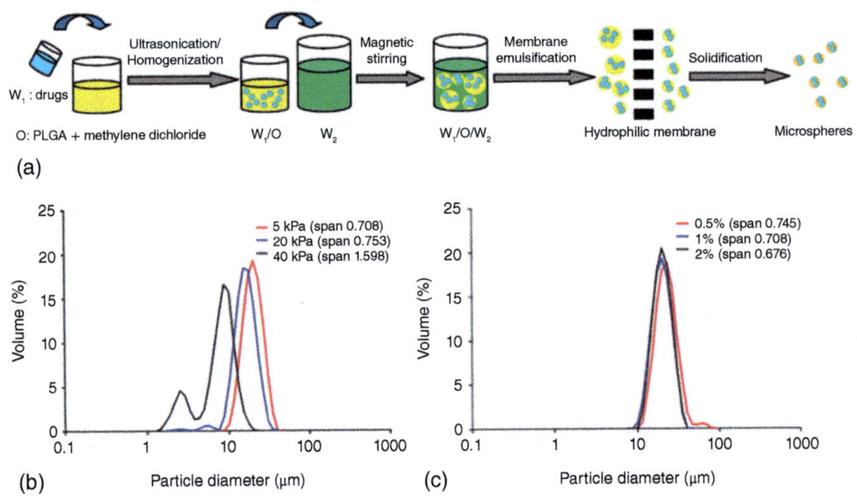

Figure 5.19 The scheme of the preparation of uniform-sized exenatide-loaded PLGA microspheres (a) and the size distribution of microspheres prepared under different transmembrane pressures (b) and PVA concentrations (c). Source: Qi et al. [27]/with permission of Elsevier.

microspheres with sizes as large as 20 μm (Figure 5.19b). For example, the average particle size of prepared microspheres got smaller (22.82, 15.99, and 10.67 μm, respectively) with increased pressure from 5 to 40 kPa, and the Span value increased from 0.708 to 1.598 (Figure 5.19c). This was due to the fact that higher pressure promoted the coarse double emulsions to cross the membrane faster, and the higher shear force caused by faster pass-through rate eventually led to the production of a large amount of smaller droplets, resulting in broad size distribution. The average particle size of the resultant microspheres was not significantly affected by PVA concentration, but the Span value was slightly decreased with increasing the PVA concentration. Moreover, the encapsulation efficiency of exenatide inside the microspheres was high (around 100%). In addition, the microspheres prepared by solvent evaporation, solvent extraction, and the cosolvent method presented different release behaviors *in vitro* and *in vivo* but exhibited a similar trend of PLGA degradation [29].

5.4.1.3 Preparation of Hollow Particles for Encapsulating Antigen/Drug

The physicochemical characteristics of particles dictate whether they are suitable for further applications. Although PLGA particles are noted for their good biocompatibility, it is a challenge to make PLGA particles work more efficiently by enabling them with high pH sensitivity. To address this issue, our group fabricated PLGA hollow NP, which encapsulated antigen and NH_4HCO_3 together from W/O/W emulsion by rapid MET [30]. Through the process of emulsion evolution, pH-sensitive PLGA particles with one cavity and a thin shell were fabricated (Figure 5.20). This structure enabled the microspheres with a larger inner space and a coencapsulation of NH_4HCO_3 as an antigen release promoter to interact with protons in endosomes/lysosomes of cells (pH ~ 5.0 and ~6.5).

In detail, pH-responsive PLGA NP with rapid antigen release behavior was prepared using the W/O/W emulsion–solvent extraction method combined with the rapid MET, after adjusting the ratio of the inner W phase (with the addition of 2.5 mg ml^{-1} NH_4HCO_3) and O phase. For PLGA polymers, a lactic/glycolic molar ratio of 75 : 25 was chosen. NPs prepared from copolymers with high lactic acid ratio degraded more slowly than those with high glycolic acid ratio. The designed pH-responsive NPs showed pH-dependent antigen release behavior (minimal at pH 7.4 and 10% within 24 hours; more than 85% antigen release at pH 5.0 and 6.5 for 24 hours; Figure 5.20d). This behavior is quitted different from that for the conventional PLGA NPs, which offered a controlled release curve at pH 5.0 within 24 hours, but the cumulative release efficiency was less than 10% (Figure 5.20e).

5.4.2 Preparation of Microspheres with Unique Structure via W/O/W Emulsion

5.4.2.1 Preparation of PLA/PLGA Microspheres with Single-Core Structure

From W/O/W complex emulsion systems, divergent structures of PLA/PLGA particles can be finely tuned. For example, a scalable and versatile strategy has been developed for the fabrication of double emulsion-templated single-core PLGA microcapsules with narrow size distribution and controllable structure [31]. A single-core water-in-oil-in-water (W_1/O/W_2) double emulsion was generated from two-step premix membrane emulsification, which guaranteed the narrow size distribution and size controllability of emulsion. As shown in Figure 5.21a, the internal aqueous phase W_1 consisted of aqueous NaCl solution, the oil phase consisted of PLGA and EA, and PELA was added as a stabilizer. A uniform W_1/O pre-emulsion was prepared with a homogenization followed by a rapid MET (using a hydrophobic membrane). Next, this emulsion was dispersed into the outer aqueous phase W_2 (containing PVA stabilizer) via a secondary emulsification process.

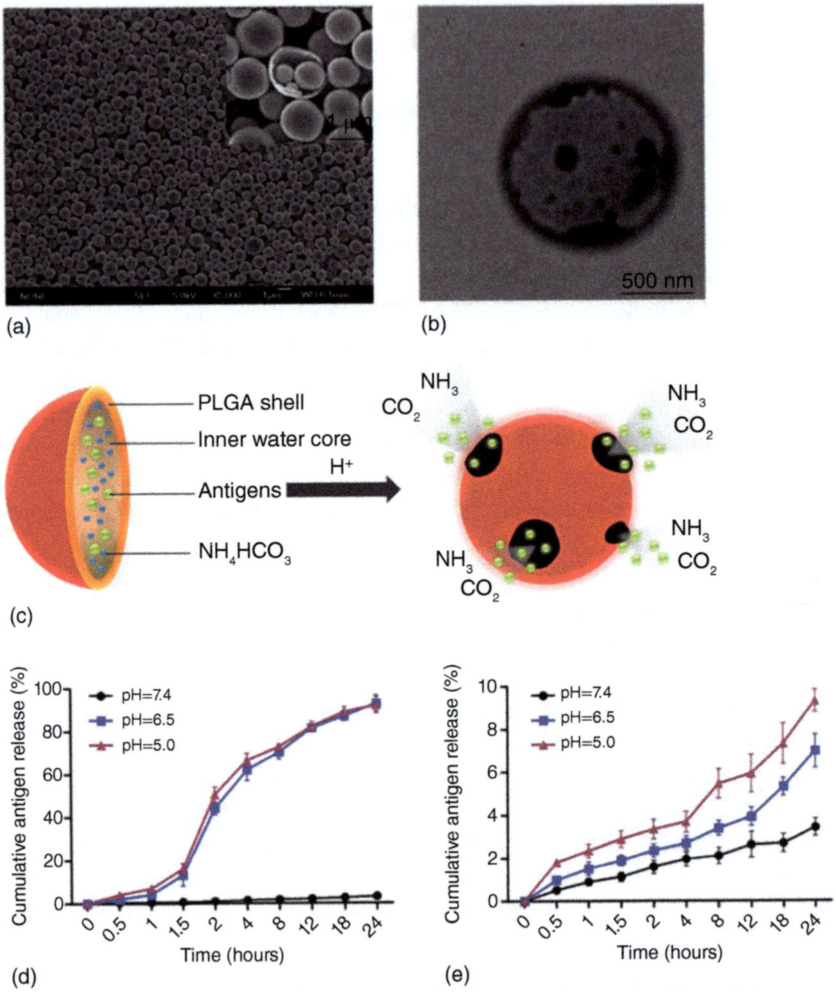

Figure 5.20 Characteristics and *in vitro* antigen release behavior of pH-responsive PLGA NP. (a) SEM and (b) TEM micrographs of pH-responsive PLGA NP; (c) schematic illustrating the composition/structure of the as-constructed pH-responsive PLGA NP and their working mechanism. *In vitro* release of antigens from (d) pH-responsive, PLGA NP and (e) normal PLGA NP incubated in test media with different pH values to mimic the cytoplasm (pH 7.4), early endosomes (pH 6.5), and late endosomes/lysosomes (pH 5.0) at 37 °C. Source: Liu et al. [30]/American Chemical Society.

During this process, the coalescence of W_1 droplets within oil droplets allowed multicore $W_1/O/W_2$ double emulsion to spontaneously transform into a single-core one. The osmotic pressure and solvent diffusion rate were used to control the shell-thickness of single-core PLGA microcapsules. The key to obtaining a single-core lies in the fusion regulation of the inner aqueous phase, especially for the regulation via tuning the concentration of PEG–PLA and NaCl. On the one hand, the inner-core number of PLGA microcapsule was dependent on the PEG–PLA concentration in the emulsion coalescence stage. Multicores structure appeared with 0.3% PEG–PLA

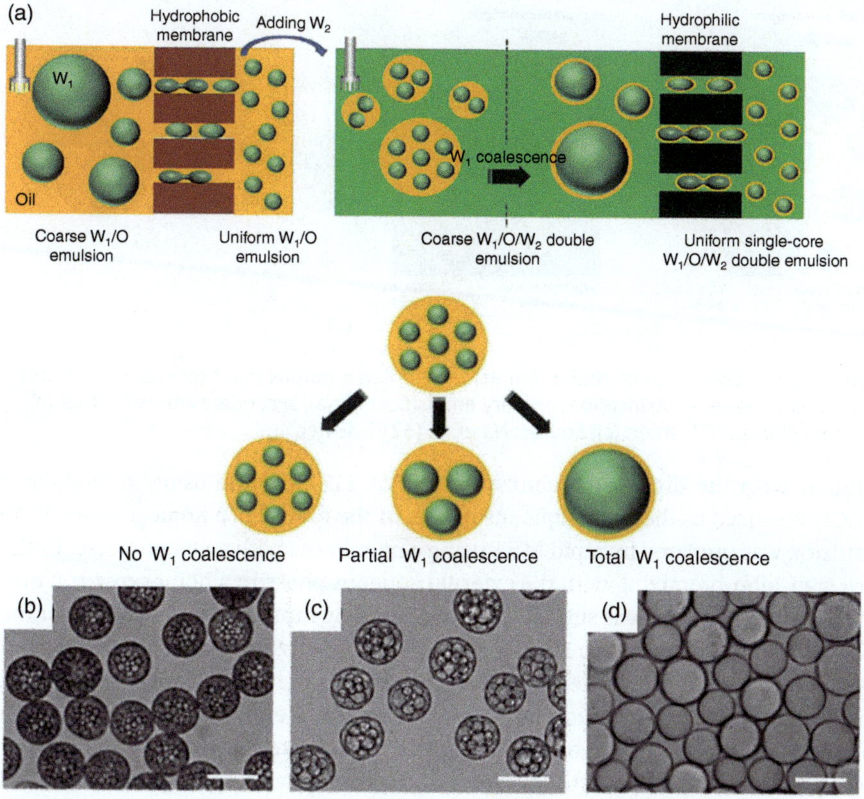

Figure 5.21 A schematic process of preparing single-core PLGA microcapsules by using a two-step premix membrane emulsification (a) and the influence of inner water droplets coalescence (PEG–PLA concentration) on morphology (b–d). (b) 0.3% (w/w); (c) 0.1% (w/w); and (d) 0.03% (w/w). Scale bar 50 μm. Source: Na et al. [31]/John Wiley & Sons.

in the oil phase after 20 minutes' emulsion coalescence (Figure 5.21b). And the lower PEG–PLA concentration, 0.1% (w/w) in the oil phase, led to the partial coalescence of W1 droplets (Figure 5.21c). In the case of 0.03% (w/w) PEG–PLA, total coalescence of W1 droplets happened, and multicores were transformed into single-core completely (Figure 5.21d). Alternatively, NaCl 0.1% (w/w) could be added in the out or inner water phase to regulate an optimum osmotic pressure and control the wall shell thickness (0.4, 1.8, and 3.1 μm).

5.4.2.2 Preparation of PMMA/PLGA Microspheres with Gigaporous Structures

Besides the polylactide-based hollow particles, other framework materials can also be employed to prepare microspheres with unique structures via W/O/W emulsion. We prepared poly(methyl methacrylate-glycidyl methacrylate) (P(MMA-GMA)) microspheres with gigaporous structure using a two-step rapid membrane emulsification method [32]. In detail, a certain amount of NaCl solution was selected as the inner aqueous phase, GMA, MMA, oil-soluble initiator (e.g. benzoyl peroxide), and stabilizer (mPEG–PLA) was dissolved in EA as the oil phase. The aqueous

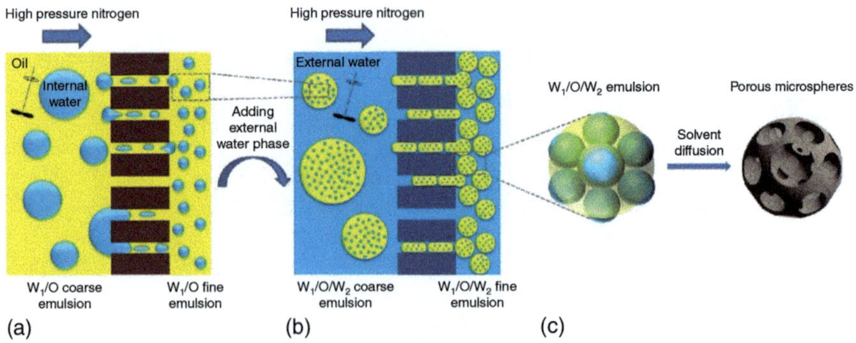

Figure 5.22 Fabrication of double-emulsion-templated porous microspheres by two-step premix membrane emulsification: primary emulsification (a), secondary emulsification (b), and emulsion solidification (c). Source: Na et al. [32]/Elsevier.

solution with the dissolved stabilizer (e.g. PVA-217) and the osmotic modulator (NaCl) was used as the outer aqueous phase. In the first step, a homogeneous W/O emulsion was prepared by rapid MET using a membrane with small pore size. In the next step, after premixing with the external aqueous phase in a homogenizer, a uniform double emulsion was subsequently prepared by rapid MET using a membrane with a large pore size (Figure 5.22). Finally, the polymerization was carried out at an elevated temperature, and the internal aqueous phase was fused to form penetrating pores, leading to the formation of gigaporous microspheres. During this emulsion solidification process, internal water droplets were packed densely together, forming a thin oil film between the internal and external water phases. After solvent diffusion, the thin film could be ruptured and the pores could be templated from the internal water droplets to form interconnected porous structures.

Since both the primary emulsion and double emulsion were prepared by rapid MET, the pore size and microsphere size were controllable and uniform. With the membrane pore size in the first MET process that was selected from 0.8, 1.4 to 2.8 μm, the microsphere pore size increased from 0.52, 0.98 to 2.10 μm, respectively. Furthermore, P(MMA–GMA) microspheres with different particle sizes could be obtained by secondary emulsification. When the membranes with the pore sizes 8, 20, and 40 μm (Figure 5.23) were used, respectively, the average particle sizes of the corresponding P(MMA-GMA) microspheres were 4.2, 15.3, and 28.6 μm; the particle size distributions were narrow with Span values at 0.23, 0.28, and 0.31, respectively. To control the porosity and pore size precisely, the osmotic pressure and Laplace pressure balance could be tuned. This method enabled the fabrication of functional polymeric microspheres with uniform and controllable porous structures and particle sizes, which improved their performance and broadened the scope of their applications, especially in chromatographic separation.

Such a gigaporous attribute has also been endowed to the biodegradable PLGA particles. In particular, a novel strategy, which is called "post-loading mode," could be integrated with the gigaporous PLGA particles for improving the encapsulation efficiency of insoluble drug. For example, ropivacaine hydrochloride (ROP), a hydrophilic drug used in the field of anesthesia and analgesia, was encapsulated

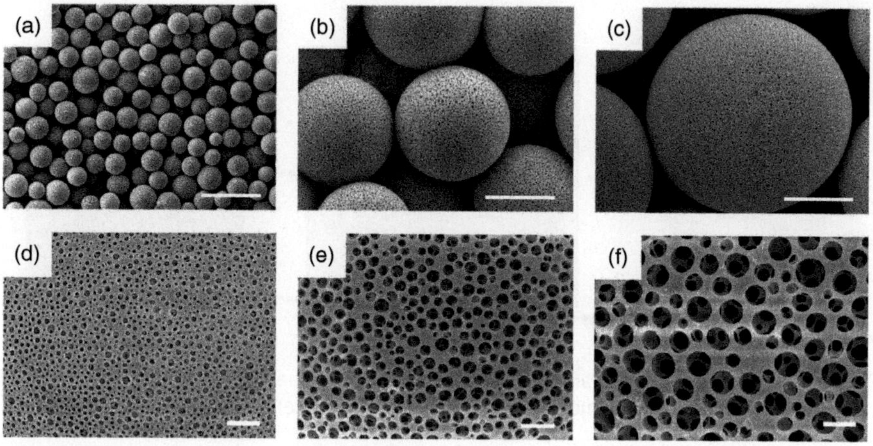

Figure 5.23 (a)–(c) Effect of membrane pore size on the particle size of microspheres in secondary premix membrane emulsification: (a) 8 μm, (b) 20 μm, and (c) 40 μm. Scale bars represent 10 μm. (d)–(f) Effect of membrane pore size on the surface morphology of microspheres in primary premix membrane emulsification: (d) 0.8 μm, (e) 1.4 μm, and (f) 2.8 μm. Scale bars represent 5 μm. Source: Na et al. [32]/Elsevier.

into the pre-prepared porous PLGA microspheres [33, 34]. Primary emulsion (W_1/O) was prepared by mixing PLGA/PELA dissolved EA (O) and 0.05–0.10% NaCl solution (W_1). The double emulsion (W_1/O/W_2) was obtained after stirring the primary emulsion with 1.5% PVA solution (W_2) and then uniformed by rapid MET under 20 μm membrane pore at 10 kPa of transmembrane pressure. These PLGA microspheres were then suspended with a saturated ROP solution for 4–12 hours, where ROP solution could be fully diffused into the porous microspheres. By using a self-healing strategy that developed in our group, the surface pores of microspheres melted during the PLGA molecular reassemble after 2.5 hours incubation in the thermostat around the glass translation temperature (42–46 °C) of PLGA, and the drug was then softly locked into the microspheres. This strategy facilitated the ROP–PLGA–Ms to achieve a drug loading (DL) at 18.0 ± 0.5 μg mg^{-1} and the EE at (55.7 ± 2.69)% [33]. The prepared ROP–PLGA microspheres had a stable sustained release effect which showed great potential in the field of sustained release of local anesthetics.

5.4.2.3 Preparation of PLGA Microspheres with Nonspherical Structures

Most polymer particles are prepared in a spherical shape because the minimization of interfacial energy leads to the spontaneous formation of spherical emulsions by using the conventional emulsion-based method. Recently, nonspherical particles are highly desired for investigating the influence of different physical properties on their biological functions. To endow the particles with a specific shape without losing uniformity, we developed a method to fabricate rod-shaped PLGA particles with uniform size (Figure 5.24) [35]. This method employed a common agent, phosphate buffer saline (PBS), as the deformation initiator and combined the rapid MET with the solvent evaporation method. The preparation of double emulsion (W_1/O/W_2)

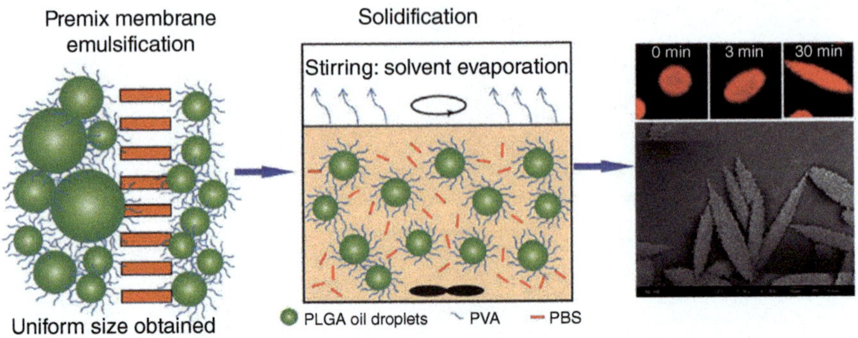

Figure 5.24 Schematic diagram and images of the rod-shaped polymer particles prepared via microporous membrane emulsification technology. Source: Fan et al. [35]/Elsevier.

and the subsequent evaporation procedure was similar to the aforementioned ones. The concentration of PBS and PVA in the external water phase was tuned to prepare uniform nonspherical particles of various shapes with aspect ratios from 2 to 40 and various sizes ranging from 800 nm to 60 μm (long axis).

The interfacial tension of the droplet and the external driving force from solidification stirring were also demonstrated to play synergistic effects on the particle deformation. For example, the pH in the aqueous phase was adjusted between 5.7 and 9.4 by modifying the ratio of Na_2HPO_4–NaH_2PO_4 at a fixed PVA concentration (0.1%). As pH increased, particles with different shapes and surface morphology (with thorn) were obtained (Figure 5.25), which was attributed to the decrease of interfacial tension. In addition, when the stirring rate was increased from 100 to 500 rpm for droplets solidification, the tendency for particle stretching became obvious, and even a filamentous shape was shown at 500 rpm condition (Figure 5.26). Thus, by introducing the rapid MET, particle size and uniformity were easily controlled, which could be beneficial for demonstrating the deformation mechanism and effectively investigating biological functions.

5.4.3 Preparation of Microspheres via O/W/O Emulsion

Particles prepared by a single W/O emulsion were successfully utilized to encapsulate or adsorb soluble drug/proteins. In response to diverse requirements for drug delivery or administration routes, particles that are sourced from different materials or with specific physiochemical properties are being perused and exploited. To resolve this problem, O/W/O double emulsions rather than single emulsion is preferred and developed with the rapid MET. On the one hand, this technique can be extended to chitosan derivate (HTCC or CMCC), alginate, and agarose materials. On the other hand, despite the conventional delivery need for subcutaneous injection and intravenous injection, the anticipation for oral administration is ardent, which raised more challenges for the particle capacity.

In the O/W/O emulsion method, the oil phase is usually composed of drug/DCM, therefore, the DCM will be evaporated simultaneously during cross-linking. The solidification process of O/W/O is crucial to drug loading and efficacy, which

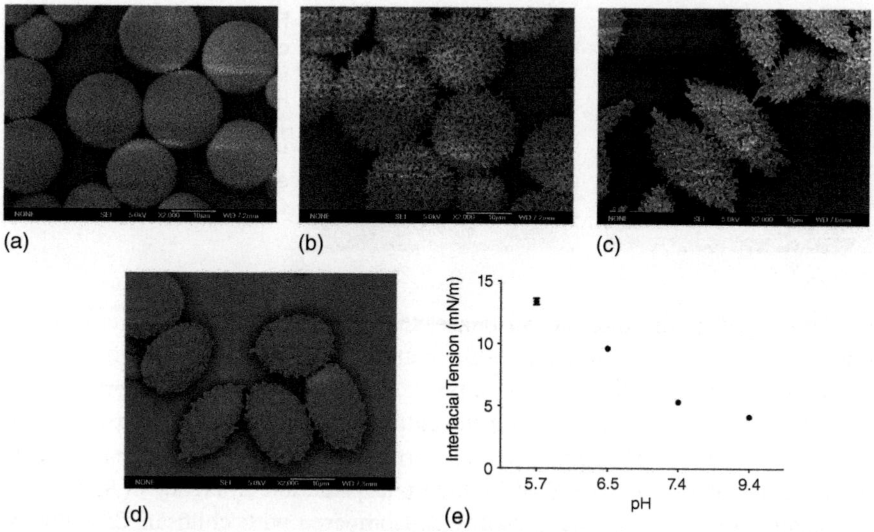

Figure 5.25 SEM photos of PLGA particles prepared at pH (a) 5.7, (b) 6.5, (c) 7.4, and (d) 9.4; (e) interfacial tension at different pH (membrane pore size was 34.8 μm). Source: Fan et al. [35]/Elsevier.

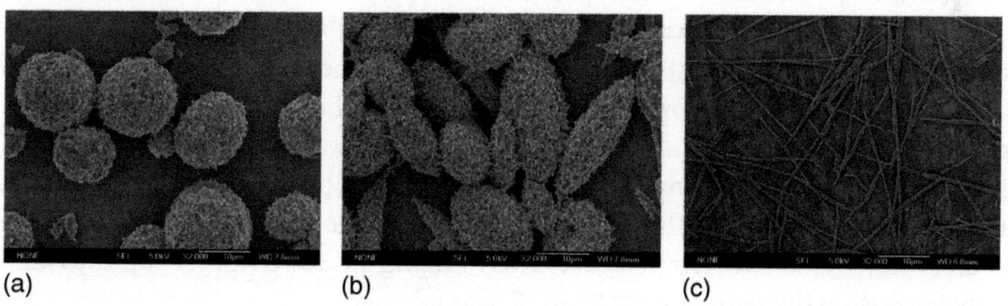

Figure 5.26 SEM photos of PLGA particles prepared at different stirring rates (a) 100 rpm, (b) 250 rpm, and (c) 500 rpm (membrane pore size was 34.8 μm). Source: Fan et al. [35]/Elsevier.

should be optimized. By choosing different membrane pore sizes and other conditions, particle size could be controlled from micro to nanosize (e.g. 130 nm), with the CV value below 15%. Herein, representative microspheres that are prepared with the material of chitosan or chitosan derivate are introduced.

5.4.3.1 HTCC Chitosan Microspheres for Oral Administration via O/W/O

Clinical application of PTX is limited because of its poor solubility in aqueous media. Commercial formulation (Taxol®) is usually administered by intravenous injection after dissolving PTX using ethanol and polyoxyethylene castor oil, which may cause allergy and significant side effects. To overcome this hurdle, our team [36] devised an oral delivery system by encapsulating PTX into quaternized HTCC chitosan NPs. The PTX encapsulated NPs were prepared using rapid MET combined with O/W/O emulsion, aiming to develop oral anticancer drug formulations. The PTX-dissolved

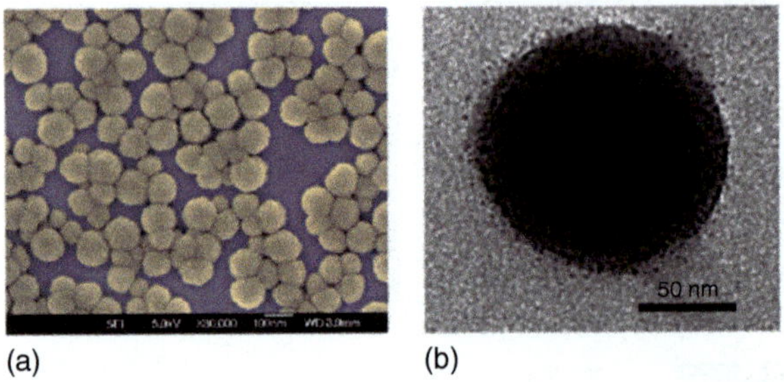

Figure 5.27 The characteristics and the *in vitro/vivo* performance of HTCC-NP:PTX. SEM (left) and TEM (right) images of HTCC-NP:PTX. Source: Fu et al. [37]/Royal Society of Chemistry.

DCM was used as the internal oil phase, the chitosan solution was used as the aqueous phase, and the liquid paraffin petroleum ether dissolved with PO-500 emulsifier was used as the external oil phase. The O/W emulsion was prepared by ultrasonication and then mixed with the external oil phase to prepare the O/W/O double emulsion. The PTX encapsulated particles were successfully obtained with a particle size of 130 nm, a PDI of 0.03, an encapsulation efficiency of 83%, and a drug loading efficiency of more than 35%. Compared with chitosan CS NP:PTX, the quaternized chitosan nanocapsules (HTCC NP:PTX) had porous structure and higher positive charge, which enhanced the intestinal permeability, facilitated the release of insoluble drugs and resulted in superior oral adsorption (Figure 5.27).

The advantage of the drug encapsulation technique via the O/W/O emulsion process was that the PTX was confined to precipitate within the internal water phase to form drug nanocrystals instead of large agglomerate, which avoided rapid escape from the particles. However, the rate of chemical cross-linking reaction and removal of DCM must be coordinated. When the DCM was volatilized too fast, PTX would form large crystals and escape from the nanocapsules before the chitosan was cross-linked. To handle this problem and remove the solvent, while continuing the cross-linking, a low-temperature cross-linking at 25 °C for 30 minutes and slowly increasing the temperature to 50 °C was adopted.

5.4.3.2 CMCC Chitosan Microspheres for Encapsulating Water-Insoluble Drugs

In spite of the HTCC chitosan particles for oral administration, we also developed a tumor-targeting delivery system for water-insoluble drug (PTX) by PEGylated O-carboxymethyl-chitosan (CMC) NPs grafted with cyclic Arg–Gly–Asp (RGD) peptide [38]. To improve the loading efficiency (LE), we combined O/W/O double emulsion method with temperature-programmed solidification technique as described above, and controlled PTX within the matrix network as in situ nanocrystallite form (Figure 5.28). The preparation of O/W/O emulsion for PTX-loaded CMC NPs (CNP:PTX) was similar to that for HTCC NP:PTX. Uniform-sized nanodroplets were achieved by extruding the coarse emulsion through the hydrophobic membrane (pore size was 0.5 μm) under high pressure (2.0 MPa). Solidification of CNP:PTX was performed by a two-step procedure, which was similar to that for encapsulating the water-insoluble drugs in HTCC nanoparticles (HNPs).

The CNP:PTX was functionalized by Maleimide PEG2000 amine with heteroterminal bifunctional groups after the COOH on CNP was activated with the assistance of 1-Ethyl-3-(3-dimethyl aminopropyl) carbodiimide (EDC) and

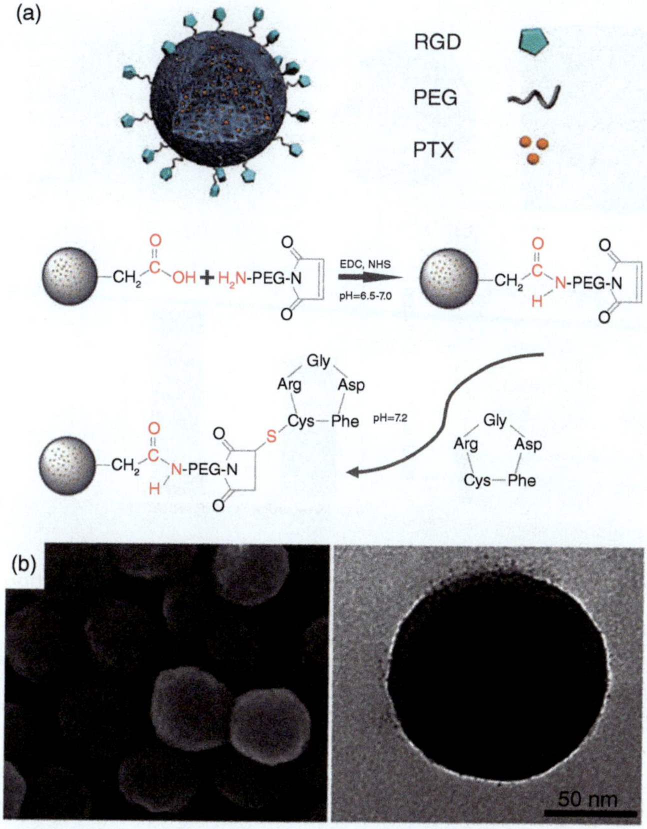

Figure 5.28 The conjugation strategy and characterizations of RGD–PEG–CNP:PTX. (a) Schematic illustration of RGD–PEG–CNP:PTX. (b) SEM (left) and TEM (right) images of RGD–PEG–CNP:PTX. PTX nanocrystals were successfully encapsulated into uniform-sized RGD–PEGCNP. Source: Lv et al. [38]/American Chemical Society.

N-hydroxysuccinimide (NHS). The amino group can be coupled to the COOH on CNP through an amide bond, and the maleimide can be used for the grafting of the subsequent cyclic RGD peptide (via the additive reaction between cysteine Cys and hydrosulfide group –SH). During this process, about 64% of carboxyl groups were substituted by the PEG chain, while 56% PEG chain was coupled to cyclic RGD peptide. These CMC NPs were PEGylated, which could reduce recognition by the reticuloendothelial system (RES) and prolong the circulation time in blood. Meanwhile, the modification of cyclic RGD peptide at the terminal of the PEG chain endowed these NPs with a higher affinity to carcinoma cells, which could be exploited in antitumor targeting.

5.4.3.3 Chitosan Microspheres for Combined Drug Delivery and Specific Administration

Apart from the hurdle faced by water-insoluble drugs, siRNA-based clinical applications are being hindered by poor intracellular uptake and enzymatic degradation.

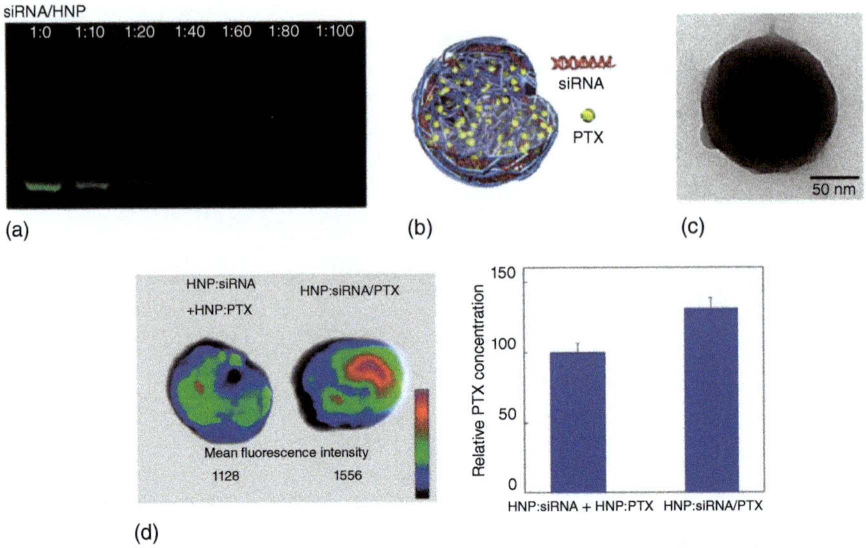

Figure 5.29 The characteristics and *ex vivo* performance of HNP:siRNA. (a) The evaluation of the siRNA-binding ability to HNP: electrophoretic mobility of free siRNA escaped from the complex. When the ratio of NPs to siRNA was above 80 (mg/mg), no significant band was observed. The structure diagram (b) and corresponding TEM image (c) of HNP:siRNA/PTX. (d) Quantitative determination of siRNA (left) and PTX (right) in tumor tissue by *In vivo* Imaging System and HPLC, respectively. Source: Wei et al. [39]/Elsevier.

To address these problems, we devised an oral delivery system for telomerase reverse transcriptase siRNA as well as PTX using HNP [39]. HNP–PTX with a narrow size distribution (PDI < 0.1) were prepared by the aforementioned method in Section 5.3.3.1, and a "two-in-one" nanocomplex (HNP:siRNA/PTX) was formed via a facile mixing for siRNA and HNP–PTX in RNase free dH_2O (Figure 5.29). In particular, another HTCC layer was further coated onto the particle surface to shield siRNA from degradation. Both the porous structure (pore size: 13.98 nm) and positive charge (zeta potential: +21.02 mV) of HNP (130–145 nm) facilitated successful siRNA loading.

It was demonstrated that HNP:siRNA/PTX could simultaneously ferry siRNA and PTX into tumor cells and increase drug concentration. The concentrations of siRNA and PTX for the "two-in-one" HNP:siRNA/PTX group were about 1.38-folds and 1.31-folds for the mixture of mono-drug-loaded HNP (HNP:siRNA+ HNP:PTX), respectively (Figure 5.29f). That is to say, the as-designed HNP:siRNA/ PTX could use fewer particle carriers but achieve a higher delivery efficiency than the mixture delivery, leading to much more effective tumor suppression. Meanwhile, the outer coating of HTCC not only protected siRNA from enzymatic degradation but also improved siRNA permeability in the intestine tract. These results suggested that the HNP, as a powerful delivery system for both siRNA and chemotherapeutic drugs might have a far-reaching application in human cancer therapy.

5.4.3.4 Preparation of Biomimetic Chitosan Microsphere with Cell Membrane

Combination chemotherapy has been proven promising for cancer treatment, but unsatisfactory therapeutic data and side effects slow down the development in the clinic. Our group developed an effective approach to coencapsulate a hydrophilic–hydrophobic chemotherapeutic drug pair (PTX and doxorubicin) into magnetic *O*-carboxymethyl–chitosan NPs (Figure 5.30) [37]. To endow them with the ability of programmed delivery, these carriers were further camouflaged with an anchored erythrocyte membrane. In brief, a DCM solution containing PTX and Fe_3O_4 NPs was mixed with an acetic acid buffer solution containing 0.5 wt% CMC and 2 wt% Tween-60 to obtain a primary emulsion (O/W) through ultrasonic emulsification. The primary emulsion was then added into the oil phase, a mixture of liquid paraffin and petroleum ether containing PO-500, to obtain a coarse double emulsion (O/W/O). Uniform-sized nanodroplets were achieved by extruding the coarse emulsion through the microporous membrane (pore size 0.5 μm) under 2.0 MPa pressure. By optimizing the CMC content, the CNPs:Fe/P (PTX) was endowed with a porous structure (pore size: 10.35 nm) and negative charge (zeta potential: −30.26 mv), which paved the way for subsequent DOX (D) loading through electrostatic adsorption. Subsequently, these particles were camouflaged with erythrocyte membrane using repeated extrusion with hemosome. During this process, DSPE–PEG-conjugated RGD can be inserted into the phospholipid layer of the erythrocyte membrane spontaneously to form the expected particle formulation (RH-CNP:Fe/P/D) (Figure 5.30).

The aforementioned particles were prepared using the O/W/O double emulsion and temperature-programmed solidification methods, which facilitated a very high loading capacity and the integration of three drugs/elements. Meanwhile, compared with the traditional polyethylene glycol coating method, this biomimetic decoration strategy was demonstrated to be superior in prolonging circulation time, improving tumor accumulation, facilitating tumor uptake, and tuning intracellular fate. These outstanding properties might enable the as-designed nanodevice to exhibit greater carrier merit and much lower side effects than the combined use of commercial formulations.

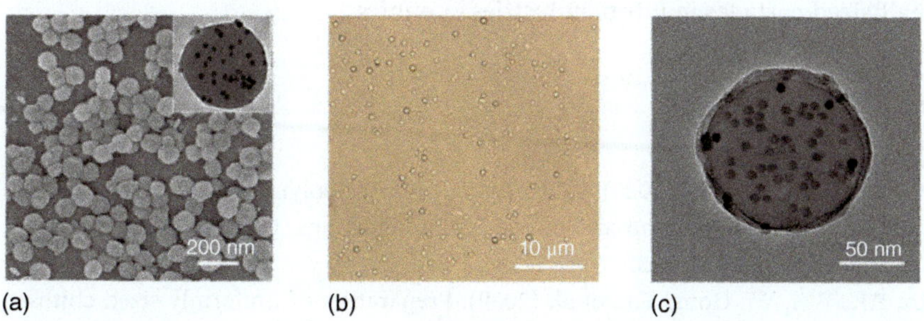

Figure 5.30 Schematic depicting the erythrocyte-membrane-camouflaged programmed codelivery system and its fabrication process. The corresponding characterizations are displayed below: (a) SEM and TEM (insert) images of CNP:Fe/P/D, (b) optical microscopy image of hemosome, and (c) TEM image of RH-CNP:Fe/P/D. Source: Fu et al. [37]/Royal Society of Chemistry.

5.5 Summary

Rapid MET overcomes the limitations of preparing small-sized particles (e.g. nanosize) or particles from preformed polymer solution with high viscosity, which is faced by conventional techniques. In particular, particle size can be controlled between 0.1–30 μm with a CV value of around 15%, by using rapid MET with optimum membrane pore size. It has been developed into a universal particle preparation method covering various material types not only for monomers but also for polymers and different emulsion types. To adapt for different material/drug types, we have developed this MET process from O/W (O phase for hydrophobic PLGA/PLA, etc.), W/O (W phase for hydrophilic chitosan, PNIPAM, agarose, alginate, glucomannan, etc.) to double emulsion (e.g. W/O/W for amphiphilic PELA and O/W/O for HTCC/CMCC). To cater for the requirements of drug/antigen loading or biological performance, divergent strategies regarding adsorption/encapsulation (especially for water-insoluble drugs), solidification (e.g. droplet swelling method), emulsification steps (two-step rapid MET for particular structures), and modification (e.g. PEG/membrane coating), that along with rapid MET have also been successfully explored.

The capabilities of rapid MET on aspect of defined particle size (micro to nano), various structures (hollow, nonspherical, one core–shell, and gigaporous), optimum biological merit (e.g. high adsorption/encapsulation efficiency, long circulation, and controlled release profile) have thus been extensively demonstrated [40]. By utilizing these uniform particles, not only fundamental research (size, shape effect on the biological response) can be studied precisely but also the reproducibility of products is guaranteed, which is especially important for biomedical applications. Correspondingly, the micro/nanospheres prepared via rapid MET could be explored in the biochemical engineering area (e.g. bioseparation media) and biomedical engineering area (e.g. antigen delivery, anticancer treatment, and bioimaging). Moreover, the rapid MET is easy to scale up and has low energy consumption, which paves the way for further practical use. It can be expected that rapid MET would become a potent technique that largely facilitates the application of small-sized particles in future industries or clinics.

References

1 Ma, G.H. and Yue, H. (2020). Advances in uniform polymer microspheres and microcapsules: preparation and biomedical applications. *Chinese Journal of Chemistry* 38 (9): 911–923.

2 Lv, P.P., Wei, W., Gong, F.L. et al. (2009). Preparation of uniformly sized chitosan nanospheres by a premix membrane emulsification technique. *Industrial and Engineering Chemistry Research* 48 (19): 8819–8828.

3 Zhou, Q.Z., Wang, L.Y., Ma, G.H., and Su, Z.G. (2008). Multi-stage premix membrane emulsification for preparation of agarose microbeads with uniform size. *Journal of Membrane Science* 322 (1): 98–104.

4 Zhou, Q.Z., Ma, G.H., and Su, Z.G. (2009). Effect of membrane parameters on the size and uniformity in preparing agarose beads by premix membrane emulsification. *Journal of Membrane Science* 326 (2): 694–700.

5 Wei, Q., Wei, W., Lai, B. et al. (2008). Uniform-sized PLA nanoparticles: preparation by premix membrane emulsification. *International Journal of Pharmaceutics* 359 (1–2): 294–297.

6 Li, X., Wei, Y., Wen, K. et al. (2021). Novel insights on the encapsulation mechanism of PLGA terminal groups on ropivacaine. *European Journal of Pharmaceutics and Biopharmaceutics* 160: 143–151.

7 Zhu, L., Gong, F., Yuan, Q., and Ma, G. (2013). Preparation of uniform and heterogeneous structure polystyrene particles by premix membrane emulsification. *The Chinese Journal of Process Engineering* 13 (2): 275–280.

8 Zhu, L., Li, Q., Gong, F.L. et al. (2013). An insight into structure regulation of uniform polystyrene micro/nano-particles by porogen in premix membrane emulsification process. *Journal of Membrane Science* 448: 248–255.

9 Lu, T., Hu, F., Yue, H. et al. (2020). The incorporation of cationic property and immunopotentiator in poly(lactic acid) microparticles promoted the immune response against chronic hepatitis B. *Journal of Controlled Release* 321: 576–588.

10 Zeng, Y.J., Wang, L.Y., Guang-Hui, M.A., and Run-Yu, M.A. (2010). Preparation of microspheres of paclitaxel-loaded PLA, PLGA and PELA by premix membrane emulsification. *The Chinese Journal of Process Engineering* 10 (3): 568–575.

11 Li, X., Wei, Y., Lv, P.P. et al. (2019). Preparation of ropivacaine loaded PLGA microspheres as controlled-release system with narrow size distribution and high loading efficiency. *Colloids and Surfaces A: Physicochemical and Engineering Aspects* 562: 237–246.

12 Tian, Q., Zhou, W., Cai, Q. et al. (2022). In situ complex coacervation supported by self-coated polydopamine interlayer on uniform-sized essential oils droplet. *Journal of Colloid and Interface Science* 623: 1027–1038.

13 Wei, W., Wang, L.Y., Yuan, L. et al. (2007). Preparation and application of novel microspheres possessing autofluorescent properties. *Advanced Functional Materials* 17 (16): 3153–3158.

14 Yue, H., Wei, W., Yue, Z. et al. (2010). Particle size affects the cellular response in macrophages. *European Journal of Pharmaceutical Sciences* 41 (5): 650–657.

15 Wu, J., Wei, W., Wang, L.Y. et al. (2007). A thermosensitive hydrogel based on quaternized chitosan and poly(ethylene glycol) for nasal drug delivery system. *Biomaterials* 28 (13): 2220–2232.

16 Wu, J., Wei, W., Wang, L.Y. et al. (2008). Preparation of uniform-sized pH-sensitive quaternized chitosan microsphere by combining membrane emulsification technique and thermal-gelation method. *Colloids and Surfaces B: Biointerfaces* 63 (2): 164–175.

17 Wang, Y.Q., Wu, J., Fan, Q.Z. et al. (2014). Novel vaccine delivery system induces robust humoral and cellular immune responses based on multiple mechanisms. *Advanced Healthcare Materials* 3 (5): 670–681.

18 Wang, Y.Q., Fan, Q.Z., Liu, Y. et al. (2016). Improving adjuvanticity of quaternized chitosan-based microgels for H5N1 split vaccine by tailoring the

particle properties to achieve antigen dose sparing effect. *International Journal of Pharmaceutics* 515 (1–2): 84–93.

19 Wang, Y.X., Qin, J., Wei, Y. et al. (2013). Preparation strategies of thermo-sensitive P(NIPAM-*co*-AA) microspheres with narrow size distribution. *Powder Technology* 236: 107–113.

20 Zhang, Y.L., Wei, W., Lv, P.P. et al. (2011). Preparation and evaluation of alginate-chitosan microspheres for oral delivery of insulin. *European Journal of Pharmaceutics and Biopharmaceutics* 77 (1): 11–19.

21 Liu, Y., Yin, Y., Wang, L.Y. et al. (2013). Engineering biomaterial-associated complement activation to improve vaccine efficacy. *Biomacromolecules* 14 (9): 3321–3328.

22 Nan, F.F., Wu, J., Qi, F. et al. (2014). Uniform chitosan-coated alginate particles as emulsifiers for preparation of stable Pickering emulsions with stimulus dependence. *Colloids and Surfaces A: Physicochemical and Engineering Aspects* 456: 246–252.

23 Nan, F.F., Wu, J., Qi, F. et al. (2014). Preparation of uniform-sized colloidosomes based on chitosan-coated alginate particles and its application for oral insulin delivery. *Journal of Materials Chemistry B* 2 (42): 7403–7409.

24 Ao, Z., Yang, Z., Wang, J.F. et al. (2009). Emulsion-templated liquid core-polymer shell microcapsule formation. *Langmuir* 25 (5): 2572–2574.

25 Mi, Y., Li, J., Zhou, W.Q. et al. (2016). Improved stability of emulsions in preparation of uniform small-sized konjac glucomanna (KGM) microspheres with epoxy-based polymer membrane by premix membrane emulsification. *Polymers-Basel* 8 (3): 53.

26 Wei, Q., Wei, W., Tian, R. et al. (2008). Preparation of uniform-sized PELA microspheres with high encapsulation efficiency of antigen by premix membrane emulsification. *Journal of Colloid and Interface Science* 323 (2): 267–273.

27 Qi, F., Wu, J., Fan, Q.Z. et al. (2013). Preparation of uniform-sized exenatide-loaded PLGA microspheres as long-effective release system with high encapsulation efficiency and bio-stability. *Colloids and Surfaces B: Biointerfaces* 112: 492–498.

28 Qi, F., Wu, J., Hao, D.X. et al. (2014). Comparative studies on the influences of primary emulsion preparation on properties of uniform-sized exenatide-loaded PLGA microspheres. *Pharmaceutical Research* 31 (6): 1566–1574.

29 Qi, F., Wu, J., Yang, T.Y. et al. (2014). Mechanistic studies for monodisperse exenatide-loaded PLGA microspheres prepared by different methods based on SPG membrane emulsification. *Acta Biomaterialia* 10 (10): 4247–4256.

30 Liu, Q., Chen, X.M., Jia, J.L. et al. (2015). pH-Responsive poly(D,L-lactic-*co*-glycolic acid) nanoparticles with rapid antigen release behavior promote immune response. *ACS Nano* 9 (5): 4925–4938.

31 Na, X.M., Guo, J.J., Li, T. et al. (2020). Double emulsion-templated single-core PLGA microcapsules with narrow size distribution and controllable structure by using premix membrane emulsification. *CHEMNANOMAT* 6 (7): 1059–1062.

32 Na, X.M., Zhou, W.Q., Li, T. et al. (2019). Preparation of double-emulsion-templated microspheres with controllable porous structures by premix membrane emulsification. *Particuology* 44: 22–27.

33 Wen, K., Wei, Y., and Ma, G. (2021). Analysis of preparation factors of ROP-PLGA microspheres with high encapsulation efficiency based on response surface method. *The Chinese Journal of Process Engineering* 21 (1): 83–91.

34 Wen, K., Na, X.M., Yuan, M.M. et al. (2022). Preparation of novel ropivacaine hydrochloride-loaded PLGA microspheres based on post-loading mode and efficacy evaluation. *Colloids and Surfaces B: Biointerfaces* 210.

35 Fan, Q.Z., Qi, F., Miao, C.Y. et al. (2016). Direct and controllable preparation of uniform PLGA particles with various shapes and surface morphologies. *Colloids and Surfaces A: Physicochemical and Engineering Aspects* 500: 177–185.

36 Lv, P.P., Wei, W., Yue, H. et al. (2011). Porous quaternized chitosan nanoparticles containing paclitaxel nanocrystals improved therapeutic efficacy in non-small-cell lung cancer after oral administration. *Biomacromolecules* 12 (12): 4230–4239.

37 Fu, Q., Lv, P.P., Chen, Z.K. et al. (2015). Programmed co-delivery of paclitaxel and doxorubicin boosted by camouflaging with erythrocyte membrane. *Nanoscale* 7 (9): 4020–4030.

38 Lv, P.P., Ma, Y.F., Yu, R. et al. (2012). Targeted delivery of insoluble cargo (paclitaxel) by PEGylated chitosan nanoparticles grafted with Arg-Gly-Asp (RGD). *Molecular Pharmaceutics* 9 (6): 1736–1747.

39 Wei, W., Lv, P.P., Chen, X.M. et al. (2013). Codelivery of mTERT siRNA and paclitaxel by chitosan-based nanoparticles promoted synergistic tumor suppression. *Biomaterials* 34 (15): 3912–3923.

40 Yue, H. and Ma, G.H. (2020). Applications of polymeric micro/nanoparticles in engineered vaccines. *Acta Polymerica Sinica* 51 (2): 125–135.

6

Applications of Uniform Particles in Sustained Release of Drugs

6.1 Introduction

Protein and polypeptide drugs are mainly administrated by injection because they are hydrophilic biological macromolecules, which are easily degraded by gastrointestinal protease in the gastrointestinal environment and cannot effectively penetrate the physiological barrier composed of the intestinal mucus layer and epithelial layer [1]. Furthermore, protein and polypeptide drugs are required to be injected frequently due to the short circulation time in the blood, resulting in poor compliance of patients. In addition, injection of insoluble drugs easily causes pain at the injection site because oil is usually used as the solvent and it is difficult to maintain a stable blood concentration in the body [2]. In order to overcome these problems, microspheres/microcapsules have been considered as being the most promising sustained-release systems, because they not only can prolong the half-life time of a drug but also potentially favor an increase in the bioavailability of a drug by controlling the release rate of the drug from the matrix [3]. Manufacture of microspheres with uniform and controlled size plays a key role in microsphere formulations because the uniform microsphere formulations have the advantages of green production, low cost, good batch-to-batch repeatability, and beneficial to scale-up. At present, uniform microsphere formulations have been successfully prepared to apply in drug delivery systems, such as triptorelin acetate, exenatide, recombinant human growth hormone (rhGH), and other peptides, protein, and insoluble drugs [3–5].

An essential component in preparing a long-acting injectable drug release system is the biodegradable polymer material. At present, microsphere matrix materials can be divided into two types: natural and synthetic polymers. Polymers from natural sources are cheap and own a wide range of sources, such as chitosan, dextran, alginate, starch, and gelatin. Compared with other natural biodegradable polymers, chitosan has two main advantages in the drug delivery field. First of all, the preferable characteristics of chitosan can be determined by the degree of deacetylation (DDA), which considerably affects physical properties. Secondly, amino and hydroxyl groups in chitosan allow them to become amenable to various chemical modifications and form chitosan derivatives with definite functional

Novel Membrane Emulsification: Principles, Preparation, Processes, and Bioapplications,
First Edition. Guanghui Ma.

groups, which show better performance than chitosan itself due to the increase in bioadhesion and permeability. Synthetic polymers can be divided into polyester, polyanhydrides, polyphosphate, and so on. Because the degraded small molecules from polymers can be metabolized by the body, the molecular weight, polymerization mode, polymerization ratio, hydrophilicity, and hydrophobicity of the polymer can be adjusted according to the requirements of sustained-release microspheres. Polylactic acid-based biodegradable polymers are the most common synthetic materials utilized for designing microparticles. Among them, poly(lactic acid) (PLA) and poly(lactic-*co*-glycolic acid) (PLGA) are few polymers approved by Food and Drug Administration (FDA) with good biodegradability and biocompatibility and have been extensively employed in medical and pharmaceutical areas [6].

Albeit promising, the microsphere-related research area was still in its infancy. The most common problems are the low encapsulation efficiency (EE) and poor bioactivity preservation during microencapsulation, storage, and release under physiological conditions [3]. For example, the conventional double emulsion method used for the rhGH encapsulation by PLGA seemed to provide unsatisfactory rhGH profiles with insufficient rhGH activity after release. Because an aqueous protein solution was apt to be dispersed into an organic polymer solution during water/oil (W/O) emulsification, using a homogenizer or sonicator, the exposure of the protein to the W/O interface might affect stability. In addition, the hydrophobic nature of the PLGA and PLA matrix and acidification arising from degradation of the polymer matrix will easily lead to protein aggregation and inactivation, and these aggregates may result in a serious immunogenic response [3]. Therefore, in this chapter, strategies to improve EE and maintain bioactivity in terms of both chitosan and polylactic acid-based biodegradable polymers will be summarized and reviewed for the encapsulation of drugs by the membrane emulsification technique.

6.2 Synthetic Polymer (PLA, PLGA, and PELA)

As shown in Figure 6.1, Boswell and Scribner applied the first patent for the PLA-based drug delivery system, including long-acting release microspheres in 1973. Later, at the beginning of the 1980s, Debiopharm, a Swiss-headquartered biopharmaceutical company, developed PLGA microparticles for the delayed controlled release of triptorelin. In 1986, Debiopharm commercialized Decapeptyl® SR into the market; it is a triptorelin-loaded long-acting release (LAR) system to treat prostate cancer. After three years, Lupron Depot®, a formulation of leuprolide acetate microspheres for prostate cancer treatment, was approved, followed by Parlodel LA® (bromocriptine mesylate), Sandostatin LAR® (octreotide acetate), and many others [7].

6.2.1 PLA

PLA is part of aliphatic polyesters and is generally synthesized from lactic acid or lactide, which is considered a renewable resource. As shown in Figure 6.2, PLA isomers

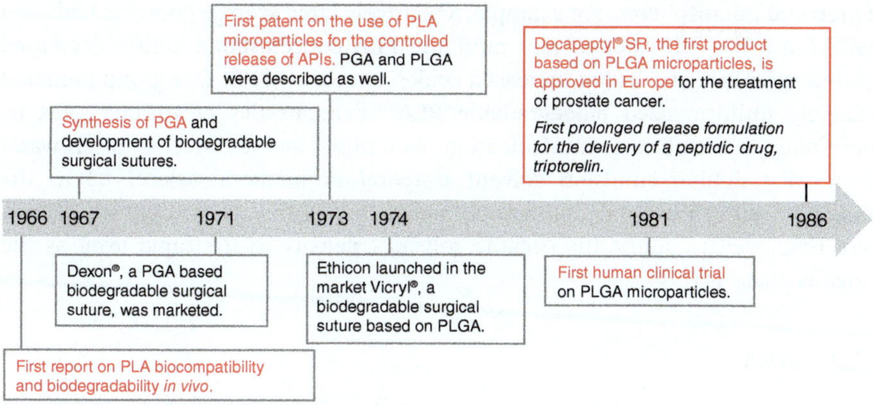

Figure 6.1 Timeline (1966–1986) indicating crucial scientific breakthrough of injectable products based on synthetic polymers (API: active pharmaceutical ingredient; PGA: polyglycolic acid).

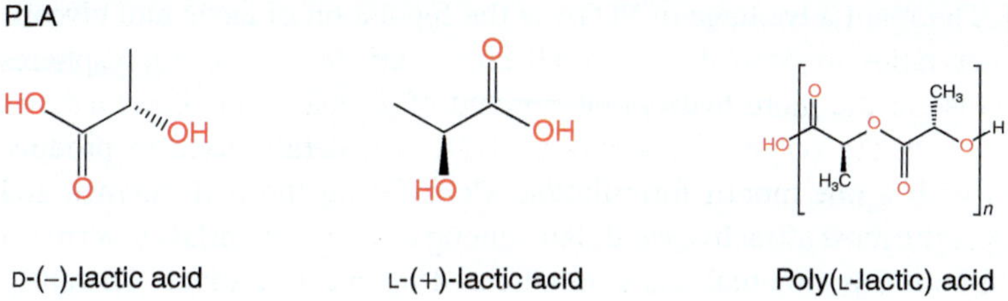

Figure 6.2 Molecular monomers D-(–)-lactic acid and L-(+)-lactic acid of the homopolymers poly(L-lactic acid).

exist as two enantiomers: poly(L-lactic acid) (PLLA) and poly(D-lactic acid) (PDLA). The ratio between L- and D-isomers mainly affects the degree of crystallinity of PLA. Some literature reported that a minimum of 85% of PLLA was required for developing the crystallinity of PLA, while its ratio PLLA/PDLA of 1/1 formed an amorphous polymer structure [8]. Its stereochemical structure can easily be changed by the polymerization reaction of a mixture of L- and D-isomers where high crystalline and amorphous polymers can be obtained, respectively [9]. The amorphous PLA is soluble in the following organic solvents, such as ethyl acetate, tetrahydrofuran (THF), chlorinated solvents, benzene, acetonitrile, and dioxane.

PLA is degraded by hydrolysis reaction by breaking the ester bond. The rate of reaction is affected by isomer ratio, temperature, time, and molecular weight. In general, PLA has been proven to have long degradation time and applied in the microspheres production of the 3- and 6-months administration. One example of this formulation is Lupron Depot, lyophilized microspheres loaded with leuprolide acetate initially launched in the USA in 1989. Hong et al. produced PLA microspheres with interconnected open-pore matrices employing the emulsion-solvent evaporation method based on solution-induced phase separation. It was reported that process parameters such as oil/water solvent ratio, the concentration of PLA, and solidification rate considerably impacted the particle and pore size distribution of prepared microspheres. For example, a relatively large average pore size had been yielded at a low oil/water solvent ratio and PLA concentration within decreased solidification rate because of increased coalescence time [10]. Our group prepared relatively uniform-sized biodegradable PLA microcapsules of various sizes by combining a membrane emulsification technique and a water-in-oil-in-water ($W_1/O/W_2$) double emulsion solvent evaporation method. Arlacel 83 as the oil-soluble emulsifier and dichloromethane (DCM)/toluene as mixture solvent were employed to adjust the organic solvent's density to the same level as the aqueous phase [11].

6.2.2 PLGA

PLGA is a copolymer of PLA and PGA and therefore consists of three different hydroxyl acid monomers (D-lactic, L-lactic, and glycolic acids). It is produced by a random ring-opening copolymerization reaction of two mentioned monomers, using tin(II) 2-ethyl hexanoate, tin(II) alkoxides, or aluminum isopropoxide as the catalyst [12]. The main advantage of PLGA is the regulation of lactic and glycolic acid copolymer ratios to control the overall characteristics of the microspheres [13]. An increase in the more hydrophilic content of glycolic acid allows a faster degradation rate of the polymer. Therefore, PLGA is generally used to produce microspheres with a one-month formulation. Considering the performance and design, PLGA is the most attractive candidate among other biomaterials in terms of application in both experimental and commercial drug encapsulations. Currently, PLGA has been already applied in numerous marketed products, such as Lupron Depot and Trelstar [14, 15].

By modifying the monomer ratio, it is possible to tune the polymer's physical properties, such as degradation time. The polymer can be processed to be a highly crystalline structure (as PLLA) or a completely amorphous structure (as PDLA). Therefore, the degree of crystallinity of the PLGA (Figure 6.3) is controlled by the ratio and type of individual monomers unit in the polymer, and it determines physical characteristics such as swelling behavior, mechanical property, and biodegradation time (Figure 6.4) [14]. The degree of crystallinity can be decreased by copolymerization of highly crystalline PGA with PLA, resulting in increase of hydrolysis rate [16]. PLGA is soluble in various volatile organic solvents, including DCM, THF, chloroform, acetone, and ethyl acetate. While in water, PLGA degrades into monomers by breaking their ester linkages. Consequently, these monomers are metabolized and converted into simple CO_2 and H_2O substances through the Krebs cycle. As the methyl side in PLA is more hydrophobic than PGA, the higher the percentage of lactide units, the less hydrophilic PLGA is; hence, less water is absorbed, and it degrades more gradually. Conversely, a polymer with a larger glycolide content is more hydrophilic, and subsequently, it tends to absorb a significant amount of water, which results in a

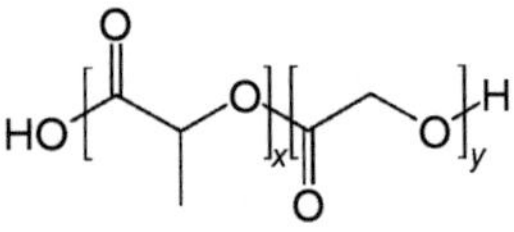

Figure 6.3 Structure of PLGA (*x* is the number of lactic acid units and *y* is the number of glycolic acid units).

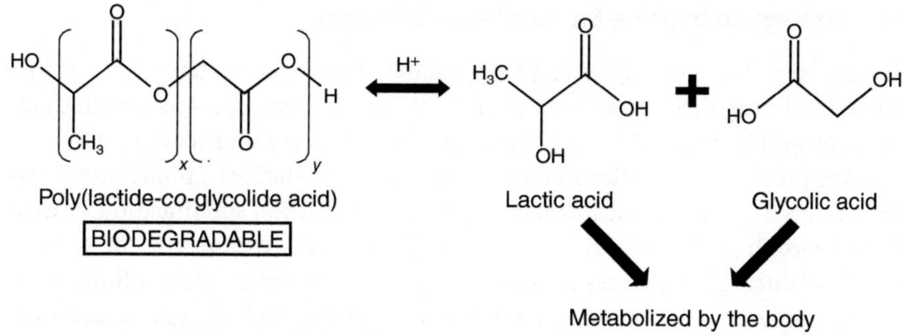

Figure 6.4 Hydrolysis of PLGA molecules.

short degradation time. This phenomenon influences the degradation rates of the microsphere matrix and the release rate of the drug [17]. In general, the physical property of PLGA is affected by different factors, including molecular weight, glass transition temperature, the ratio of lactide to glycolide, and contact with water [5].

6.2.3 PELA

To overcome the disadvantages resulting from the hydrophobic nature of PLA, a hydrophilic component poly(ethylene glycol) (PEG), which has been widely used to improve the biocompatibility of blood-contacting materials, is introduced to form amphiphilic copolymer poly(D, L-lactide) and PEG (PELA). It illustrates high potential because it enhances the properties of the microsphere's products. It is proved that the introduction of the PEG into the PLA matrix could reduce the additive amount of surfactant during the fabrication process. This can be explained by the fact that poly(monomethoxypolyethylene glycol-co-D,L-lactide) (PELA) polymer exhibits emulsifier properties as the PEG component is hydrophilic, and the PLA is hydrophobic [18]. When PELA is used in microsphere preparation with the double emulsion method, the hydrophilic unit of PELA polymer behaves as a protein-stabilizing agent or surface modifier of PLA polymer, which improves the stability of the drug thus increasing the EE of the drug [19].

In comparison with a hydrophobic polymer, such as PLA and PLGA, PELA has several advantages in the drug delivery system. First of all, PELA microparticles enhance the drug EE. Secondly, PELA microspheres decrease the protein aggregation and promote its stability. Lastly, PELA microparticles could reduce the initial burst effect of the drugs [5]. In addition, it has been reported that during the protein release, hydrophobic PLGA prevented the penetration of water into the center of the microspheres, thus forming an acidic microenvironment as a result of the accumulated acidic products, such as lactic and glycolic acid end groups [20]. The encapsulated proteins can be severely hydrolyzed or aggregated because of the acidic microenvironment. Incorporating a hydrophilic sequence PEG into relatively hydrophobic PLA block can promote water uptake and swelling of the microsphere matrix and modulate the diffusion of proteins from the carrier system, resulting in a more stabilizing environment for proteins [21].

6.2.4 Strategy to Improve Encapsulation Efficiency

Although great breakthroughs in biodegradable microspheres have been gained in the field of drug delivery and some of them have become commercial products, many hydrophilic drugs still have the problem of low EE and loading efficiency (LE) in the preparation of microspheres. This is because the hydrophilic proteins or peptide drugs are usually encapsulated by W/O/W emulsion solidification process, and the hydrophilic drugs in the inner water phase easily escape into the external water phase during the process of emulsion preparation and droplets solidification. The main strategies to improve encapsulation efficiency and LE will be described in more detail below. Before introduction of strategies, the definition of EE and LE is explained briefly. The EE was calculated by the following equation:

$$\mathrm{EE} = \frac{m}{m_0} \times 100\%$$

where m_0 is total mass of drug added and m is the mass of drug loaded in the microparticles.

The LE was calculated by the following equation:

$$\mathrm{LE} = \frac{a}{b} \times 100\%$$

where b is total mass of microparticles (including mass of polymer and drug loaded) and a is the mass of drug loaded in the microparticles.

6.2.4.1 Effect of Additives on Encapsulation Efficiency

Oil-soluble emulsifier usually is added in oil phase (polymer solution) to keep W/O emulsion stable, and coalescence of the inner and external aqueous phases is avoided. As a result, drug EE is improved by utilizing an emulsifier in the oil phase. We explored the effect of Arlacel 83 on lysozyme-loaded PLA microcapsules, and results showed that EE was increased by increasing emulsifier concentration when using the conventional stirring method. As shown in Figure 6.5, however, an opposite trend was observed when microcapsules were prepared with direct membrane emulsification technology, and EE was decreased from 77.83% to 56.24%. This is because Arlacel 83 has hydrophilic ends, and it can contact and wet the membrane's pores. Hence, the polydispersed double emulsion was yielded, which continually caused coalescence of droplets and breakup of large droplets [11].

Additives in the internal aqueous phase also have a significant effect on EE. First of all, it can contact with oil emulsifiers to upgrade the interfacial layer's strength between the oil and internal water phase. Thus, it decreases the chance of coalescence between the inner and outer aqueous phases. Secondly, it interacts with the active ingredient, preventing drug diffusion to the external phase. Lastly, the additive added in the internal water phase is relatively viscous, which can retard drug escape from the internal to the external aqueous phase. Common additives during the microsphere preparation are poly(vinyl alcohol) (PVA), polyvinylpyrrolidone (PVP), Tween 20, and gelatin. We found out that these inner water additives impacted the EE with a positive trend, and among them, the most effective one was PVA, which increased EE from 62.22% to 77.83% (Figure 6.6) [11].

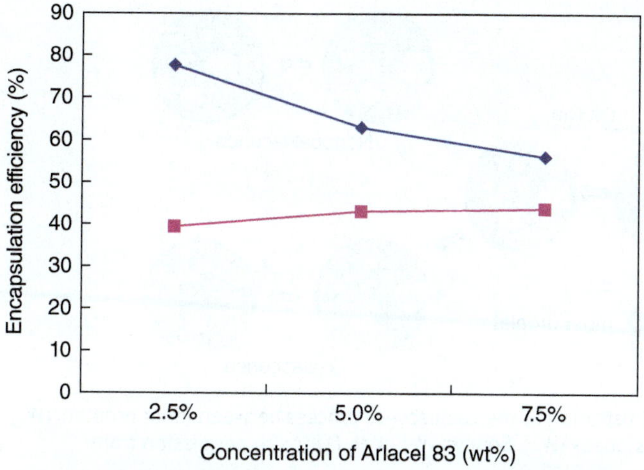

Figure 6.5 Effect of the emulsifier concentration in the oil phase on the drug encapsulation efficiency by different preparation methods. —◆—: membrane emulsification; —■—: stirring method. Source: Liu et al. [11]/with permission from Elsevier.

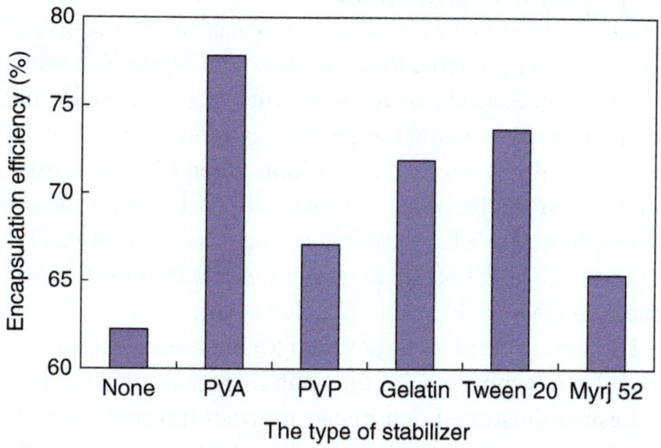

Figure 6.6 Effect of different additives on drug encapsulation efficiency. Source: Liu et al. [11]/with permission from Elsevier.

6.2.4.2 Effect of pH in the External Phase

Concerning the pH value of the external aqueous phase, when it was proximate to the isoelectric point of the encapsulated active ingredient, the solubility of the active ingredient in the outer aqueous phase decreased. Therefore, the drug escape into the outer aqueous phase was restricted. Significant enhancement of the drug loading efficiency was observed when the pH value of the external aqueous phase was 6.0 by adjusting the molar ratio of HCl and NaOH [22]. We also found that when the external water phase's pH was raised from 5.2 to 9.0, the loading efficiency (LE%) increased from 42.13% to 49.06%. The alkaline environment reduced the drug base's solubility in the external water phase, further reducing the drug's escape [17].

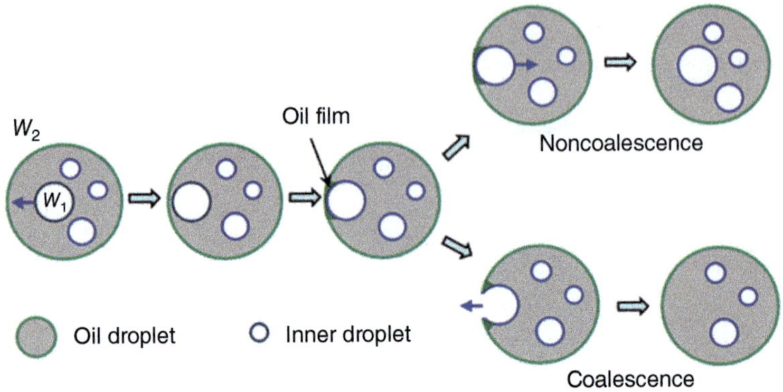

Figure 6.7 Schematic illustration of the coalescence process between inner droplets (W_1) and the external aqueous phase (W_2). Source: Wei et al. [18]/with permission from Elsevier.

6.2.4.3 Effect of Polymer

In order to overcome the problem that the protein in the inner droplets tends to merge with external water phase, resulting in the loss of encapsulated protein during the fabrication, the amphiphilic PELA (PLA-mPEG) was used as microcapsules matrix polymer to compare with hydrophobic PLA. As shown in Figure 6.7, when PLA was used as polymer, the inner droplets moved to the interface of oil phase and external water phase, the inner droplets would merge into the external water phase rapidly and disappear once the oil film was evacuated. While, when PELA was used, the PLA segment was anchored at the interface, the evacuation of oil film therein would be restrained, resulting in higher EE. According to our research results, PLA microcapsules encapsulated 61.3% of the active ingredient, while there was a dramatic increase in EE (almost 90%) with PELA microcapsules [18].

The molecular weight of polymer also plays important role in protein drug encapsulation. Our group prepared microspheres with four different molecular weights: 10, 20, 100, and 300 kDa. Results illustrated that higher polymer molecular weight resulted in the highest EE (71.6%), and it was achieved with a PLA of 300 kDa. High molecular weight formed a primary emulsion with increased viscosity of the oil phase, and its stability was enhanced. As a result, the drug's escape from the viscous dispersed phase to the external aqueous phase became difficult [22].

However, we found the opposite result when we used single emulsion method to encapsulate a lipid-soluble drug that had interaction with the polymer material. Our group encapsulated ropivacaine (RVC) in PLGA microsphere from O/W emulsion by membrane emulsification technology, where the drug was dissolved in PLGA solution directly. We found that the EE of ropivacaine decreased with the increase of the molecular weight of the polymer. This was because the amino group in the ropivacaine molecule interacted with the carboxyl group in the PLGA molecule. When the molecular weight of PLGA decreased, more carboxyl groups interacted with ropivacaine to produce stronger electrostatic attraction, so the number of drugs that escaped in the preparation process can be reduced, which

Table 6.1 Characterization of OH-RVC-MS, COOH-RVC-MS, and COOR-RVC-MS (results are mean ± standard deviation (SD), $n = 3$).

	Average diameter (μm)	Span	LE%	EE%
OH-RVC-MS	17.01 ± 0.67	0.572	33.52 ± 0.87	83.8 ± 2.03
COOH-RVC-MS	17.15 ± 0.38	0.553	37.84 ± 1.23	94.6 ± 3.075
COOR-RVC-MS	17.49 ± 0.71	0.538	32.2 ± 0.49	80.5 ± 1.23

led to a higher EE. Based on the narrow particle size distribution, we investigated in detail the influences and mechanisms of the terminal groups on the properties of ropivacaine-loaded microspheres. It was discovered that the terminal groups of PLGA showed a significant effect on ropivacaine incorporation and content in the microspheres. The results are shown in Table 6.1, the 94.6% of EE% was obtained by using PLGA with the COOH terminal group (COOH-RVC-MS), as compared to 83.8% with the OH terminal group (OH-RVC-MS) and 80.5% with the COOR terminal group (COOR-RVC-MS). Figure 6.8 shows the changes in frequency (ΔF) and dissipation (ΔD) for the different terminal groups of PLGA film during the ropivacaine adsorption and subsequent desorption process. As shown in Figure 6.8b, PLGA-COOH resulted in a decrease in frequency (ΔF) of 41.5 Hz and an increase in dissipation (ΔD) of 5.96. For PLGA-OH and PLGA-COOR ones (Figure 6.8a,c), a smaller decrease in frequency (10.16 and 0.24 Hz, respectively) and smaller increases in dissipation (3.89 and 0.26, respectively) were observed. These results demonstrated that PLGA-COOH can absorb more ropivacaine than PLGA-OH and PLGA-COOR. Furthermore, they observed the topographical changes after the desorption of ropivacaine by QCM-D. Figure 6.9 shows that the amount of ropivacaine adsorbed on the surface of PLGA-COOH was much larger than that on the PLGA-OH and PLGA-COOR. This experiment phenomenon also identified the results of the QCM-D test. Therefore, it can be concluded that the different terminal group of PLGA leads to different absorption ability of ropivacaine, resulting in different EE [23].

6.2.4.4 Effect of Solidification Technique

The solidification of the microsphere is a process in which the oil phase (organic solvent) volatilizes and the film material is deposited to cover the core. In this process, the droplets of the emulsion gradually form into microspheres. Several commonly used solidification methods, such as solvent extraction method, solvent evaporation method, or cosolvent method, are essential to obtain drug-loaded microspheres with different properties. For the emulsion–solvent evaporation method, the solidification time plays an important role in drug encapsulation in microspheres [24, 25].

In order to control the solidification rate, the process of solidification was modified with a procedure of additional presolidification. The double emulsions obtained by rapid membrane emulsification were preliminarily partially solidified by a little amount of aqueous solution and then completely solidified by adding a large

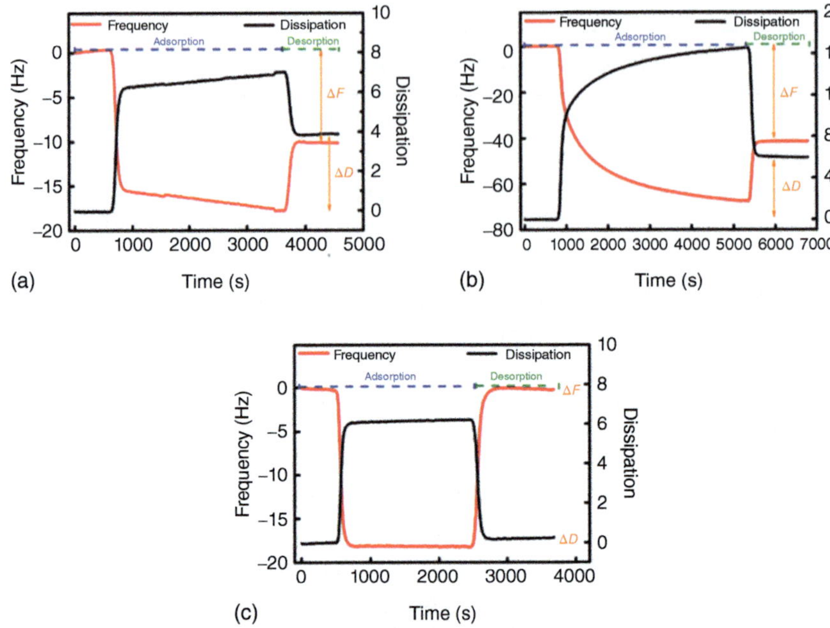

Figure 6.8 Diagram of frequency (red line) and dissipation (black line) vs. time for exposure of the (a) PLGA-OH, (b) PLGA-COOH, and (c) PLGA-COOR surfaces to ropivacaine solution, followed by the exchange of the ropivacaine solution with buffer solution. ΔF and ΔD represent the changes in frequency and dissipation, respectively, when desorption becomes saturated. Source: Li et al. [23]/with permission from Elsevier.

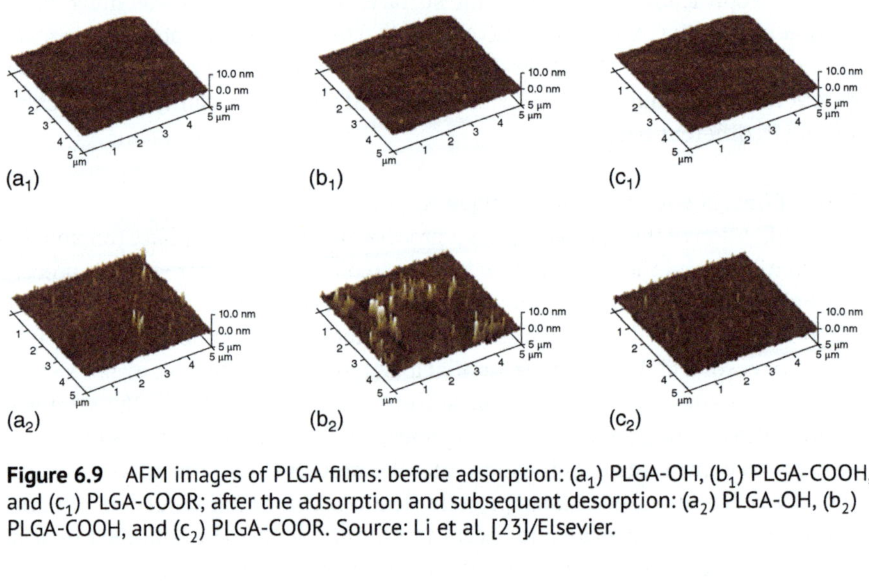

Figure 6.9 AFM images of PLGA films: before adsorption: (a_1) PLGA-OH, (b_1) PLGA-COOH, and (c_1) PLGA-COOR; after the adsorption and subsequent desorption: (a_2) PLGA-OH, (b_2) PLGA-COOH, and (c_2) PLGA-COOR. Source: Li et al. [23]/Elsevier.

amount of water further. The solidification rate of microspheres fabricated by direct solidification technique is faster than those of step step-wise solidification technique. In step-wise solidification technique, the solidification rate of microspheres would be raised with the increase of presolidification aqueous solution volume. PLA-mPEG was chosen as the polymer and three presolidification solution volumes of microspheres were prepared, with batch C volume of 0 ml (direct solidification technique), batch D of 40 ml, and batch E of 20 ml. Thus, batch C > batch D > batch E in terms of solidification rate of obtained microspheres. As shown in Figure 6.10, the surface morphologies of microspheres were observed by SEM. All the microspheres revealed rough appearances, which were more developed with the increase in solidification rate. The EEs of batches C, D, and E, were 90.4%, 83.9%, and 77.3%, respectively. The EE was improved with the increase in the solidification rate of microspheres [18].

The solidification rate can also be controlled by the vacuum degree. We investigated the impact of two different solidification methods, namely: solvent evaporation under ambient pressure evaporation (ASE) and vacuum negative pressure evaporation (NSE) (Figure 6.11). The NSE method resulted in higher drug loading efficiency (37.28%), while the relatively lower LE (28.19%) resulted in applying the ASE method [17].

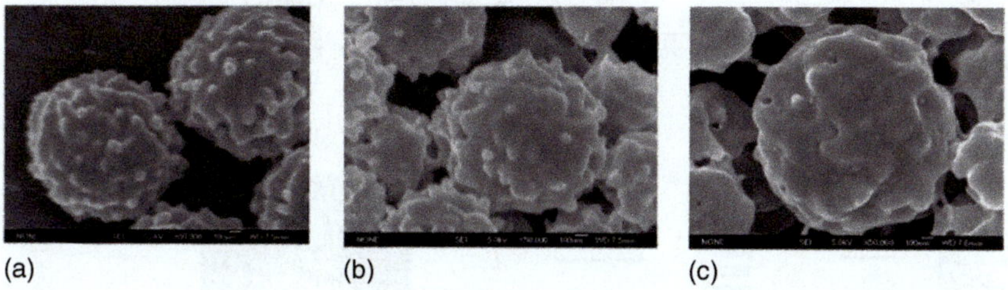

Figure 6.10 SEM photographs of microspheres prepared by different solidification techniques: (a) batch C, (b) batch D, and (c) batch E. Source: Wei et al. [18]/Elsevier.

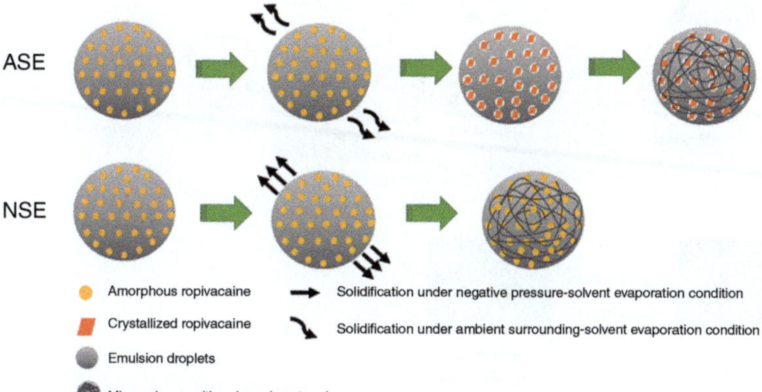

Figure 6.11 The schematic diagrams of microspheres formation process for different solidification methods. Source: Li et al. [17]/with permission from Elsevier.

6.2.4.5 Using Post-loading Mode Instead of Pre-loading Mode

Our group developed a new strategy to prepare porous microspheres with narrow size distribution for drug encapsulation, involving the fabrication of porous microspheres followed by a pore closing process. We used this porous microsphere to encapsulate ropivacaine (ROP) hydrochloride in post-loading mode, and compared it with the pre-loading mode. The flow diagram of the pre-loading mode is shown in Figure 6.12a. ROP was dissolved in the internal aqueous phase (W_1) during the preparation of microspheres. The drug loading efficiency of microspheres prepared by this method reached 1.85% and the EE reached 55.7% [27]. However, in the post-loading mode, the porous microspheres were prepared firstly, and then the porous microspheres were resuspended in the ROP solution. The ROP was passively diffused to fill the porous microspheres. Subsequently, the pores of microspheres were allowed to heal by increasing the temperature to achieve drug encapsulation, as shown in Figure 6.12b. The drug loading efficiency of ROP microspheres prepared by this method can reach up to 8.72% [26].

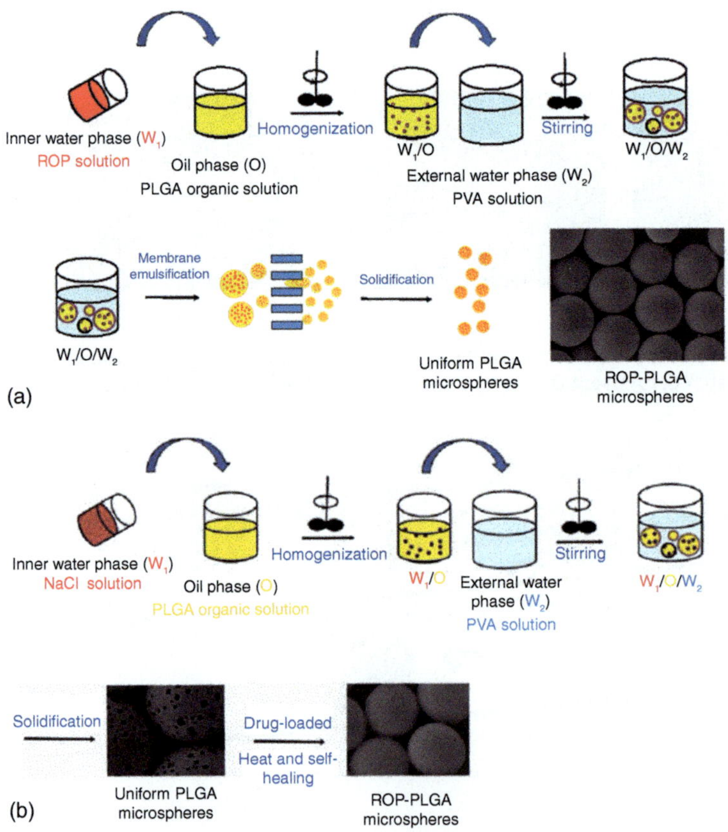

Figure 6.12 Schematic diagram of the preparation of ropivacaine hydrochloride sustained-release microspheres in pre-loading mode (a) and post-loading mode (b). Source: Wen et al. [26]/Elsevier.

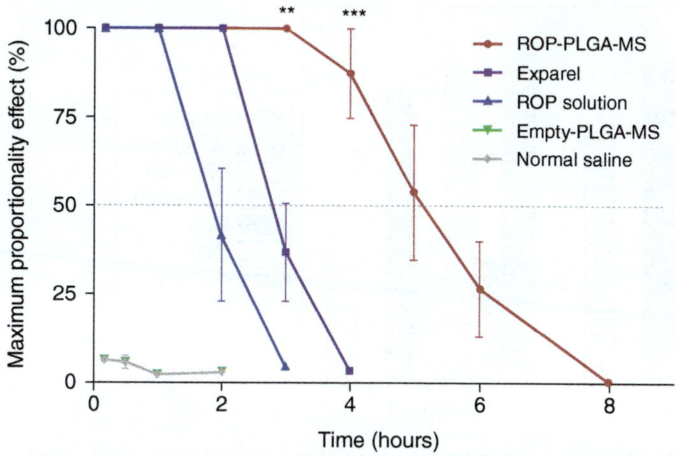

Figure 6.13 The result of a sensory nerve block test: the dotted line is the control line (50%), and the value above which is considered as effective analgesia. Source: Wen et al. [26]/with permission from Elsevier.

Using post-loading mode instead of pre-loading mode for ROP encapsulation can increase the drug loading efficiency by nearly five times. Thus, the higher LE means the less microspheres injected, which can relieve the pain of patients. As shown in Figure 6.13. The effective analgesia time of ROP sustained-release microspheres (ROP-PLGA-MS) prepared in post-loading mode was up to five to six hours, which was much higher than the commercially available products (Exparel) and ROP solution.

6.2.5 Strategy to Maintain Bioactivity of Drugs

The main challenge in the encapsulation of protein/peptide drugs is that the drug can easily lose its bioactivity during the preparation, storage, and release process, which has not been encountered in conventional chemical drugs. This is because the proteins/peptides have three-dimensional structures and some of them have four-dimensional structures, and protein unfolding and protein aggregation are easy to occur when they contact with the oil phase, oil/water interface, or hydrophobic surface [28, 29]. Besides, the shear force during emulsion preparation also has a great influence on protein denaturation [30]. The inactivated protein drugs will be therapeutically inactive; furthermore, the denatured proteins may cause unpredictable side effects, including toxicity and immunogenicity. The main strategies to maintain the bioactivity of drugs will be described in more detail below.

6.2.5.1 Adding Additives to Prevent Proteins from Denaturation

During the emulsification process, especially in the double emulsion, protein interaction with the oil/water phase is a significant reason for drugs' deactivation. The addition of stabilizers or additives into the primary water phase can avoid the unfolding or aggregation of active ingredients. We investigated the impact

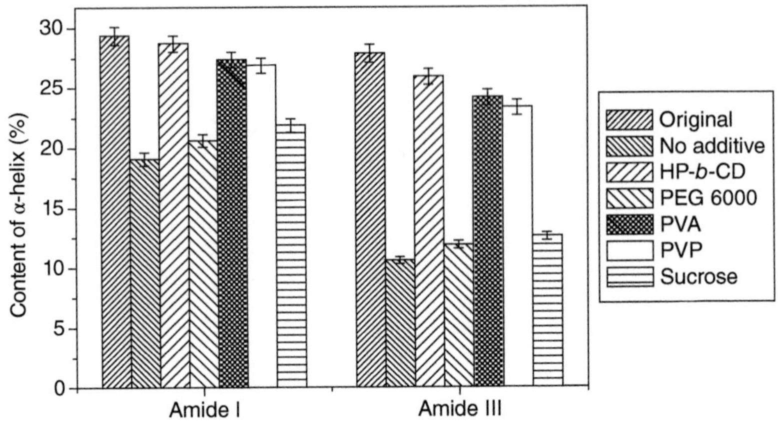

Figure 6.14 Impact of additives in the inner aqueous phase on the bioactivity of a drug. Source: Ma et al. [2]/with permission from Elsevier.

of five different additives in the inner aqueous phase: PVA, PVP, Sucrose, PEG 6000, and HP-β-CD (hydroxypropyl-β-cyclodextrin). Results shown in Figure 6.14 demonstrated that PVA, PVP, and HP-β-CD assisted in maintaining the bioactivity of the protein drug. When the additive was HP-β-CD, the α-helix content of lysozyme was hardly affected by the oil–water interface, and the α-helix content of the amide I band only decreased from the original 29.4–28.7%, and the amide III bands only decreased from 27.9% to 25.9%. When PVA was added, the α-helix content of amide I and amide III bands in lysozyme decreased from 29.4% and 27.9% to 27.4% and 24.1%, respectively. When PVP was added, the α-helix content of the amide I and amide III bands in lysozyme decreased from 29.4% and 27.9% to 26.9% and 24.1%, respectively, 27.9% to 26.9% and 23.9%, respectively. The working principle of additives can be elucidated for numerous reasons. First of all, additives are adsorbed preferentially on the interface between oil and water, hence protecting active ingredients from denaturation. Secondly, a specified interaction of additive with protein surface can cover a protein's hydrophobic site and reduce contact between drug and W/O interface [31].

Acylation of some peptides during the encapsulation by PLA or PLGA is a big problem for sustained-release formulation, because it may potentially result in structural changes, loss of activity, and even producing immunogenic response. In general, acylation normally occurs on lysine or arginine residues and the N-terminus of the peptides [32]. In our studies, we selected exenatide as a model peptide drug and used PLGA to encapsulate it. Besides, we also prepared three systems by adding different metal cations Ca^{2+}, Mn^{2+}, and Zn^{2+} in inner water phase with the exenatide drug together, and studied their effect on the acylation of drug. Among them, Zn^{2+} has the most potent effect, as it produced complex with exenatide drug. The schematic diagram for a mechanism of acylation inhibition by Zn^{2+} cation is shown in Figure 6.15. The steric hindrance of the obtained complex restricted the nucleophilic attack of amine groups of exenatide to PLGA. Simultaneously, it also prevented the adsorption of free exenatide to the binding site of the PLGA. Hence, the acylation could

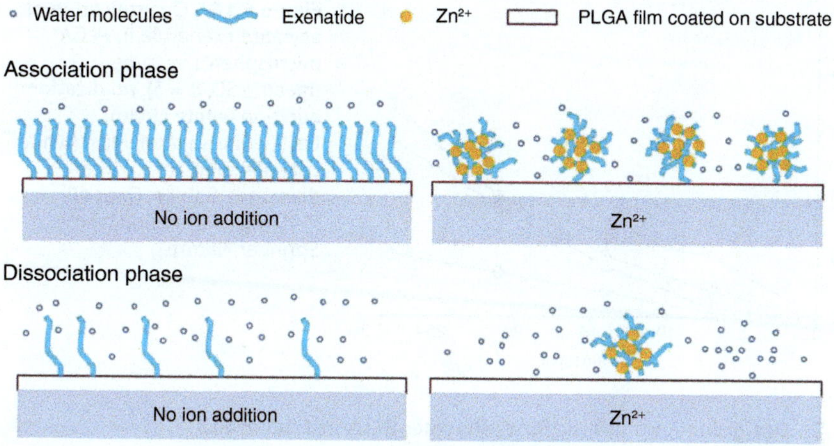

Figure 6.15 The schematic diagram for the mechanism of acylation inhibition by Zn^{2+} cation. Source: Qi et al. [33] /With permission from Springer Nature.

Table 6.2 Characteristics of PLGA microspheres prepared by different dications.

Dication	Size (μm)	Span	Encapsulation efficiency of exenatide (%)	Encapsulation efficiency of dication (%)
None	20.41	0.676	90.34 ± 5.10	—
Ca^{2+}	21.67	0.656	95.69 ± 1.13	55.7 ± 4.24
Mn^{2+}	22.09	0.645	92.31 ± 2.65	66.0 ± 3.37
Zn^{2+}	22.12	0.677	94.68 ± 2.54	64.7 ± 7.93

Source: Qi et al. [33]/with permission from Springer Nature.

be considerably hindered when metal cation was incorporated into the system. As shown in Table 6.2, after the addition of different dications in the formulation, all the microspheres with the size of 20 μm possessed high EE of peptide, and the EEs of dication were also at the same level. Exenatide was extracted from the microspheres after incubation. As shown in Figure 6.16, in the early phase (0 ~ 7 days), Mn^{2+} and Zn^{2+} could greatly inhibit acylation, and Ca^{2+} showed weak inhibition. With incubation further proceeding, the inhibition effect of both Ca^{2+} and Mn^{2+} gradually got weaker, because the dications were also released, accompanied with exenatide release. At the late phase (14 ~ 28 days), especially for Ca^{2+}, there was no difference in the ratio of formed acylated derivatives compared with the no-addition group. Mn^{2+} still presented a weaker effect, and Zn^{2+} could significantly inhibit the acylation. Therefore, it was concluded that Zn^{2+} possessed a high acylation inhibition effect for exenatide in PLGA microspheres [33].

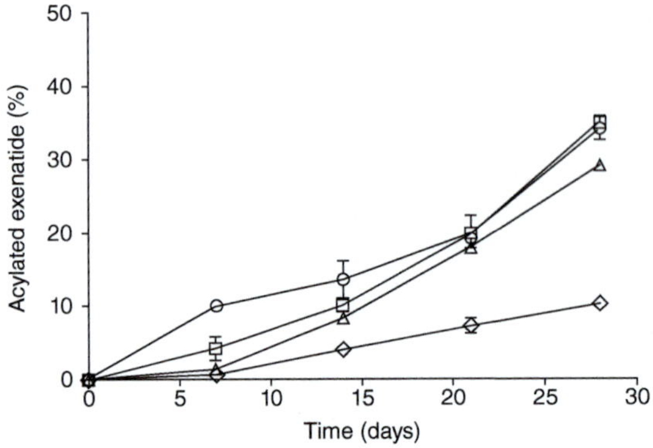

Figure 6.16 Quantification of acylated exenatide in PLGA microspheres vs. time (mean ± SD, $n = 3$), no dication addition (white circle), Ca^{2+}(white square), Mn^{2+}(white triangle), and Zn^{2+}(white diamond). Source: Qi et al. [33]/with permission from Springer Nature.

6.2.5.2 Designing Amphiphilic Block Polymer PELA Instead of PLA

The hydrophobic interaction between the drug and the polymers may lead to protein denaturation. Additionally, during the sustained-release stage, it forms an acidic environment because of formation of acidic products such as lactic acid and glycolic acid. Hence, the entrapped proteins are subjected to this harsh environment and can be easily hydrolyzed or agglomerated [34]. Therefore, PELA polymer can prevent these incidents as it has the following advantages: (i) self-association of PELA at the W/O interface and PEG part of the polymer surrounds the inner aqueous phase, avoiding the contact of protein with hydrophobic PLA or PLGA polymers. (ii) PELA can be employed as a stabilizer to maintain primary emulsion, and extra additives are not required. (iii) PELA promotes water uptake, which causes faster degradation of carrier material and accelerates the flux of the drug, resulting in a stabilizing condition for proteins. When PELA possessed the higher ratio of hydrophilic mPEG, which can be hydrated more easily in the releasing medium, the microcapsules can be swelled more readily and the water-soluble acids produced by polyester hydrolysis can diffuse out more easily, resulting in less acidic environment. As shown in Figure 6.17, by comparing three types of PELA microcapsules with

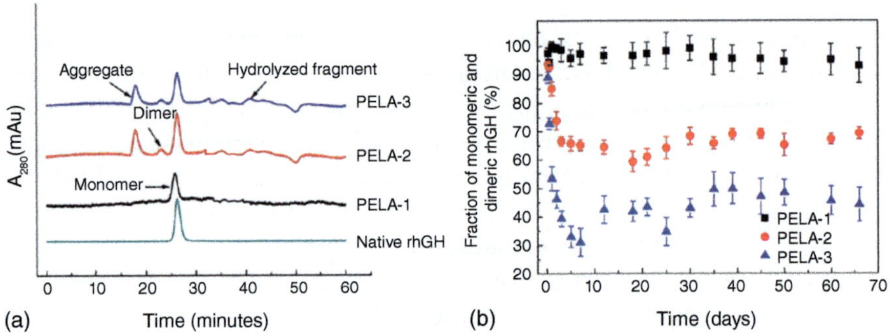

Figure 6.17 (a) Size exclusion chromatography (SEC-HPLC) of rhGH after recovery from microcapsules on the second day. (b) Fraction of monomeric and dimeric rhGH released from microcapsules (mPEG-PLA with number-average molecular weights (Mn) of 20982 (PELA-1), 39878 (PELA-2), and 45912 kDa (PELA-3)). Source: Wei et al. [34]/with permission from American Chemical Society.

different hydrophilicity degrees, the drug (rhGH) entrapped in the most hydrophilic microcapsules (PELA-1) was exposed to the less acidic environment, which had a smaller deteriorating effect on protein stability. The confocal laser scanning microscope (CLSM) images of the acidic microenvironment created inside the microcapsules are shown in Figure 6.18. When the strongest fluorescence emission of the dye is near 580 nm, the dye shows a red light and the local pH is close to 6.0. When the strongest fluorescence emission is near 640 nm, the dye shows a green light and the local pH is close to 9.0. Therefore, if the dye is yellow, the local pH is close to neutral [34].

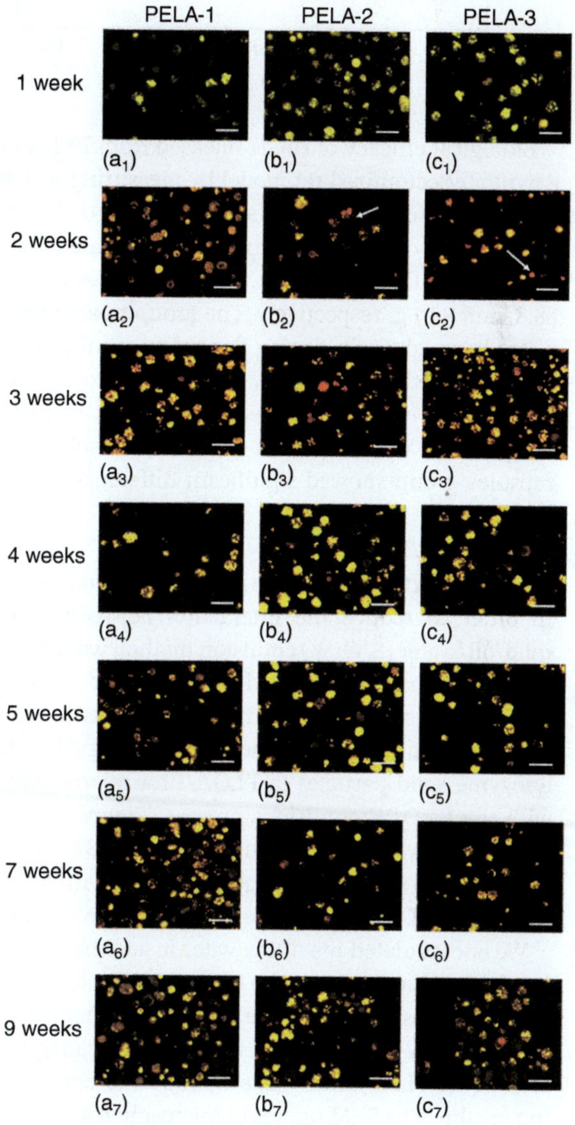

Figure 6.18 CLSM images of pH-sensitive dye-loaded PELA-1 (a_1–a_7), PELA-2 (b_1–b_7), and PELA-3 (c_1–c_7) microcapsules at different incubation times (1–5, 7, and 9 weeks). The scale bar is 10 μm. Source: Wei et al. [34]/American Chemical Society.

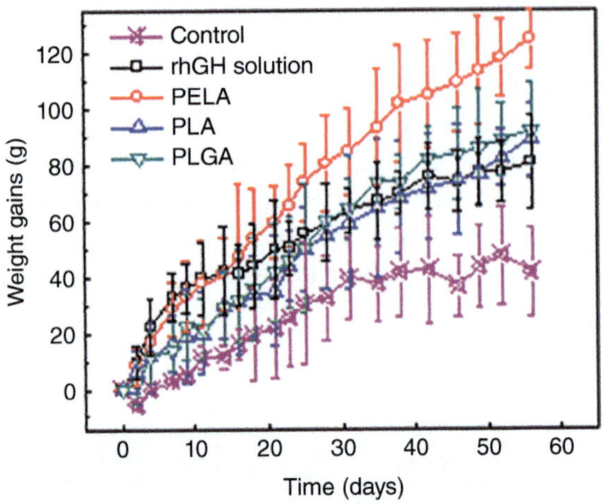

Figure 6.19 The weight gains in hypophysectomized rats treated with rhGH solution and different rhGH-loaded microcapsules. Data are means ± SD for six rats per group. Source: Wei et al. [34]/with permission from American Chemical Society.

Biological efficacy of rhGH released from PELA microcapsules was examined in a hypophysectomized rat model by measuring weight gain. PLA and PLGA microcapsules loading rhGH were also prepared. As shown in Figure 6.19, the cumulative weight gains of hypophysectomized rats treated by negative control, daily injection of rhGH solution, and PELA, PLA, and PLGA microcapsules were 41.4, 80.8, 124.7, 88.8, and 92.1 g, respectively. The group treated with PELA microcapsules showed a steady weight gain during the entire monitoring period of 56 days, whereas the groups treated with the PLA or PLGA microcapsules showed a significantly lower weight gain in the same period than that of PELA microcapsules. In addition, compared to daily injection of the rhGH solution group, only the PELA microcapsules group showed significant differences based on cumulative weight gains ($p < 0.05$).

6.2.5.3 Effect of Preparation and Solidification Method

In order to reduce the interaction between a protein with the interface, the solid/oil/water (S/O/W) emulsion method was applied by straightly dispersing solid powder of protein into the dispersed phase, instead of using the protein aqueous solution. As a result, the protein molecules inside the powder will be prevented from contacting the interface. The strategy has been employed to encapsulate lysozyme solid particles in PLGA/PLA microcapsules and results were compared with the W/O/W emulsions process (Figure 6.20). The bioactivity of lysozyme in the S/O/W system was maintained at 94.48% (D sample) of native lysozyme, and much higher than protein aqueous phase without additive (78.97%, B sample) and with HP-β-CD as a protective additive (89.05%, C sample) [31].

We encapsulated insulin powder inside the PLGA microcapsules just by dispersing the insulin powder in the PLGA/DCM solution to form a S/O emulsion [22]. The S/O was pressed through the pores of the membrane into the outer water phase to form a S/O/W emulsion with a uniform droplet by premix membrane emulsification technique. Then DCM was evaporated to obtain a uniform microsphere containing insulin. The SEM photos of microspheres prepared by employing the membrane

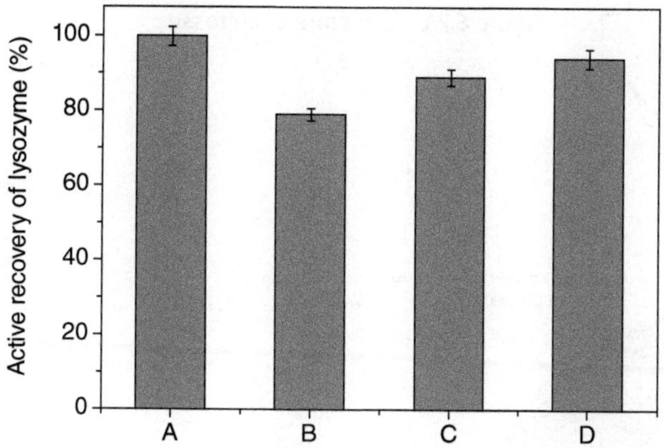

Figure 6.20 Active recovery of lysozyme in the PLA/PLGA microsphere/microcapsules. (A) native lysozyme; (B) and (C) via W/O/W emulsion without additive and with HP-β-CD as additive in inner water phase; (D) via S/O/W emulsion by dispersing lysozyme solid particle in oil phase. The data are presented as mean ± SD ($n = 3$). Source: Ma et al. [2]/with permission from Elsevier.

with different pore sizes are shown in Figure 2.14 [2]. It can be seen that the uniform microsphere can be obtained by this technique, which guaranteed the reproducibility of the PLGA microsphere.

6.3 Natural Polymer (Polysaccharide) Chitosan

Chitosan is a natural linear polysaccharide consisting of randomly distributed D-glucosamine and *N*-acetyl-D-glucosamine. It has a similar structure to mucopolysaccharide, which is extensively distributed in connective tissues and is one of the organic constituents of plasma membranes. It exhibits favorable biological characteristics such as nontoxicity, biocompatibility, biodegradability, film-forming, and antibacterial properties [35]. In addition, chitosan can be degraded into nontoxic components and quickly cleared from the body and there is no irritation and antigen immunity. Chitosan is usually synthesized by alkaline deacetylation of chitin, which is composed of β-(1,4)-linked *N*-acetyl-glucosamine. Chitin is the second most abundant organic polymer and central component in fungal cell walls such as aspergillus, mucor, and protective exoskeletons of crustaceans such as crabs, shrimps, prawns, and lobsters [36].

Chitosan has a complex two-helix structure, and each of them comprises one principal amino and two free hydroxyl groups for each C_3 and C_6 building unit (Figure 6.21). Because of plenty of available free amino groups in chitosan, it carries a positive charge and interacts with substances with negatively charged surfaces. As a result, chitosan has been exhibited to possess mucoadhesive properties because of strong electrostatic interaction with sialic acid residues on negatively charged mucosal layers. Hence, it promoted the drug or microparticle itself to be absorbed from the gastrointestinal tract [37]. Furthermore, chitosan is insoluble in water and

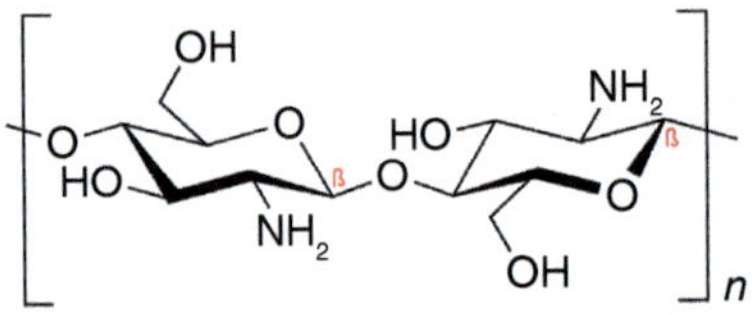

Figure 6.21 Structure of chitosan.

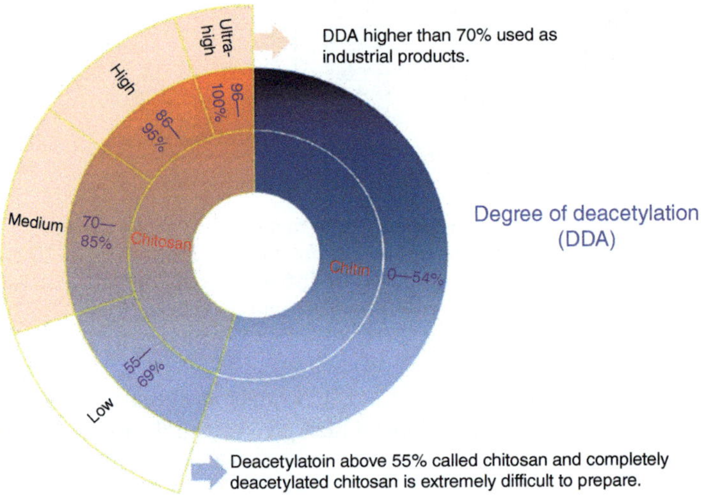

Figure 6.22 The DDA classification standard of chitosan.

most organic solvents, except dilute acid, acetic acid, and formic acid [38]. Properties of chitosan are mainly dependent on two characteristics: molecular weight and degree of acetylation. The DDA is determined by the percentage of deacetylated glucosamine units concerning the total number of monomers (Figure 6.22). In general, the DDA considerably affects solubility, viscosity, crystallinity, ion exchange capacity, and chitosan flocculation.

The -NH_2 unit in the chitosan is protonated to produce NH_3^+ ion in an acidic environment and results in better solubility in the acid condition. The higher the DDA, the greater number of –NH_2 groups in chitosan, and it dissolves better in the acid. Past studies revealed that a decrease in deacetylation degree is favorable for cell proliferation and adhesion [39, 40]. Other physicochemical characteristics are also impacted by the molecular weight of the chitosan, such as viscosity, solubility, adsorption on solids, elasticity, tear strength, biofunctional activities, crystal size, and morphological character. The higher the relative molecular weight, the solubility is lower and the degree of entanglement between molecules becomes larger [38].

6.3.1 Strategies to Improve Encapsulation Efficiency

The W/O emulsion-crosslinking method using glutaraldehyde as chemical crosslinker is widely used for the preparation of chitosan microspheres. And, hydrophilic protein/peptide drugs can be dissolved in chitosan aqueous phase and then can be encapsulated in chitosan microsphere easily. However, the EE of chitosan microspheres prepared by conventional method was low because a lot of water was squeezed out of the droplets during crosslinking by glutaraldehyde and the shrinkage occurred from droplets to microspheres, resulting in drug leakage during this process. In addition, protein/peptide drugs usually have amino groups, which also will be crosslinked by glutaraldehyde to induce denaturation of drugs and to form protein aggregates [2, 41]. Therefore, researchers tried to use mild methods to adsorb protein on the prepared blank chitosan microspheres to maintain the activity of protein drugs. Unfortunately, the drug loading and EEs of chitosan–glutaraldehyde (CG) microspheres prepared by the adsorption method are very low because the drugs can only be adsorbed on the surface of the microspheres due to the smooth surface and dense structure. The main strategies to improve the EE are described in more detail below.

6.3.1.1 Using Step-wise Crosslinking Method to Avoid Shrinkage Stage

Based on the conventional crosslinking method, we developed a step-wise crosslinking method combined with membrane emulsification technique to prepare insulin-loaded chitosan microsphere [42]. That is, the W/O chitosan emulsion containing insulin was initially hardened by tripolyphosphate (TPP) ionic gelation; then, it was crosslinked by a glutaraldehyde agent (Figure 6.23). In order to compare the difference and show the advantages of the step-wise crosslinking method (Method A), another two insulin-loading methods were used in this study: (i) direct encapsulation method, where W/O emulsion was crosslinked by glutaraldehyde directly (Method B); (ii) adsorption insulin method (Method C), where insulin was adsorbed on blank microsphere prepared by step-wise crosslinking method (TPP ionic gelation combined with glutaraldehyde crosslinking). The uniform size

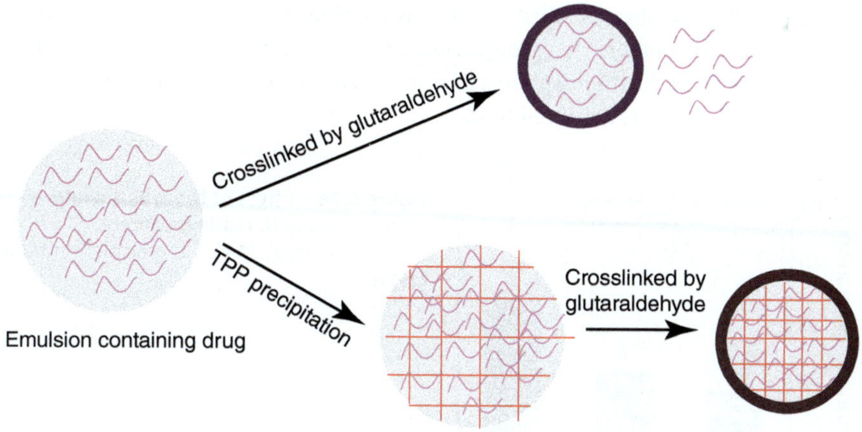

Figure 6.23 The schematic explanation for chitosan microspheres formation by different crosslinking methods, showing the drug (red short wavy line) leakage during the crosslinking process. Source: Wang et al. [43]/with permission from Elsevier.

of microsphere allowed it to investigate the relationship between EEs and drug loading methods [43].

The EEs of insulin by Methods A, B, and C were 76%, 40%, and 22%, respectively. The EE of insulin by Method C was the lowest. In this case, the insulin was loaded into the microspheres mainly by surface adsorption and diffusion into the microspheres during the swelling of microspheres. Most of insulin was adsorbed near the surface of microspheres. Because the structure of microspheres was dense and the swelling degree of crosslinked microspheres was limited, the saturated adsorption amount was lower. In the case of direct encapsulation, the EE of insulin by Method B was lower than that of Method A. A lot of water was squeezed out of the droplets during crosslinking by glutaraldehyde because chitosan concentration was very low (1.5 wt%); therefore, shrinkage occurred from droplets to microspheres and it led to drug leakage during this process. The method A developed in this study can overcome the problem of drug leakage, because the insulin was interlaced in the gel matrix made by TPP and chitosan during the gelation step, inhibiting the drug's outflow during the following glutaraldehyde crosslinking process [43]. Therefore, the EE was the highest by method A.

6.3.1.2 Using Chitosan Derivatives as Polymer to Adjust Microsphere Structure to Avoid Drug Crosslinking and Leakage

Based on crosslinking method, our group also developed hollow quaternized chitosan microspheres. The mixture of chitosan and a derivative *N*-[(2-hydroxy-3-trimethylammonium) propyl] chitosan chloride (HTCC) (chitosan:HTCC = 1 : 1) was used to prepare uniform hollow microspheres (CH microspheres) by combining the direct membrane emulsification technique and glutaraldehyde crosslinking method. The insulin analog (IA) was used as model drug and added in the chitosan mixture before emulsification. The substitution of HTCC was 40% and 80% in CH40 and CH80, respectively. We found that the loading content of IA increased when HTCC with a high degree of substitution was employed (Figure 6.24), because the steric hindrance by quaternized groups would reduce the crosslinking IA in particle and its leakage from the microspheres during the crosslinking reaction and a higher loading content could be achieved [44].

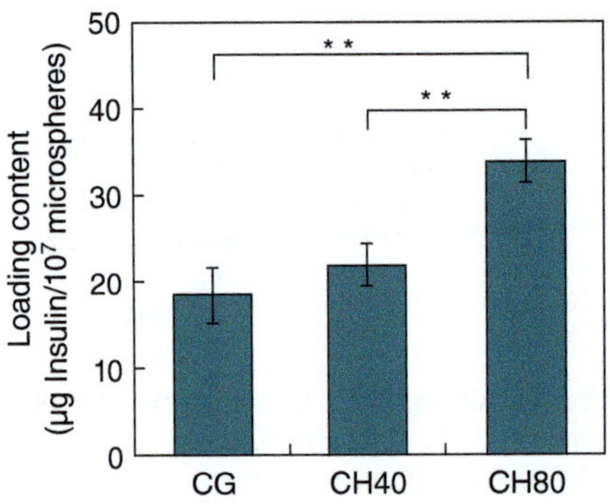

Figure 6.24 Effect of the crosslinking method on the IA loading content. Source: Wei et al. [44]/with permission from Elsevier.

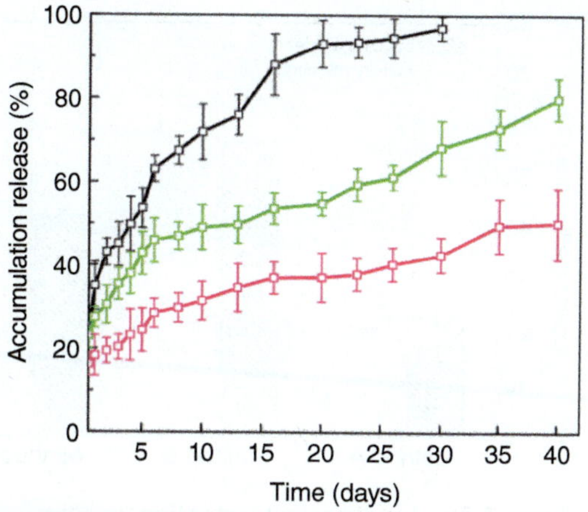

Figure 6.25 IA release profiles from CG (pink), CH40 (green), and CH80 (black) microspheres *in vitro* ($n = 3$).

The hollow structure displayed a significant role in encapsulation, as a microsphere with a hollow structure has a greater available interior space for drug storage than an ordinary rigid structure. Furthermore, the CH80 microspheres with a higher quaternization degree resulted in a lower crosslinking degree; as a result, the release rate of encapsulated IA increased compared with the traditional CG microspheres with a compact structure (Figure 6.25) [44].

6.3.1.3 Preparing Chitosan/Alginate Complex Microsphere by Two-step Solidification Method to Avoid Drug Leakage

Chitosan can form complex with alginate; therefore, chitosan/alginate microsphere can be prepared without chemical crosslinking. We also developed a two-step solidification process for preparing alginate/chitosan microspheres. The uniform W/O alginate emulsion was prepared by membrane emulsification technology. At the first solidification step, $CaCl_2$ W/O miniemulsion prepared by sonification was mixed with alginate W/O emulsion to solidify alginate droplets to obtain the alginate particles. At the second solidification step, after the particle was washed and redispersed in the aqueous phase, chitosan/insulin aqueous phase was added to crosslink alginate particle further and load insulin simultaneously [45]. The whole process was mild and without using chemical crosslinker. The alginate–chitosan microspheres demonstrated homogeneous gel network structure and insulin was embedded in the network. The alginate–chitosan microsphere was uniform with diameter of 7.5 μm and Span value of 0.731. In comparison with other drug loading methods, insulin was loaded in the microsphere by three methods: (i) Method A: 1.5 mg insulin was dissolved in the 1.5 ml $CaCl_2$ solution, the insulin was loaded in microsphere during the first step solidification by $CaCl_2$ miniemulsion; (ii) Method B: 1.5 mg insulin was dissolved in 1.5 ml chitosan solution at the second solidification step as mentioned above; and (iii) Method C: combining Method A and Method B, that is, 1.5 mg insulin in the 1.5 ml $CaCl_2$ solution (the first solidification step) and 1.5 mg insulin in the 1.5 ml chitosan solution (the second solidification step). As shown in Figure 6.26, Method B had the highest LE (EE as defined in Section 6.2.1) of about 57%. The results show that the network formed between chitosan and alginate–Ca

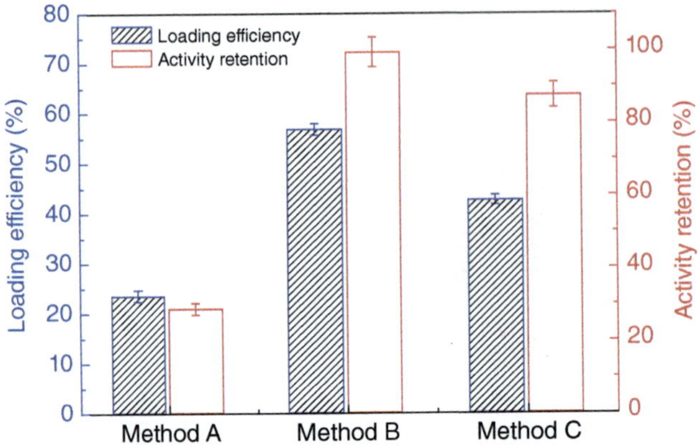

Figure 6.26 Insulin activity retention was obtained with different preparation methods. Source: Zhang et al. [45]/with permission from Elsevier.

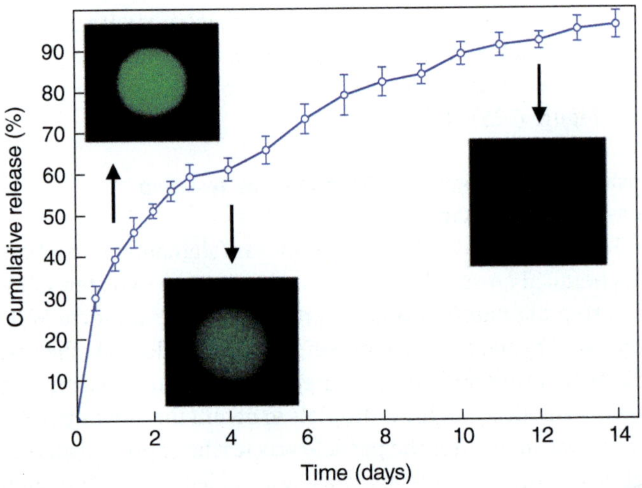

Figure 6.27 Cumulative release of insulin and corresponding LSCM images of insulin (green fluorescence) release from alginate–chitosan microspheres prepared by the Method B. Source: Zhang et al. [45]/Elsevier.

microspheres significantly decreased the porosity and minimized the leakage of insulin [45].

The alginate–chitosan microspheres prepared by two-step solidification method not only owned the high LE but also showed the ideal controlled-release profile. Figure 6.27 shows the cumulative release curve of insulin in PBS (pH 7.4), and the fluorescence images of corresponding stages were observed. Approximately, 91% of the total insulin was released after 14 days. In detail, the release process comprised two parts. First, about 40% of insulin was released in one day at a faster rate. This revealed that most of the insulin near the surface of microsphere was released at the first stage. Then, a decreased release rate followed, and about 60% of insulin was released in the later three days. The fluorescence became weaker, which suggested insulin was released from the framework of the whole microsphere. When 90% of insulin was released, the fluorescence almost disappeared. The result indicated that alginate–chitosan microspheres system was suitable for the controlled release of insulin.

6.3.1.4 Controlling Morphologies of Microspheres to Increase Drug Loading

In order to overcome the problem of low adsorption efficiency of drug on blank chitosan microspheres, we attempted to prepare chitosan microsphere with various morphologies to improve the LE and drug release behavior. Compared with the traditional compact C–G microsphere crosslinked by glutaraldehyde, we prepared hollow (C–PG), hollow-porous (CH–G), and macroporous (CH–PG) morphologies by utilizing a special crosslinking technique and mixture of chitosan and quaternized chitosan. As shown in Figure 6.28a, CH–G microspheres yielded the highest LE, reaching about 390 μg BSA/10^6 microspheres. The LEs of CH–PG and C–PG microspheres were also higher than those of traditional C–G microspheres. A possible explanation for the very high LE can be proposed, based on the role of electrostatic effects. The isoelectric point of BSA is 4.7, and the BSA molecules are negatively charged in phosphate-buffered saline (PBS). As a result, strong electrostatic interactions are established between BSA and microspheres containing positively charged HTCC, resulting in an enhanced LE. Furthermore, spheres with a hollow structure have a relatively greater amount of accessible space for BSA storage, as compared to their conventional solid counterparts. In addition, the porosity of the microsphere walls not only facilitates the migration of BSA into the lumen, but also creates a higher surface area and therefore adds to the number of sites available for BSA loading by absorption.

The LE can be increased by adjusting morphologies of microspheres, and the release behavior was also affected (Figure 6.28b). The macroporous (CH–PG) microsphere showed the highest initial burst and rapid release rate due to its largest pores, although it showed higher LE. The conventional solid (C–G) microsphere also showed a high initial burst due to its adsorption on the surface. BSA release in hollow (C–PG) microsphere presented a triphase release behavior with an initial burst followed by a platform for about 60 hours and completed 85% release after a second burst. The first burst resulted from the release of BSA loaded in the shell, and the second was caused by the BSA in the cavity, which diffused outward much slower than the BSA in the outer shell. Compared with the microspheres above, the BSA initial burst was well controlled within 20% in hollow porous (CH–G) microsphere, which was followed by a slower linear release. This was because most of BSA molecules entered the interior of CH–G microsphere through the pore channels and were released later by diffusion due to the concentration gradient. The structural properties, as well as the uniform size, enabled the development of oral delivery of insulin because it was necessary to maintain insulin at a specific therapeutic serum concentration. CH–G microspheres with a minimal initial burst and relatively well-controlled release seem to be ideal carriers for insulin [46].

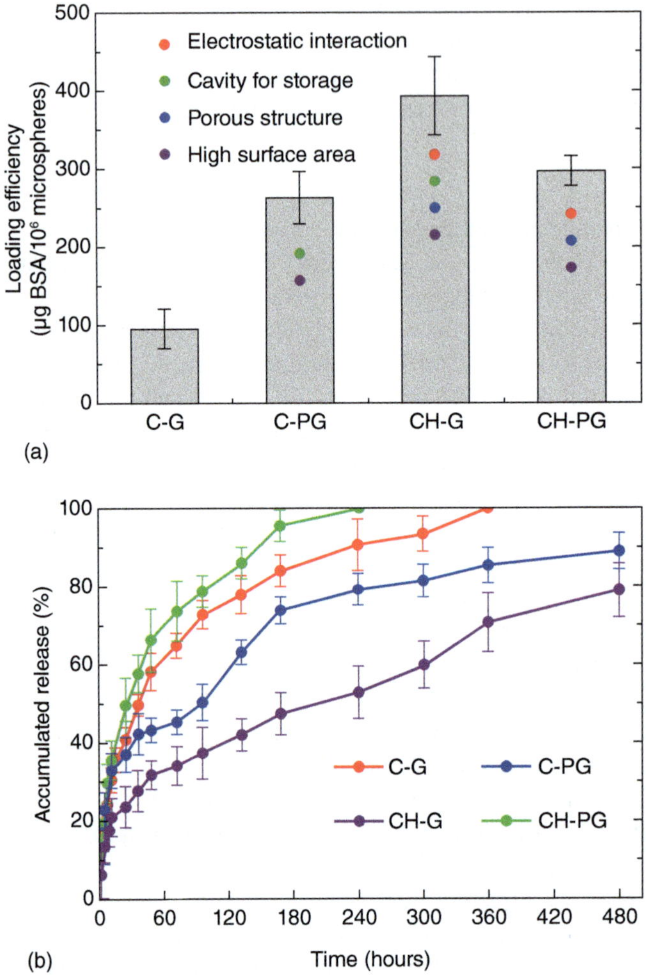

Figure 6.28 *In vitro* BSA (a) loading efficiency and (b) release profiles measured for the different types of microspheres. Source: Wei et al. [46]/with permission from JOHN WILEY & SONS, INC.

6.3.2 Strategies to Maintain Bioactivity of Drugs

Chitosan is a hydrophilic material; the problem encountered in hydrophobic PLA material can be overcome by using chitosan to encapsulate protein/peptide drugs. However, as mentioned above, usually water-soluble drugs are simply dispersed in chitosan solution and encapsulated by an emulsion-crosslinking process using chemical crosslinker, such as glutaraldehyde, to obtain solid chitosan microcapsule. The peptide and protein drugs will lose their bioactivity significantly during this process, due to reaction between amino groups of the drugs and aldehyde groups of the crosslinker [47]. In the following part, some of above strategies to increase the drug loading efficiency also can improve the bioactivity. And the uniform size of microspheres was expected to increase bioavailability of drug, and repeatability of release behavior, and it is also become possible to investigate the relationship between the properties of microspheres and bioavailability [48].

6.3.2.1 Step-wise Crosslinking Method

As mentioned in Section 6.3.1.1, we developed a step-wise crosslinking strategy of insulin-loaded chitosan microspheres by sodium tripolyphosphate (TPP) ionic gelation combined with glutaraldehyde crosslinking. This method decreased the interaction between the drug and chemical crosslinking agent, which in turn prevented the drug from being inactivated due to the crosslink. In order to show the advantages of the step-wise crosslinking method (Method A), another two insulin-loading methods were used, Methods B and C as described in Section 6.3.1.1. Method D is another adsorption method where insulin was adsorbed on blank chitosan microspheres prepared by glutaraldehyde crosslinking. The integrity of model drug insulin released from chitosan microspheres is shown in Figure 6.29, which was characterized by HPLC, and compared with conventional methods. There was no loss of insulin activity in the case of adsorption method. However, this caused a lot of active loss of insulin in the method of direct crosslinking of chitosan microsphere by glutaraldehyde (Method B). The main reason was that glutaraldehyde produced crosslinking reaction between insulin and chitosan. In the step-wise crosslinking method, TPP reacted with chitosan to form gel to keep water inside of the microsphere, which weakened the dehydration in the subsequent crosslinking by glutaraldehyde. Moreover, the gel formed between TPP and chitosan would reduce the contact between glutaraldehyde and insulin, so the crosslinking reaction between insulin and chitosan was greatly avoided. Hence, the step-wise crosslinking method provided the microspheres with higher chemical stability of insulin (>95%) [43].

The *in vitro* release profile of insulin was also affected by solidification and loading method. As shown in Figure 6.30b, the obvious burst release of more than 70% was observed in the case of adsorption method (Methods C and D) although the insulin activity can be maintained well. This was because most of insulin was adsorbed on the surface of the chitosan microspheres. In the case of direct crosslinking by glutaraldehyde (Figure 6.30a, Method B), the release rate was extremely slow; only 10% of insulin was released after 80 days because most of insulin was crosslinked into the microspheres as described above. On the other hand, the release profile was greatly improved when step-wise crosslinking method was used (Figure 6.30a, Method A), no burst effect was observed, and the release rate was relatively stable within 80 days. This was because the addition of TPP could reduce the reaction degree of glutaraldehyde, so the structure of insulin-loaded microsphere was loose, this was favorable for the diffusion of insulin [43].

6.3.2.2 Using Chitosan Derivatives as Polymer to Protect Protein from the Crosslinking Process

As described in Section 6.3.1.2, we have developed a method to prepare the hollow quaternized chitosan microspheres (CH microspheres) using a mixture of chitosan and derivative HTCC as material [44]. In this case, the crosslinking reaction by glutaraldehyde was limited near the surface of the microsphere due to the steric

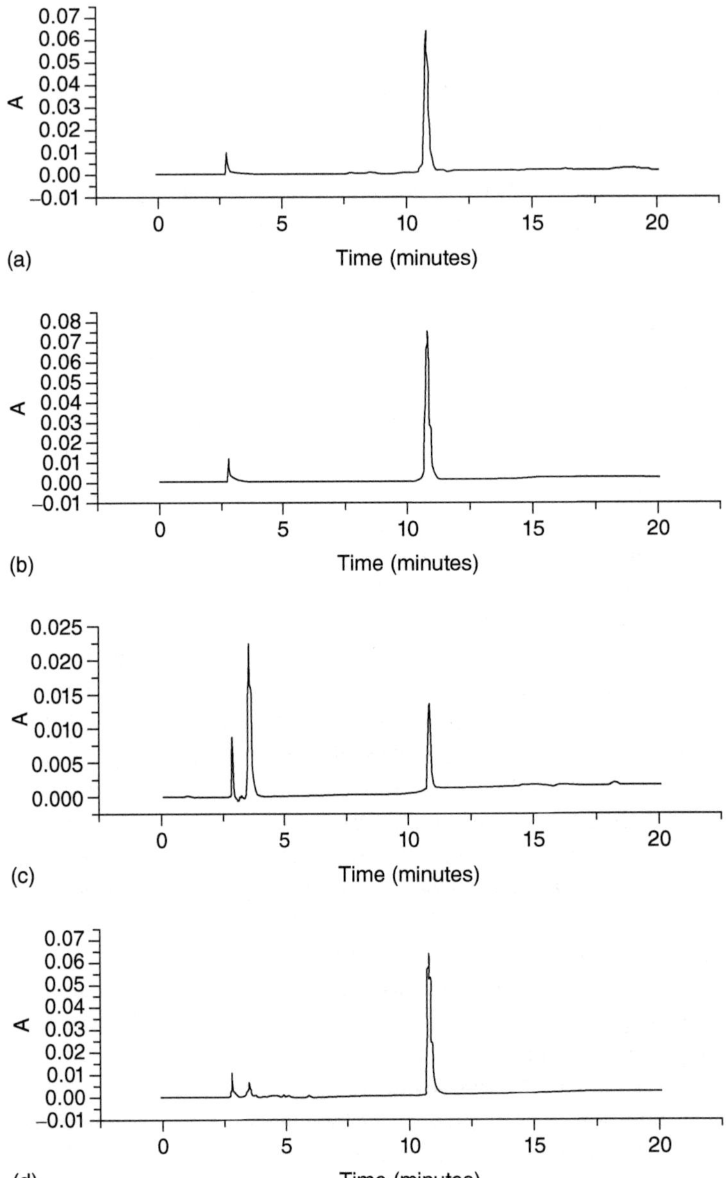

Figure 6.29 The integrity of insulin released from chitosan microspheres by different insulin-loading methods. (a) Standard insulin, (b) Insulin loaded by adsorption on blank particle, (c) Insulin loaded by direct encapsulation and crosslinking by glutaraldehyde (conventional), and (d) Insulin loaded by direct encapsulation and TPP ionic gelation combined with glutaraldehyde crosslinking (step-wise). Source: Wang et al. [43]/with permission from Elsevier.

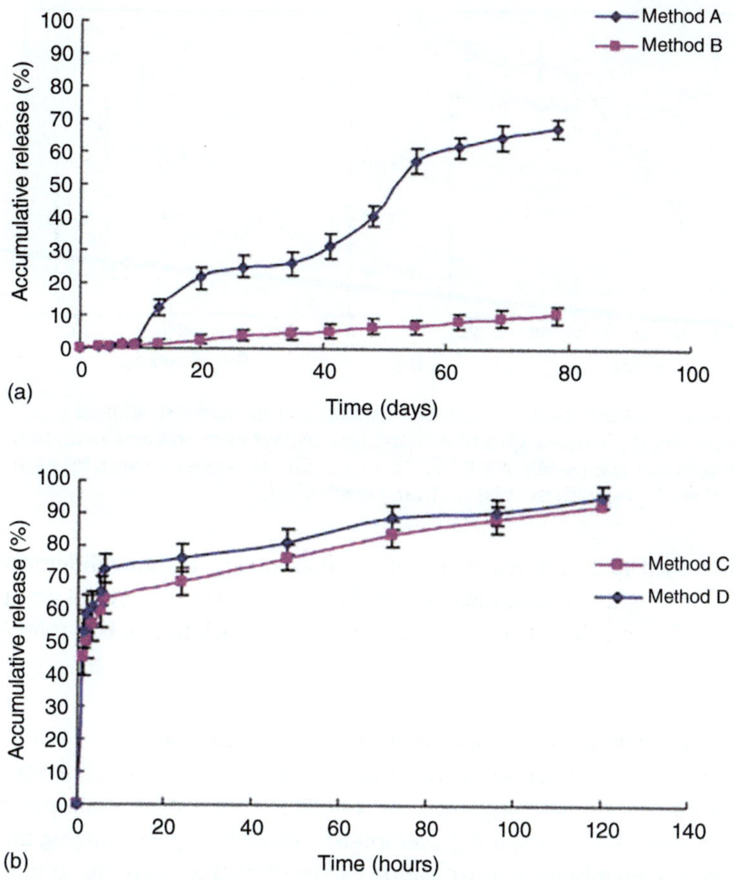

Figure 6.30 The effect of insulin-loading methods on *in vitro* release profile of insulin (a) direct encapsulation in chitosan microspheres with different crosslinking process and (b) adsorption (Methods C and D). Each point represents the mean ± SD ($n = 6$). Source: Wang et al. [43]/with permission from Elsevier.

hindrance of quaternary ammonium groups. As a result, the internal drug IA was free from the destruction of the crosslinking agent, and the activity was maintained well. In the traditional CG microspheres, the activity was only 29.8% of the IA standard. When we increased the quaternization degree of chitosan in the matrix from 40% to 80%, the activity of the embedded insulin increased from 43.2% to 63.5%.

In order to evaluate the potential use of the prepared CH microspheres for oral delivery, the therapeutic effects of CH microspheres loaded with IA were tested in diabetic rats. Obvious differences in plasma glucose reduction between the microspheres were observed (Figure 6.31a). A small hypoglycemic effect was found in rats treated with traditional CG microspheres. The hypoglycemic effect of the CH40 microspheres was moderate and lasted for about 20 days. It was notable that the blood glucose level decreased significantly after administration of IA-loaded CH80 microspheres. The maximal hypoglycemic effect occurred on day 4, then the blood

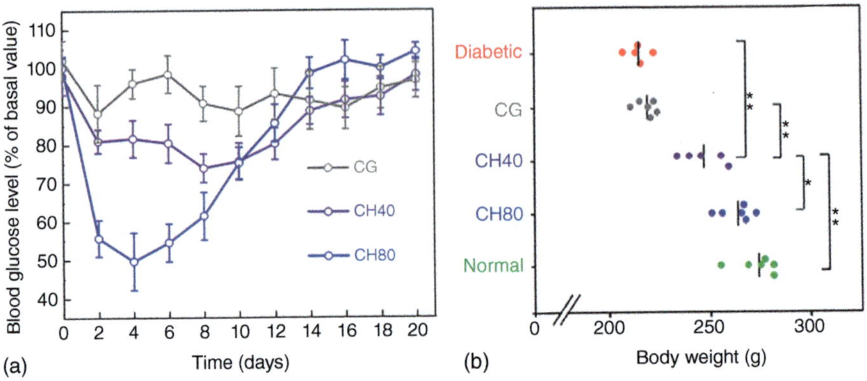

Figure 6.31 (a) Blood glucose levels of diabetic rats after an oral administration of IA-loaded microspheres. (b) Body weights of diabetic rats 14 days after oral administration. Results are expressed as means ± SD, $^{*}p < 0.05$; $^{**}p < 0.01$. Six rats were prepared for each group. Source: Wei et al. [44]/with permission from Elsevier.

glucose level increased again and reached control values on day 14. Little difference was found in body weight between the CH80 group and the healthy control group (Figure 6.31b), confirming the satisfied therapeutic effect of IA-loaded CH80 microspheres [44].

6.3.2.3 Self-solidification System Instead of Using Chemical Crosslinker

We have developed a temperature-sensitive chitosan system, which is solution state at low temperature, and become a gel at 37 °C [49, 50]. This system is composed of quaternized chitosan HTCC and α-β-glycerophosphate (α-β-GP). By utilizing this system, therefore, the membrane emulsification can be carried out at room temperature to obtain uniform W/O emulsion, and chitosan droplets can be solidified just by increasing the temperature above 37 °C without using chemical crosslinker [51]. As a result, the bioactivity of protein/peptide drugs can be maintained. Figure 6.32 shows the optical microscopic photo of W/O emulsion (a), SEM of solidified microsphere (b), and CLSM photo of chitosan microsphere containing FITC-labeled BSA

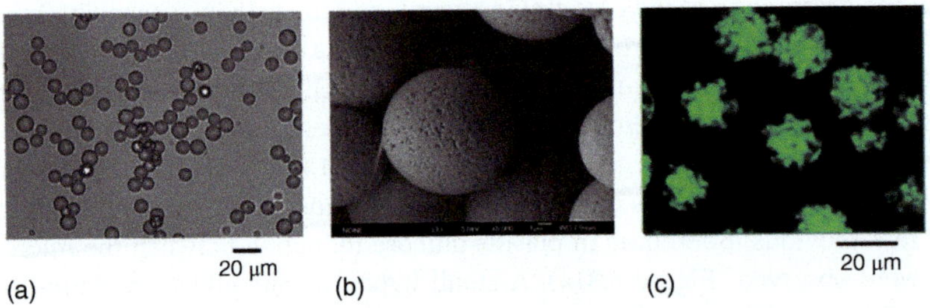

Figure 6.32 Optical microscopic photo of uniform chitosan W/O emulsion (a), SEM photo of solidified chitosan microsphere (b), and CLSM of chitosan microsphere containing FITC-labeled BSA (c) prepared by combining membrane emulsification process and self-solidification system. Source: Wu et al. [51]/Elsevier.

(c) by utilizing the above self-solidification system. It can be seen that the uniform droplet and microsphere containing BSA can be successfully obtained.

6.3.2.4 Preparing Chitosan/Alginate Complex Microsphere Instead of Chemical Crosslinking of Chitosan

As described in Section 6.3.1.3, we have developed a two-step solidification process to prepare alginate/chitosan microsphere loading insulin [45]. The whole process was mild and without using chemical crosslinker, so the bioactivity of protein drug can be maintained.

The results showed Method B maintained bioactivity of 99.4%. The denaturation of the insulin in Method A was probably caused by the ultrasonication during the preparation of the miniemulsion where insulin was added in the water phase. In Method B, insulin was loaded in the microspheres with chitosan together, which was a mild process and effectively maintained the activity of insulin. Method C was the combination of Methods A and B; it was not surprising that the EE and activity retention decreased to 42.6% and 87.4%, respectively, accordingly, Method B was the best way for insulin loading [45].

The chemical stability of insulin was evaluated after it was released from the microspheres, which were treated with simulated gastric fluid containing pepsin. The result shown in Figure 6.33b and c demonstrated that the chemical stability of insulin released from the microspheres was well preserved. The release profile of Figure 6.33a showed that the alginate/chitosan microspheres had a small quantity of release in stomach.

The insulin-loaded and insulin-free alginate–chitosan microspheres were orally administrated to the streptozotocin (STZ)-induced diabetic rats, and changes in plasma glucose are shown in Figure 6.34. A significant difference between insulin-loaded and empty microspheres was observed. The hyperglycemia was sustained in the group administrated of insulin-free microspheres, while the blood

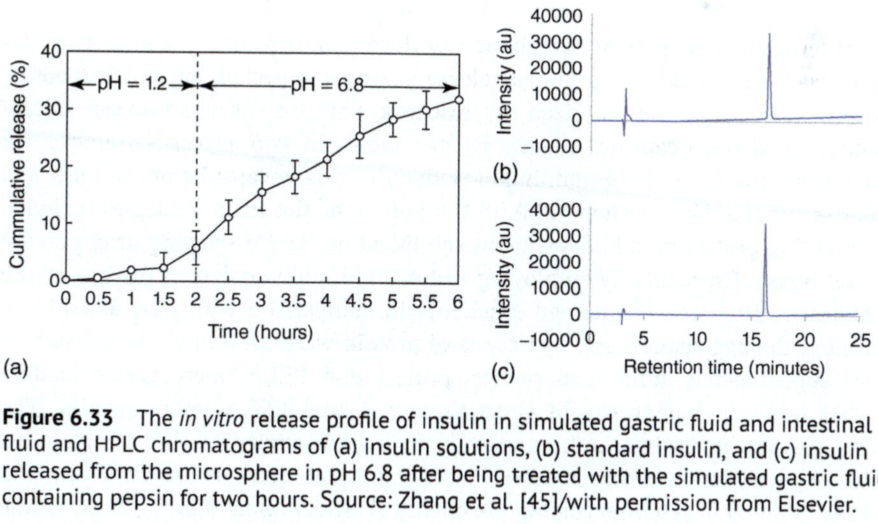

Figure 6.33 The *in vitro* release profile of insulin in simulated gastric fluid and intestinal fluid and HPLC chromatograms of (a) insulin solutions, (b) standard insulin, and (c) insulin released from the microsphere in pH 6.8 after being treated with the simulated gastric fluid containing pepsin for two hours. Source: Zhang et al. [45]/with permission from Elsevier.

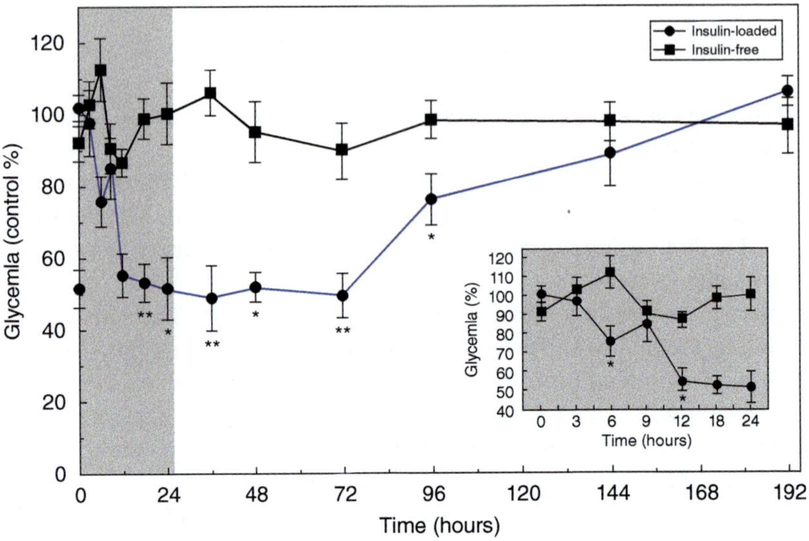

Figure 6.34 The serum glucose concentration after oral administration of insulin-free and insulin-loaded (100 IU kg^{-1}) alginate–chitosan microspheres to STZ-induced diabetic rats. Statistically significant difference from insulin-free microspheres: *$p < 0.05$; **$p < 0.01$. Source: Zhang et al. [45]/with permission from Elsevier.

glucose concentration of the diabetic rats in the treatment group was decreased evidently after oral administration, which confirmed that the bioactivity of insulin in alginate/chitosan microspheres could be retained. The result suggested the alginate/chitosan microspheres could be orally administrated only once in 3 to 4 days, which will bring convenience to diabetic patients.

6.4 Summary

In order to improve patient compliance, uniform microspheres and microcapsules have been developed for sustained release of drugs, especially of protein/peptide drugs. This chapter summarized the research work on sustained-release microspheres and microcapsules mainly by our group in two aspects: strategies for increasing the LE and maintaining bioactivity during encapsulation, storage, and release process. For synthetic PLA/PLGA polymers, the main strategies included the adding protection additives, rapid solidification, and dispersing drug powder in oil phase. Especially by employing PELA amphiphilic polymer, the bioactivity of protein drug was maintained efficiently in comparison with PLA and PLGA, because the aggregation and hydrolysis of protein were prevented. As a result, *in vivo* subcutaneous administration test proved that PELA microcapsule loading rhGH provided higher weight gains than PLA and PLGA microcapsules. The strategies for increasing the LE and maintaining bioactivity for chitosan included the step-wise crosslinking method, crosslinking of quaternized chitosan, and self-solidification system, avoiding the chemical crosslinking of drugs in the inside of chitosan microsphere. And, we also prepared blank chitosan microsphere with hollow, hollow-porous, and macroporous morphologies instead of conventional solid microsphere, then used them to adsorb protein/peptide drugs. Hollow-porous microspheres showed the highest protein loading efficiency, lowest burst effect, and relatively constant release behavior, which had potential to be an ideal carrier for insulin delivery.

Compared with other traditional drug formulations, the commercialization of sustained-release microspheres is relatively slow. The following bottlenecks have to be solved from the laboratory to the clinic stage: (i) the scale-up production of microspheres is difficult. Especially for conventional mechanical stirring method, the reproducibility is poor; (ii) the types and specifications of biodegradable polymers approved by FDA for clinical use are limited; and (iii) after drug-loaded microspheres enter human body, the mechanism of degradation, release, and the interaction between microspheres with tissues have not been clarified completely. The toxicity and safety evaluation need to be further verified.

In recent years, the investment and support for the research and development of new drug delivery systems have increased year by year, and the research on new sustained-release injection drug delivery systems for protein and peptide drugs has also become a hot topic. From laboratory technology to industrialization, there are many obstacles to be overcome. How to reduce the cost and industrialize technology rapidly is an urgent and important task.

References

1 Czuba, E., Diop, M., Mura, C. et al. (2018). Oral insulin delivery, the challenge to increase insulin bioavailability: influence of surface charge in nanoparticle system. *International Journal of Pharmaceutics* 542 (1–2): 47–55.

2 Ma, G.H. (2014). Microencapsulation of protein drugs for drug delivery: strategy, preparation, and applications. *Journal of Controlled Release* 193: 324–340.

3 Wei, Y., Wang, Y.X., Wang, W. et al. (2011). mPEG-PLA microspheres with narrow size distribution increase the controlled release effect of recombinant human growth hormone. *Journal of Materials Chemistry* 21: 12691–12699.

4 Wei, Y., Wang, Y.X., Kang, A.J. et al. (2012). A novel sustained release formulation of recombinant human growth hormone and its pharmacokinetic, pharmacodynamic and safety profiles. *Molecular Pharmaceutics* 9 (7): 2039–2048.

5 Qi, F., Wu, J., Hao, D.X. et al. (2014). Comparative studies on the influences of primary emulsion preparation on properties of uniform-sized exenatide-loaded PLGA microspheres. *Pharmaceutical Research* 31 (6): 1566–1574.

6 Qi, F., Wu, J., Yang, T.Y. et al. (2014). Mechanistic studies for monodisperse exenatide-loaded PLGA microspheres prepared by different methods based on SPG membrane emulsification. *Acta Biomaterialia* 10 (10): 4247–4256.

7 Blasi, P. (2019). Poly(lactic acid)/poly(lactic-*co*-glycolic acid)-based microparticles: an overview. *Journal of Pharmaceutical Investigation* 49 (4): 337–346.

8 Liu, R., Ma, G.H., Wan, Y.H., and Su, Z.G. (2005). Influence of process parameters on the size distribution of PLA microcapsules prepared by combining membrane emulsification technique and double emulsion-solvent evaporation method. *Colloids and Surfaces B: Biointerfaces* 45 (3–4): 144–153.

9 Garlotta, D. (2001). A literature review of poly(lactic acid). *Journal of Polymers and the Environment* 9 (2): 63–84.

10 Hong, Y., Gao, C., Shi, Y., and Shen, J. (2005). Preparation of porous polylactide microspheres by emulsion-solvent evaporation based on solution induced phase separation. *Polymers for Advanced Technologies* 16 (8): 622–627.

11 Liu, R., Ma, G.H., Meng, F.T., and Su, Z.G. (2005). Preparation of uniform-sized PLA microcapsules by combining Shirasu porous glass membrane emulsification technique and multiple emulsion-solvent evaporation method. *Journal of Controlled Release* 103 (1): 31–43.

12 Sharma, S., Parmar, A., Kori, S., and Sandhir, R. (2016). PLGA-based nanoparticles: a new paradigm in biomedical applications. *TrAC Trends in Analytical Chemistry* 80: 30–40.

13 Mao, S., Guo, C., Shi, Y., and Li, L.C. (2012). Recent advances in polymeric microspheres for parenteral drug delivery–part 1. *Expert Opinion on Drug Delivery* 9 (9): 1161.

14 Makadia, H.K. and Siegel, S.J. (2011). Poly lactic-*co*-glycolic acid (PLGA) as biodegradable controlled drug delivery carrier. *Polymers (Basel)* 3 (3): 1377–1397.

15 Zhou, J., Hirota, K., Ackermann, R. et al. (2018). Reverse engineering the 1-month Lupron Depot®. *The AAPS Journal* 20 (6): 105.

16 Wischke, C. and Schwendeman, S.P. (2008). Principles of encapsulating hydrophobic drugs in PLA/PLGA microparticles. *International Journal of Pharmaceutics* 364 (2): 298–327.

17 Li, X., Wei, Y., Lv, P.P. et al. (2019). Preparation of ropivacaine loaded PLGA microspheres as controlled-release system with narrow size distribution and high loading efficiency. *Colloids and Surfaces A: Physicochemical and Engineering Aspects* 562: 237–246.

18 Wei, Q., Wei, W., Tian, R. et al. (2008). Preparation of uniform-sized PELA microspheres with high encapsulation efficiency of antigen by premix membrane emulsification. *Journal of Colloid and Interface Science* 323 (2): 267–273.

19 Zhou, S., Liao, X., Li, X. et al. (2003). Poly-D,L-lactide-*co*-poly(ethylene glycol) microspheres as potential vaccine delivery systems. *Journal of Controlled Release* 86 (2–3): 195–205.

20 Li, X., Deng, X., Yuan, M. et al. (2000). In vitro degradation and release profiles of poly-DL-lactide-poly(ethylene glycol) microspheres with entrapped proteins. *Journal of Applied Polymer Science* 78 (1): 140–148.

21 Li, X., Deng, X., and Huang, Z. (2001). In vitro protein release and degradation of poly-DL-lactide-poly(ethylene glycol) microspheres with entrapped human serum albumin: quantitative evaluation of the factors involved in protein release phases. *Pharmaceutical Research* 18 (1): 117–124.

22 Liu, R., Huang, S.S., Wan, Y.H. et al. (2006). Preparation of insulin-loaded PLA/PLGA microcapsules by a novel membrane emulsification method and its release in vitro. *Colloids and Surfaces B: Biointerfaces* 51 (1): 30–38.

23 Li, X., Wei, Y., Wen, K. et al. (2021). Novel insights on the encapsulation mechanism of PLGA terminal groups on ropivacaine. *European Journal of Pharmaceutics and Biopharmaceutics* 160: 143–151.

24 Katou, H., Wandrey, A.J., and Gander, B. (2008). Kinetics of solvent extraction/evaporation process for PLGA microparticle fabrication. *International Journal of Pharmaceutics* 364 (1): 45–53.

25 Falke, L.L., Van Vuuren, S.H., Kazazi-Hyseni, F. et al. (2015). Local therapeutic efficacy with reduced systemic side effects by rapamycin-loaded subcapsular microspheres. *Biomaterials* 42: 151–160.

26 Wen, K., Na, X.M., Yuan, M.M. et al. (2022). Preparation of novel ropivacaine hydrochloride-loaded PLGA microspheres based on post-loading mode and efficacy evaluation. *Colloids and Surfaces B: Biointerfaces* 210: 12215.

27 Wen, K., Wei, Y., and Ma, G.H. (2021). Analysis of preparation factors of ROP-PLGA microspheres with high encapsulation efficiency based on response surface method. *The Chinese Journal of Process Engineering* 21 (1): 83–91.

28 Weert, M.V.D., Hoechstetter, J., Hennink, W.E., and Crommelin, D.J.A. (2000). The effect of a water/organic solvent interface on the structural stability of lysozyme. *Journal of Controlled Release* 68 (3): 351–359.

29 Morlock, M., Koll, H., Winter, G., and Kissel, T. (1997). Microencapsulation of rh-erythropoietin, using biodegradable poly(D,L-lactide-*co*-glycolide): protein stability and the effects of stabilizing excipients. *European Journal of Pharmaceutics and Biopharmaceutics* 43 (1): 29–36.

30 Meng, F.T., Zhang, W.Z., Ma, G.H., and Su, Z.G. (2003). The preparation and characterization of monomethoxypoly(ethylene glycol)-*b*-poly-DL-lactide microcapsules containing bovine hemoglobin. *Artificial Cells Blood Substitutes and Immobilization Biotechnology* 31 (3): 279–292.

31 Lianyan, W., Tingyuan, Y., and Guanghui, M. (2015). Particle design of membrane emulsification for protein drug and vaccine delivery. *Current Pharmaceutical Design* 20 (19): 2563–2598.

32 Shirangi, M., Hennink, W.E., Somsen, G.W., and van Nostrum, C.F. (2016). Acylation of arginine in goserelin-loaded PLGA microspheres. *European Journal of Pharmaceutics and Biopharmaceutics* 99: 18–23.

33 Qi, F., Yang, L., Wu, J. et al. (2015). Microcosmic mechanism of dication for inhibiting acylation of acidic peptide. *Pharmaceutical Research* 32 (7): 2310–2317.

34 Wei, Y., Wang, Y.X., Wang, W. et al. (2012). Microcosmic mechanisms for protein incomplete release and stability of various amphiphilic mPEG-PLA microspheres. *Langmuir* 28 (39): 13984–13992.

35 Mitra, A. and Dey, B. (2011). Chitosan microspheres in novel drug delivery systems. *Indian Journal of Pharmaceutical Sciences* 73 (4): 355–366.

36 Sinha, V.R., Singla, A.K., Wadhawan, S. et al. (2004). Chitosan microspheres as a potential carrier for drugs. *International Journal of Pharmaceutics* 274 (1–2): 1–33.

37 Thanou, M., Verhoef, J.C., and Junginger, H.E. (2001). Oral drug absorption enhancement by chitosan and its derivatives. *Advanced Drug Delivery Reviews* 52 (2): 117–126.

38 Li, B.L., Elango, J., and Wu, W.H. (2020). Recent advancement of molecular structure and biomaterial function of chitosan from marine organisms for pharmaceutical and nutraceutical application. *Applied Sciences-Basel* 10 (14): 4719.

39 Mati-Baouche, N., Elchinger, P.H., de Baynast, H. et al. (2014). Chitosan as an adhesive. *European Polymer Journal* 60: 198–212.

40 Yuan, Y.L., Chesnutt, B.M., Haggard, W.O., and Bumgardner, J.D. (2011). Deacetylation of chitosan: material characterization and in vitro evaluation via albumin adsorption and pre-osteoblastic cell cultures. *Materials* 4 (8): 1399–1416.

41 Wang, L.Y., Ma, G.H., and Su, Z.G. (2005). Preparation of uniform sized chitosan microspheres by membrane emulsification technique and application as a carrier of protein drug. *Journal of Controlled Release* 106 (1–2): 62–75.

42 Wang, L.Y., Gu, Y.H., Zhou, Q.Z. et al. (2006). Preparation and characterization of uniform-sized chitosan microspheres containing insulin by membrane emulsification and a two-step solidification process. *Colloids and Surfaces B: Biointerfaces* 50 (2): 126–135.

43 Wang, L.Y., Gu, Y.H., Su, Z.G., and Ma, G.H. (2006). Preparation and improvement of release behavior of chitosan microspheres containing insulin. *International Journal of Pharmaceutics* 311 (1–2): 187–195.

44 Wei, W., Ma, G.H., Wang, L.Y. et al. (2010). Hollow quaternized chitosan microspheres increase the therapeutic effect of orally administered insulin. *Acta Biomaterialia* 6 (1): 205–209.

45 Zhang, Y.L., Wei, W., Lv, P.P. et al. (2011). Preparation and evaluation of alginate-chitosan microspheres for oral delivery of insulin. *European Journal of Pharmaceutics and Biopharmaceutics* 77 (1): 11–19.

46 Wei, W., Yuan, L., Hu, G. et al. (2008). Monodisperse chitosan microspheres with interesting structures for protein drug delivery. *Advanced Materials* 20 (12): 2292–2296.

47 Zhang, W., Ren, G.L., Xu, H. et al. (2016). Genipin cross-linked chitosan hydrogel for the controlled release of tetracycline with controlled release property, lower cytotoxicity, and long-term bioactivity. *Journal of Polymer Research* 23 (8): 156.

48 Wei, W., Wang, L.Y., Yuan, L. et al. (2008). Bioprocess of uniform-sized crosslinked chitosan microspheres in rats following oral administration. *European Journal of Pharmaceutics and Biopharmaceutics* 69 (3): 878–886.

49 Wu, J., Su, Z.G., and Ma, G.H. (2006). A thermo- and pH-sensitive hydrogel composed of quaternized chitosan/glycerophosphate. *International Journal of Pharmaceutics* 315 (1–2): 1–11.

50 Wu, J., Wu, Y.B., Hao, D.X. et al. (2012). Thermal-sensitive hydrogels as nasal vaccine delivery systems. *Therapeutic Delivery* 3 (10): 1151–1153.

51 Wu, J., Wei, W., Wang, L.Y. et al. (2008). Preparation of uniform-sized pH-sensitive quaternized chitosan microsphere by combining membrane emulsification technique and thermal-gelation method. *Colloids and Surfaces B: Biointerfaces* 63 (2): 164–175.

7

Applications of Uniform Particles for Targeted Delivery of Anticancer Drugs

7.1 Introduction

The incidence and mortality related to various cancers continue to increase worldwide. With rapid growth and aging of the global population, cancer has emerged as a leading cause of death in several geographic regions, and as such, cancer poses a major obstacle to the public health goal of increasing life expectancy. According to *Global Cancer Statistics 2020* [1], 19.3 million new cases of cancer and almost 10 million cancer-related deaths were reported in 2020, with 28.4 million new cancer cases (including nonmelanoma skin cancer (NMSC), except basal cell carcinoma) estimated among the global population by 2040, representing a 47% increase from that in 2020 if current rates remain constant. Cancer has thus arisen as a prominent threat to human health worldwide that limits both societal advancement and economic growth, particularly in resource-limited regions.

Historically, chemotherapy has served as the primary strategy for treating most malignancies. However, the large majority of traditional anticancer drugs lack specificity, resulting in the need to administer disproportionately large doses *in vivo* to ensure that sufficient levels of drug accumulate at the tumor site. This strategy leads to broad distribution of the drug throughout the patient, commonly leading to high toxicity and adverse effects on healthy tissues and organs, which consequently limits the overall therapeutic efficacy. Therefore, the development of delivery modes that enable specific accumulation of the anticancer drug to the targeted tumor cells, but not normal tissues to increase the treatment efficacy and reduce toxicity and side effects remains a primary and highly significant objective of the research community.

To this end, several new approaches for drug carriers, such as microcapsules, nanotubes, and hollow nanoparticles (NPs), have recently risen to prominence as potentially viable modes of cancer drug delivery that merit continued research and development [2]. In particular, tumor-targeting formulations that utilize nano/microcarriers, such as polymer particles or liposomes, have shown promise toward overcoming the long-standing disadvantages of direct administration by improving the specificity of anticancer drug delivery [3]. These strategies exploit the enhanced permeability and retention effect (EPR effect) common to tumor cells; by

Novel Membrane Emulsification: Principles, Preparation, Processes, and Bioapplications,
First Edition. Guanghui Ma.

modulating the vehicle size or deformability, the loaded drug can more efficiently enter tumor tissues via the tumor vascular endothelial system and accumulate to sufficiently toxic levels. Moreover, functional modifications to the carrier particle surface enable intelligent drug delivery through targeted localization of the particles and controlled release of the pharmacological component, thereby minimizing off-target toxicity and other side effects. In addition, water-insoluble anticancer drugs can be loaded into soluble nano/microcarriers specially designed to suit the unique structure and physicochemical properties of each drug, thus allowing delivery of insoluble anticancer drugs. For quantitative research on drug delivery systems, uniformity of particle size is of utmost importance. Our research group has developed a novel membrane emulsification strategy that allows preparation of uniformly sized micro or nanoparticles. This is already being applied, including in stimulus-response or controlled-release drug delivery methods. In this chapter, we focus on the theoretical basis and biological applications of nano/microspheres and nano/microcapsules, especially the uniform nano/microparticles via membrane emulsification technique, in the targeted delivery of anticancer drugs.

7.2 Influence of Physical and Chemical Particle Properties on Antitumor Efficacy

The physical and chemical properties of materials used in biological applications are well known to affect the biological response to a given treatment, such as drug half-life, biocompatibility of prosthetic implants or devices, as well as the targeting specificity and release rates of drug carriers [4]. Similarly, the properties of materials used for nano/microspheres or capsules can directly determine their *in vivo* behavior or function. A comprehensive understanding of the relationship between the physicochemical properties of a nano/microparticle and its biological effects, especially in tumor cells and the tumor microenvironment (TME), is of vital importance; then particles can be designed with a high likelihood of achieving the desired anticancer therapeutic effects [5]. Moreover, control of a single variable, most notably uniform particle size, can ensure consistency in the structure–activity relationship between specific particle properties and effects on the target cell/TME. A wide distribution of particle sizes may raise difficulties for precision and accuracy, and lead to ambiguous results in biological assessment. This is the case with most traditional methods, which result in a wide particle size distribution, compromising the accuracy of research findings. Our novel membrane emulsification strategy overcomes this issue by producing uniformly sized micro and nanoparticles, which will benefit future studies of the structure–activity relationship.

Drugs must overcome numerous obstacles to their delivery in tumor cells, including engulfment by macrophages, osmotic resistance and heterogeneity in dense tumor tissues, transport through the target cell membrane, and localization to off-target intracellular organelles. Thus, only a small percentage of administered drugs typically accumulate at the target site to induce therapeutic effects. A complete understanding of the structure–activity relationship of antitumor carriers is therefore essential for the rational design of polymer-based drug carriers with specific biological functions or targets.

7.2.1 Size

Among the various properties of nano/microparticles that have been studied to improve anticancer efficacy, small size and high surface area are most important for bypassing host defense, accumulation at the target site, and import into tumor cells [6]. Indeed the effects of carrier particle size have always been a hotspot in antitumor drug delivery research, especially in the context of developing new biomaterials for novel vehicles [4].

Several important *in vivo* anticancer drug carrier functions are largely dependent on particle size, such as circulation time, extravasation, targeting specificity, tumor cell internalization, intracellular trafficking, degradation rate, fluid dynamic properties, clearance, and uptake mechanisms [7]. Particle diameter can enhance or restrict their transport and adhesion through airways, the gastrointestinal tract, and most importantly, the circulatory system [8], since there are several different size-based mechanisms for clearance of foreign particles from the bloodstream [9]. For example, Kupffer cells are responsible for sequestration of microparticles in liver tissue or microparticles may be physically trapped in capillary beds [10]. By contrast, nanoparticles smaller than 100 nm can exit blood vessels through fenestrations in the endothelial lining. In tumors, vasculature is leaky compared to that in healthy tissue, resulting in leakage of diverse particles from blood. Larger particles, such as those >500 nm, are targets of phagocytosis by macrophages, while smaller particles are subject to endocytosis by either professional phagocytes or nonphagocytic cells, regardless of the means of administration [11, 12].

In addition to particle diameter, size distribution is also an important factor in evaluating potential efficacy. It is widely recognized that particles designed for extravasation into tissues should be smaller than 100 nm in diameter, and microscale particles (diameters 1–10 μm) are therefore unsuitable as vehicles for delivery by intravascular injections since they will be efficiently cleared from circulation [13]. By using O/W emulsion technique, our group prepared and reported a uniform iRGD-modified paclitaxel (PTX) nanodots (~10 nm) with superior intratumoral penetration for enhanced tumor restriction and metastasis prevention in murine model, compared with iRGD-nanoparticles (~70 nm), revealing the significant influence of particle size variation in nanoscale [14] (Figure 7.1). Size guidelines are thus generally useful in the design of anticancer delivery systems and should be carefully considered in conjunction with the other relevant parameters/strategies.

7.2.2 Surface Charge

Surface charge can be an essential determining factor in biological interactions with NPs or microspheres [15]. For instance, differences in carrier surface charge can affect both the total quantity and rate of tumor cell endocytosis, in addition, to the

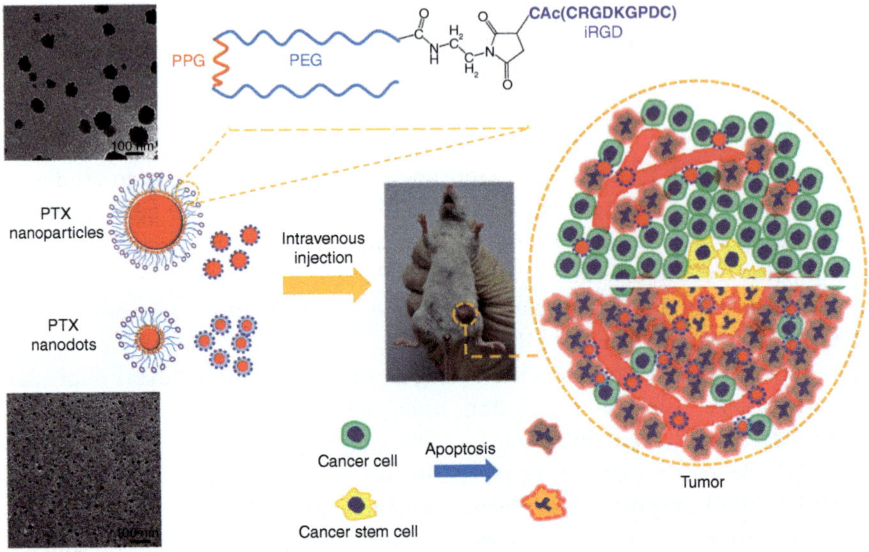

Figure 7.1 iRGD-conjugated PTX nanoparticles/nanodots and their disparate performances on intratumoral distribution. Source: Ni et al. [14]/Reproduced with permission from John Wiley & Sons, Inc.

intracellular distribution of the pharmacological agent. More specifically, positively charged particles are relatively easier to be internalized by tumor cells because of the negative charge of the cell membrane. However, the positive surface charge also facilitates interception and clearance by the reticuloendothelial system (RES), which thus negatively affects circulation time for anticancer drugs throughout the body. By contrast, negatively charged or neutral drug-loaded carriers may exhibit longer *in vivo* circulation times, but have a limited capacity for transmembrane drug delivery into target cells due to electrostatic repulsion.

Using specialized porous glass membrane emulsification technique – cross-linking method and subsequent deposition method of chitosan with different charge, we were able to eliminate variation in nanoparticle properties in order to fabricate three types of uniform nanoparticles (~215 nm) that only differed in their surface charge (Figure 7.2). These three variations on the same NP were tested in panel of eight cell lines, which revealed that both the rate of cellular uptake and the total amount of internalized particles were positively correlated with surface charge in all cell lines. Subsequent intracellular trafficking assays indicated that a proportion of positively charged nanoparticles could escape from lysosomes after internalization and localize to the perinuclear region, while NPs with a negative or neutral charge preferentially accumulated in lysosomes. These results suggest that we can overcome the barrier of intracellular delivery and targeting by preparing nanoparticles with different properties (e.g. positive charge) [15]. This finding provided a critical breakthrough in the fundamental understanding required for the design of high-specificity, high efficacy NP delivery platforms.

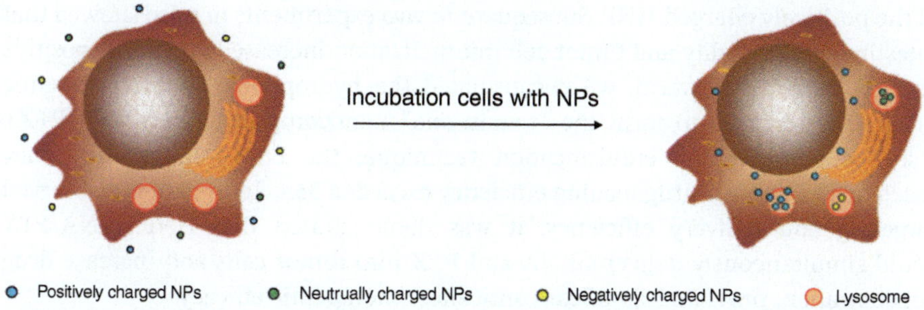

Figure 7.2 Cellular uptake and intracellular distribution of NPs with different surface charges. Source: Yue et al. [15]/Reproduced with permission from American Chemical Society.

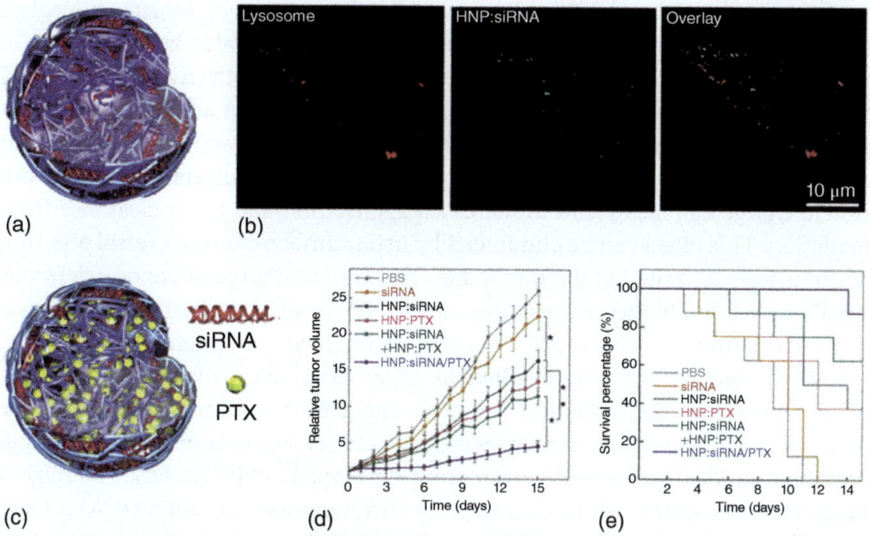

Figure 7.3 (a) The structure diagram of HNP:siRNA. (b) Confocal laser scanning microscope (CLSM) images of lysosome, HNP:siRNA, and overlay signals after incubating HNP:siRNA with LLC cells for six hours. (red: lysosomes, cyan: HNP:siRNA; scale bars: 10 μm). (c) The structure diagram of HNP:siRNA/PTX. (d) Inhibition of LLC xenograft tumor growth and (e) survival curve of tumor-bearing mice in different treatment groups. Source: Wei et al. [17]/ Reproduced with permission from Elsevier.

Based on these characteristics, we modified our design to suit oral delivery by encapsulating PTX into *N*-((2-hydroxy-3-trimethylammonium) propyl) chitosan chloride (HTCC) NPs, which increased the therapeutic efficacy toward non-small-cell lung cancer [16]. Furthermore, by taking advantage of the positive charge and structural characteristics of quaternized chitosan HTCC nanoparticles (HNP), siRNA loading was facilitated (HNP:siRNA), solving the issue of serum degradation and limited cell internalization [17] (Figure 7.3). Experiments proved that HNP could efficiently deliver siRNA to tumor, and drugs that escaped from the lysosome could be delivered to the cytoplasm, facilitated by the proton sponge effect of the positively charged HNP. Subsequent *in vivo* experiments in mice showed that intestinal permeability and tumor cell internalization increased due to the positive surface charge. Afterward, we encapsulated the hydrophobic chemotherapeutic drug PTX into NPs to form the "two-in-one" nanocomplex (HNP:siRNA/PTX). Using our membrane emulsification technique, the encapsulation efficiency reached 83% and the drug loading efficiency exceeded 35%. In light of the observed targeting and delivery efficiency, it was demonstrated that HNP:siRNA/PTX could simultaneously deliver siRNA and PTX into tumor cells and increase drug concentration, providing significant anticancer therapeutic efficacy.

The proton sponge effect exerted by the positive NP surface charge facilitated lysosomal escape and cytoplasmic translocation for drug release, leading to striking inhibitory effects on the targeted cancer cells, compared to that under mixed administration of both drugs individually loaded into the carrier. These remarkable examples provide strong evidence that particle charge strongly impacts biological and pharmacodynamic drug behavior *in vivo*, and should also be considered in carrier design. NPs that carry a positive charge are more effectively internalized by cells due to their affinity with anionic cell membranes. In addition, the proton sponge effect induced by positively charged NPs leads to lysosomal swelling and rupture, resulting in subsequent drug release in the cytoplasm [18]. However, positive charge can also result in increased aggregation and faster clearance from circulation. This effect can be eliminated by using surface charge-reversal NPs that are exclusively activated at the tumor site. The surface charge of nanocarriers was initially either uncharged or negative, and could reverse to positive in response to specific stimuli, such as pH, enzymes, light, heat, or others. This negative to positive reversal not only resulted in faster drug release in an acidic TME, but also increased the uptake of the drug into cells via electrostatic absorptive endocytosis, enhancing anticancer therapeutic effects. For example, with the rapid membrane emulsification method, our group developed a chitosan-based intelligent charge-reversed system via modification by tris(2-aminoethyl) amine (TAEA) and 2,3-dimethylmaleic anhydride (DMMA) [19]. PTX-loading charge-reversed NPs were negatively charged (−11 mV) at normal tissue circumstances (pH ~ 7.4) and were rarely internalized by macrophages. Once reaching the cancer site, these NPs could intelligently transform to a positive charge (+5 mV) by the low-pH microenvironment in tumor tissue, which could improve tumor cell uptake and increase the drug concentration at the lesion site. Compared with commercial PTX formulation, these NPs possessed superior biocompatibility and better cytotoxicity, and have significant potential as efficient anticancer drug carriers.

7.2.3 Surface Hydrophobicity

The relationship between the polarity of molecules and their internalization was described as early as 1899 by the British cell physiologist, Overton. Based on his observations of cellular uptake of different compounds, "Overton's Rule" proposed that the greater the polarity, the slower the entry into cells, while the coupling of nonpolar functional groups, such as alkyl chains, could enhance the rate of cellular uptake [20]. Subsequent studies reported that particles prepared from high molecular weight or hydrophobic materials had higher intracellular internalization rates. Coarse-grained molecular dynamics simulations by Li and coworkers examining differences in hydrophilic or hydrophobic NP interactions with the cell membrane revealed that surface hydrophobicity affected the mechanisms by which NPs interacted with the cell membrane [21]. Calculation of free energy indicated that embedding of hydrophobic particles into the cell membrane is a spontaneous behavior, whereas hydrophilic particles must overcome a considerably higher energy barrier to cross the membrane that might require assistance from external materials or energy, implicating that the cell could potentially import hydrophilic particles by phagocytosis.

For delivery of insoluble antitumor drugs, poly(lactic acid) (PLA) polymers, which have a main body comprising hydrophobic components, have fundamental advantages over other vehicles since insoluble drugs can be loaded at high capacity. However, the effects of hydrophobicity on carrier delivery efficiency are often overlooked in PLA-derivative particles. Using the membrane emulsification technique, our group fabricated a series of PLA, poly(lactide-*co*-glycolide) (PLGA), and polyethylene glycol-*b*-polylactide (PEG-*b*-PLA) NPs [22] (Figure 7.4). These NPs prepared by membrane emulsification technique had a uniform particle size (~290 nm) and a very narrow size distribution (polydispersity index <0.1), which thus allowed clear comparisons of the effects of the materials by eliminating the effects of size variability. While the original PLA nanoparticles were the most hydrophobic, hydrophilicity increased from PLGA to PEG-*b*-PLA NPs. Macrophage uptake data aligned well with established principles of NP hydrophobicity property of nanoparticles, with lower hydrophobicity leading to less macrophage internalization and consequently longer circulation time in blood. A potential mechanism for the difference in uptake could be due to large amounts of complementary components and immunoglobulins adhesion to hydrophobic NPs, leading to macrophage phagocytosis. We also found these three NPs had different intratumoral delivery capabilities. With the same particle size and exposed pressure, a higher penetration percentage reflected a better deformation performance, indicating that the deformation ability of PEG-*b*-PLA NPs exceeded those of PLA and PLGA. The ordered molecular arrangement of PEG-*b*-PLA NPs could give them more freedom for mutual slipping of molecules and lead to excellent deformability. These results indicated that less hydrophobicity leads to longer circulation time and better deformation performance leads to deeper intratumoral delivery, ultimately resulting in enhanced passive accumulation at the tumor site.

7.2.4 Morphology

Particle morphology refers to shape and surface patterning of the vehicle microstructure. During interactions between particles and tumor cells, the shape of the carrier can also affect the ability of cells to take up the drug [4]. Experimental studies have confirmed that particle shape contributes to the efficiency of drug delivery. For example, phagocytosis by macrophages, a key step in drug delivery, strongly

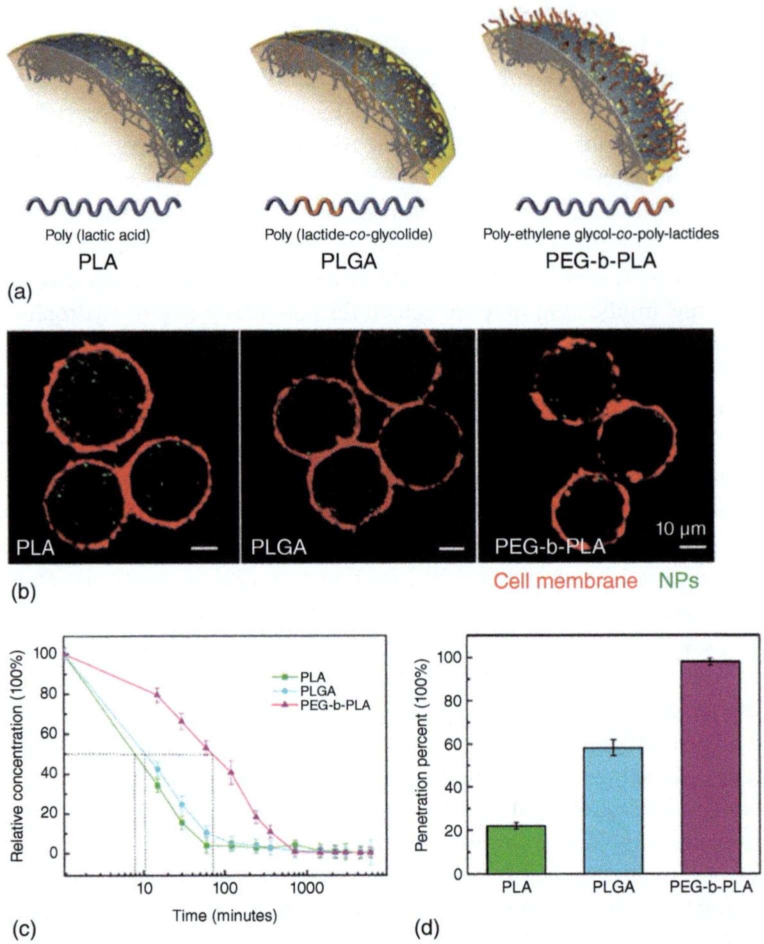

Figure 7.4 (a) Schematic diagrams of the arrangement pattern of different polymers at the interface of an oil/water emulsion droplet. (b) CLSM images of the corresponding internalization in macrophage. (c) Pharmacokinetics of PLA, PLGA, and PEG-*b*-PLA NPs after intravenous injection in tumor-bearing mice. (d) Penetration percentages of NPs through the millipore to mimic tumor interstitium. Source: Yue et al. [22]/ Reproduced with permission from Royal Society of Chemistry.

depends on particle shape [23]. For example, we used two differently shaped NPs to examine the delivery of the same drug and found that rod-shaped particles showed a higher level of delivery of insoluble antitumor drugs due to their easier uptake by tumor cells compared to spherical particles [24]. Champion and Mitragotri reported that size and shape mutually influence each other's effects in phagocytosis [25]. On a smaller scale, Gratton et al. demonstrated that the internalization of cylindrical particles strongly depended on their aspect ratio. When the aspect ratio was three, associated particle internalization was four times faster than for spherical particles of the same volume [26]. Moreover, Muro et al. compared tissue targeting efficiency of spheres with various diameters (from 100 nm to 10 μm) and microscale elliptical discs (1 × 3 μm) and found that the micrometer-scale discs accumulated to higher levels in target tissues than spherical particles of any size, even those with considerably smaller nanoscale dimensions [27]. While these studies have contributed substantial advances to the design and development of new antitumor drug carriers, further consideration is still needed to improve the uptake of these particles by tumor cells, while maintaining a sufficiently long circulation time.

Zero-dimensional (0D) particles and one-dimensional (1D) nanotubes have shown great potential in drug delivery and cancer treatments. In addition, two-dimensional (2D) materials, such as graphene oxide (GO), have become a new research focus [28, 29]. For targeting at the tissue, cell, or subcellular levels, refining the size of particles is critical. It had not yet been described how the size of particles, including GO with novel 2D structures (at the nanoscale thickness), affected cellular response. We examined the effect of the lateral dimension of GO particles, on the nano to micro scales, on cellular responses, including cellular uptake, internalization mechanisms, and intracellular trafficking. We prepared GO particles with 350 nm and 2 μm in the lateral dimension and employed two macrophages and four nonphagocytic cells to explore the intrinsic properties of these 2D materials. There was a dramatic uptake disparity between different cell types, which differed from the conventional particles (Figure 7.5). GO internalization at either size was negligible in all nonphagocytic cells, which likely resulted from the strong electrostatic repulsion from the dense carboxyl distributions and flat curvature of 2D GO. Therefore, GO needs to be functionalized on the surface to overcome this repulsion when being utilized as a drug carrier for these nonphagocytes. In sharp contrast to the nonphagocytes, GO was widely internalized into macrophages in a size-independent manner, and this was due to active Fcγ receptor-mediated phagocytosis.

Our group further explored a "sandwich" superstructure formed by 2D particles in the cell membrane to achieve drug delivery between cell membranes [30]

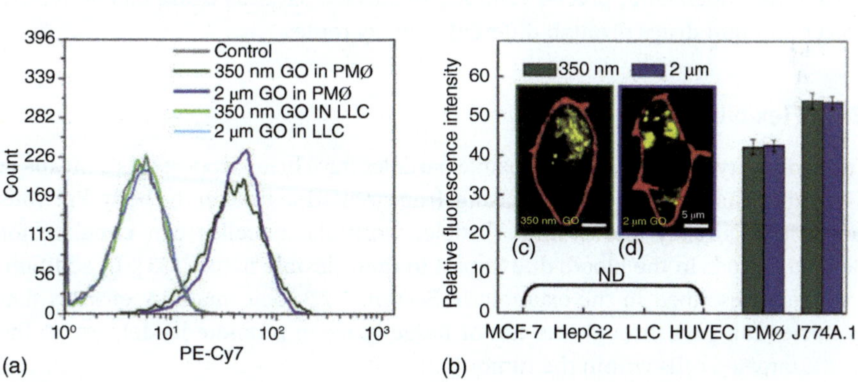

Figure 7.5 Cellular internalization of GO in macrophages and nonphagocytic cells. (a) Flow cytometry exhibiting fluorescent signals from GO-treated macrophages (PMØ) and nonphagocytic cells (LLC). (b) GO internalization by different cells. (c–d) CLSM images of differently sized GO internalization in macrophages. Source: Yue et al. [29]/ Reproduced with permission from Elsevier.

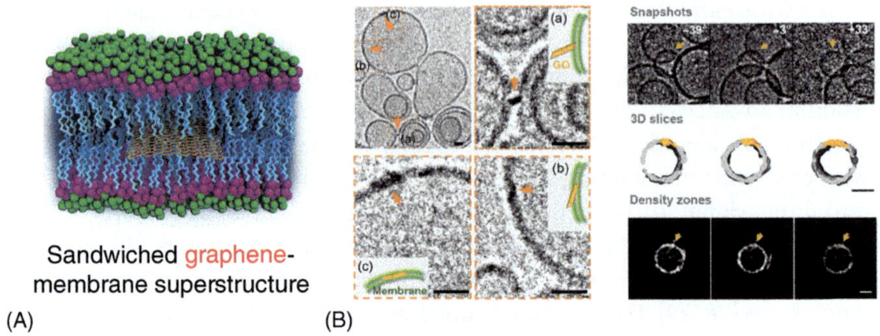

Figure 7.6 (A) Image illustrating the superstructure integrating GO inside the cell membrane. (B) Cyro-TEM images (left) and tomography views (right) of the lipid vesicles in the presence of GO (Scale bars: 20 nm). Source: Chen et al. [30]/ Reproduced with permission from American Association for the Advancement of Science – AAAS.

(Figure 7.6). This special structure allowed the drug carrier to enter the limited domain between the cell membranes to facilitate the rapid movement and diffusion of the drug in the membrane phospholipid bilayer. This "sandwich" superstructure could effectively interact with receptor targets on the membrane and significantly increase the killing ability of cancer cells, which was better than the whole-cell diffusion and delivery of liposome encapsulated drugs.

In addition to overall shape, the surface of nano and microstructures has been well established as a morphological determining factor in particle adhesion or motility [31]. Studies examining differences in uptake between surface-patterned and nonpatterned particles of the same size and overall surface composition indicated that patterning greatly impacted cellular uptake [32]. In particular, the surface patterns on NPs affected cell spreading and focal adhesion dynamics, but depended on cell-recognizable spacing [28]. Although the underlying mechanisms were poorly understood, precise tailoring of surface patterns could enable NPs to transfer different drugs through different delivery routes.

7.2.5 Flexibility

In drug delivery research, softer, flexible particles have been proposed as a means of minimizing phagocytosis and increasing drug circulation time in the body. Previous studies have already shown that flexible, wormlike micelles can circulate for extended periods in the blood, due in part to their flexible nature [33]. In addition, as we have described in the example in Section 7.2.3, deformability ensures that the carrier can pass through the tumor tissue space to promote its delivery to the specific targeted cells within the tumor [22].

Due to the highly heterogeneous and osmotic environment inside the TME, most conventional chemotherapy drugs exhibit poor penetration into the tumor core. The use of specific polymer materials in the carrier can confer specialized properties, such as deformability and hydrophilicity, to the carrier in order to improve targeting efficiency. The progress of the above system opens new avenues for research toward overcoming obstacles for drug delivery, such as tight cell packing within tumor tissues. Although comprehensive and systematic biological and immunological evaluation is necessary, full consideration of the physical and chemical properties of the carrier and ligand modification may enable the development of carrier particles that effectively penetrate the tumor to eradicate tumors.

7.3 Classical Strategies for Targeting Tumor Tissues

Early targeting strategies primarily focused on passive targeting, such as modifying the carrier with hydrophilic molecules (e.g. polyethylene glycol), avoiding nonspecific clearance, and prolonging the half-life of drugs in the blood. At the same time, the reduced size of nanocarriers exploited the high permeability and retention of tumor tissues, carrying drugs through the tumor vascular endothelial system to enter and accumulate in the tumor tissue. Through a full, mechanistic understanding of tumor development and progression, the exploration of unique structures and properties of biological carrier materials has led to a variety of strategies for targeting tumor tissues [34].

7.3.1 Ligand/Receptor-Induced Targeting

To enable rapid growth, metastasis, and infiltration, tumor cells generally overexpress several specific antigens or receptors (e.g. HER2, folate receptor, CD44, etc.) on their surface compared with normal cells. Thus, grafting ligands on the surface of the drug carrier, such as antibodies, aptamers, or transmembrane peptide RGD/iRGD, can facilitate the import of anticancer drugs into tumor tissues. Relying on the specific recognition between receptor ligands (or antigen antibodies), a large number of carriers can accumulate on the cancer cell surface, resulting in enhanced cell internalization with reduced host toxicity and side effects.

It is well known that asialoglycoprotein (ASGP) receptors are overexpressed on the surface of hepatocytes. These ASGP receptors are activated by galactose-terminated glycoproteins and mediate their internalization to the cytosol. For instance, our group described uniform 10-hydroxycamptothecin (HCPT) nanocrystallites carried in a galactosylated chitosan vehicle that exhibited enhanced internalization in human hepatocellular carcinoma HepG2 cells through an ASGP receptor-mediated pathway *in vitro* [35]. Based on this innovation, we next developed a galactosamine-conjugated albumin nanoparticle (GAL-AN) as a drug carrier for targeted liver cancer therapy [36]. Delivery of doxorubicin (DOX) in the GAL-AN particles increased its inhibitory activity on HepG2 cell growth through enhanced ASGP receptor-mediated internalization.

We also demonstrated that insoluble PTX could be loaded into hydrophilic carboxyl chitosan nanospheres (CNP:PTX) to generate PTX-loaded nanocrystals *in situ* [37]. For this innovation, an O/W/O rapid membrane emulsification method was combined with a programmed heating technique to produce uniform nanospheres. The polyethylene glycol (PEG) chain and the tripeptide sequence (RGD) targeting tumor cells were subsequently linked by carboxyl groups on the nanosphere surface, resulting in an advanced drug delivery system (RBD-PEG-CNP:PTX) that can bypass host defenses to specifically target tumor cells. This formulation effectively displayed prolonged circulation *in vivo*, enhanced affinity to tumor cells, improved drug bioavailability, and increased antitumor effects in cells *in vitro* and in tumor-bearing mice *in vivo*. In addition, compared with traditional injected formulations, this platform also resulted in lower toxicity and side effects, supporting the strong potential for clinical application of PTX.

More recently, we reported the clinical discovery that CD71 was stably expressed at high levels in leukemia cells obtained from patients with a variety of different forms of leukemia [38]. In light of this finding, we developed a nano-medicine that showed remarkably high affinity binding to leukemia cells for delivery of cytotoxic As^{III}. This ferritin-encapsulated As system, As@Fn, showed high internalization in both primary leukemia cells obtained from patients and several leukemia cell lines (Figure 7.7), while showing high efficacy against a variety of different cell line- and patient-derived xenograft models, with notably higher potency than current gold standard interventions. These results supported the considerable potential of this platform for precise and effective treatment of different leukemias.

7.3.2 Tumor Microenvironment Sensitive Targeting

The TME is widely understood to be a highly complex niche that includes a wide diversity of stromal cells (i.e. endothelial cells, fibroblasts, and immune cells) in addition to tumor cells. Despite advances in passive and active targeted-drug delivery, additional properties of drug delivery systems to enhance the bioavailability of

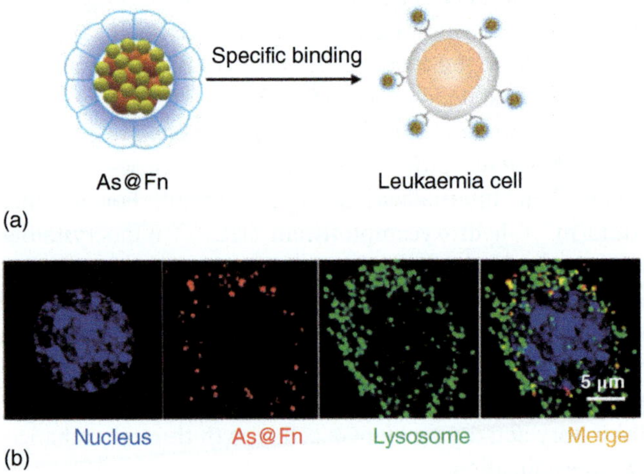

Figure 7.7 (a) Schematic illustration of As@Fn nanomedicine and the specific binding to leukemia cells. (b) Representative CLSM images and analysis of As@Fn colocalized with lysosomes after endocytosis (green, lysosome; blue, nucleus; red, As@Fn; scale bar, 5 μm). Source: Wang et al. [38]/ Reproduced with permission from Springer Nature.

drugs at tumor site and reduce adverse side effects are very important [39]. Thus, a primary objective in drug targeting is responsiveness to tumor tissue conditions to ensure that the drug is only released directly to the appropriate tumor site. Fortunately, the TME has markedly different properties than the environment of normal tissues, and tumor-targeting vectors can be constructed using materials that are sensitive to TME-specific conditions, such as low pH, aberrant enzyme activity, hypoxia, and reducing (rather than oxidizing) conditions. The design principle relies on the carrier exhibiting sufficiently high stability and structural integrity to remain intact in normal tissues, but degrade under conditions at the tumor site, thereby releasing the drug cargo, which consequently accumulates specifically at the targeted tumor site.

7.3.2.1 pH- Sensitive Drug Delivery

While the pH of normal tissue is around 7.4, tumor cells accumulate high concentrations of lactic acid produced by glycolysis under anaerobic conditions in the tumor site, resulting in a weakly acidic environment (pH 6.0–7.0) around tumor cells. Therefore, a pH-sensitive drug carrier requires sensitivity to minute changes in pH, so that the carrier remains intact in circulating blood at neutral pH, but disintegrates or undergoes conformational changes in the weakly acidic TME to release the drugs. The most common strategy for achieving this effect is by introducing "ionizable" chemical groups (e.g. amines, guanidyl, and imidazolium) into the polymer structure to make the carrier swell and release drugs at low pH. Some biocompatible inorganic substances could also act as superior carriers to achieve this effect. To this end, our group prepared hollow $CaCO_3$ nanoparticles (HP-a) using a one-pot strategy [2]. Each individual hollow HP-a comprised ~50 nm particle building units (Figure 7.8). Because the component $CaCO_3$ could dissolve in an acidic extracellular TME or within tumor cell lysosomes, HP-a showed promise as a pH-sensitive and highly biocompatible drug delivery carrier. We found that HP-a loaded with DOX released this DOX in a pH-sensitive, localized, and effective manner, and HP-a was dissembled to much smaller nanosized building units. These units could accumulate around the nucleus or directly enter it. Entry into the nucleus is attributable to electrostatic interactions between the nuclear membrane, and the positively charged DOX-loaded nano-$CaCO_3$ building units. In addition to increased uptake, these behaviors could enhance the efficiency of drug delivery to the nucleus, and increase cytotoxicity toward tumor cells.

Ulteriorly, our group also developed different biomineralized platforms with uniform carrier particles, such as a biomineralized calcium phosphate (CaP) shell encasing a ferritin payload (Fn@CaP) [40, 41] (Figure 7.9). The high pH sensitivity of the CaP shell results in its dissolution in low pH, which thus limits its accumulation in the liver but promotes Fn delivery to tumor cells for binding to its receptor CD71. Biomineralization could also be applied to other biological materials, such as bacterial outer membrane vesicles. The pH-sensitive CaP shells can also drive a buffering effect, neutralizing acidity in the TME, which subsequently promotes

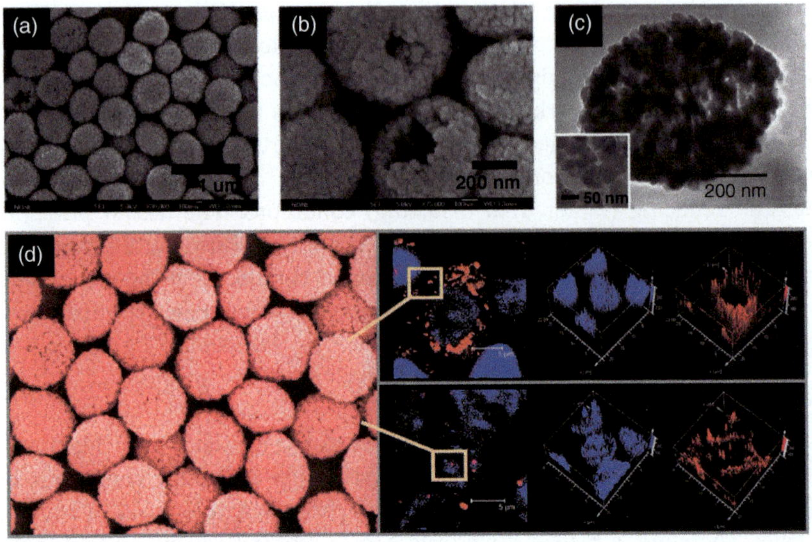

Figure 7.8 (a) SEM image of uniform particle size of HP-a. (b) SEM image of HP-a superstructures. (c) TEM images of a single HP-a, showing the hollow structure of the spheres. (d) Perinuclear accumulation and nuclear location of DOX loaded HP-a (red dot, scale bars: 5 μm). Source: Wei et al. [2]/ Reproduced with permission from American Chemical Society.

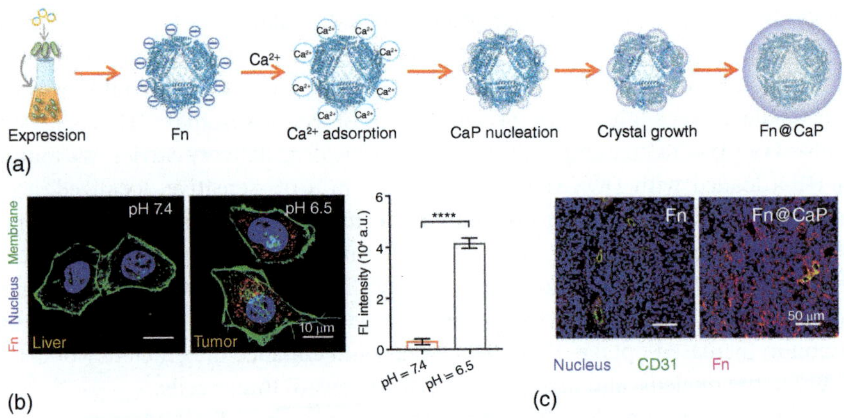

Figure 7.9 (a) Schematic illustration of the preparation of the Fn@CaP nanoplatform, comprising an inner Fn core shielded by a CaP shell. (b) Representative CLSM images of Fn@CaP endocytosed by liver cells at pH 7.4 (to simulate normal physiological condition) and by tumor cells at pH 6.5 (to simulate TME; scale bars: 10 μm). (c) CLSM images of Fn and Fn@CaP accumulated in tumor tissues (red: Fn, blue: nucleus, green: CD31; scale bars: 50 μm). Source: Wang et al. [41]/ Reproduced with permission from John Wiley & Sons, Inc.

the macrophages polarization from anti-inflammatory M_2 phenotype to proinflammatory M_1 phenotype among tumor-resident macrophages, and intratumoral calcification. These combined effects illustrate the versatility and effectiveness of biomineralized particles for antitumor drug delivery.

7.3.2.2 Enzyme Responsive Drug Delivery

Interactions between tumor cells and the extracellular matrix lead to an imbalance in extracellular matrix environment and can induce tumor cells invasion and metastasis. Matrix metalloproteinase (MMP) is a protease with high proteolytic activity that is typically highly expressed on the surface of tumor cells where it functions in the regulation of migration, epithelial-to-mesenchymal transition, cancer cell survival, activation of immune response, tumor angiogenesis, and establishment of the TME, which are all essential steps in tumorigenesis [42].

Hatakeyama et al. connected PEG to 1,2-dioleoyl-*sn*-glycero-3-phosphoethanolamine (DOPE) using peptide segments, resulting in PEG–peptide–lipid ternary conjugates (PPD) [43]. These PPDs were then tethered to a vehicle carrying a gene therapy, which exhibited MMP expression-dependent antitumor effects in different tumor tissues *in vivo*. High MMP expression at the tumor site results in cleavage of the peptide linker between PEG and DOPE, releasing the drug from PEG protection so it can accumulate at the tumor site, thus improving its targeting of tumor cells. Using an alternative strategy, Iaccarino et al. developed an MMP-2-dependent cross-linked polyelectrolyte nanocapsule for delivery of hydrophobic drugs [44] (Figure 7.10). In this case, cross-linking the nanocapsule vehicle with a peptide cleavable by MMP-2 resulted in longer drug stability and highly specific conditional release of the drug payload at the tumor site, suggesting that this platform increases the safety and control of hydrophobic drug release by exploiting endogenous enzymatic activity at the tumor site.

In addition to developing targeted vectors based on MMP, our group constructed a fluorescent probe (iprobe) activated by MMP activity that was based on two-dimensional planar particles, then loaded the probe with the insoluble antitumor drug hydroxycamptothecin (HCPT, loading rate reached 9.6%) [45]. The iProbe fluorescence could be activated at the tumor site, but blocked in the presence of MMP inhibitor. The probe not only showed stronger antitumor effects than injection with commercially available drugs but also facilitated real-time monitoring of treatment response in patients.

7.3.2.3 Hypoxia-Targeted Drug Delivery

Hypoxia in the TME is defined as an oxygen partial pressure within the tumor of 0–20 mmHg (1 mmHg = 133.322 Pa), lower than that of normal tissues, i.e.

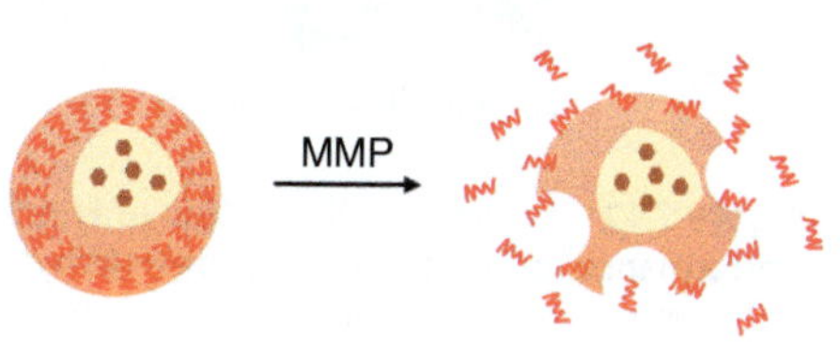

Figure 7.10 Schematic representation of cross-linked nanocapsules before and after MMP cleavage.

24–66 mmHg. Hypoxia is common in solid tumor tissues due to the development of new, aberrant blood vessels at the tumor site, which exhibit poor blood flow [46]. Tumor cells can survive and proliferate in a hypoxic environment after appropriate genetic and adaptive changes that benefit tumor cell invasion and metastasis. Two major strategies have emerged for suppressing hypoxia in the TME to reduce its positive effects on tumor growth.

On the one hand, disrupting self-regulation by tumor cells can decrease their tolerance of hypoxic environments, subsequently inducing apoptosis, which is an ideal strategy for anticancer therapy. Mechanistically, hypoxia in the TME can activate complex shifts in the transcriptome, especially in the hypoxia-inducible factor-1 (HIF-1) pathway, which plays a critical role in tumor development and chemotherapy resistance. Therefore, suppression of HIF-1 has emerged as a promising anticancer strategy [47] (Figure 7.11). To specifically suppress HIF-1 expression in hypoxic, deep tumor tissues, a platform (HM/S/MC) for delivery of small interfering RNAs (siRNAs) targeting HIF-1 was developed with a magnetic Fe_3O_4 nanocluster core in a chimeric membrane carrier with surface-tethered hyaluronidase. This system showed extended circulation time could be guided by magnetic resonance imaging, accumulated specifically in tumors through magnetic stimulation, effectively penetrated the hypoxic tissues, and displayed cytoplasmic internalization through homotypic tumor targeting. These magnetosomes effectively suppressed HIF-1, leading to significant therapeutic benefits, while also reducing chemoresistance with few detectable side effects, indicating its clinical potential as an anticancer treatment.

On the other hand, hypoxia promotes tumor deterioration and subsequent metastasis, in addition to increasing resistance to chemotherapy/radiotherapy. Therefore, rather than alleviating hypoxia via gene therapy, He et al. developed a porphyrinic metal–organic framework (MOF)-gold nanoparticle (AuNP) hybrid therapeutic

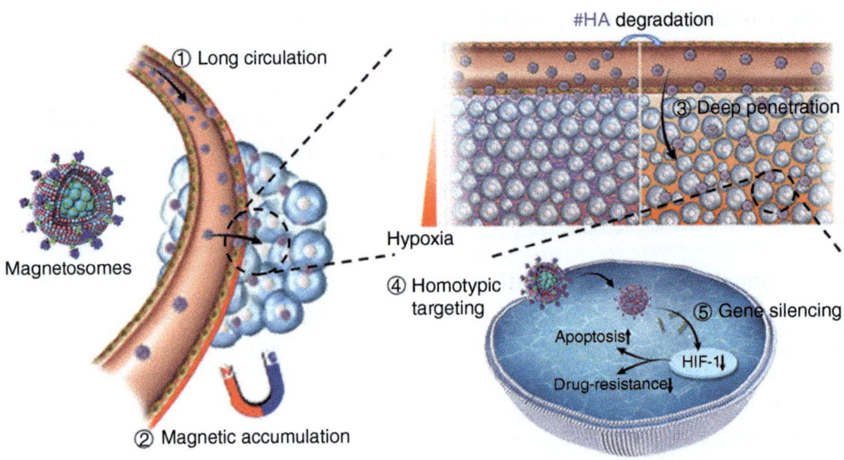

Figure 7.11 Schematic illustration of HM/S/MC with efficient HIF-1 siRNA delivery performance for antitumor therapy (HA: hyaluronic acid). Source: Lyu et al. [47]/ Reproduced with permission from Elsevier.

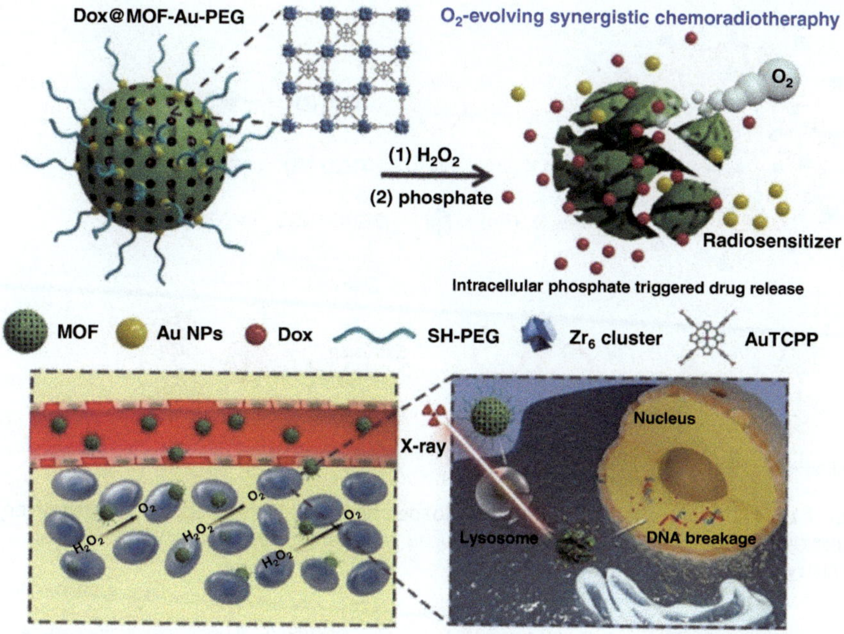

Figure 7.12 Schematic representation showing the main components of Dox@MOF-Au-PEG and the mechanism of O_2 self-supplying synergistic chemoradiotherapy. Source: He et al. [48]/ Reproduced with permission from John Wiley & Sons, Inc.

platform (Dox@MOF-Au-PEG) that relies on catalyzing the oxygenation of tumor tissues *in situ,* i.e. an O_2-evolving chemoradiotherapy [48] (Figure 7.12). In this system, AuNPs bound to the MOF surface act as radiosensitizers while also stabilizing the composite particles. The MOF scaffold serves as the carrier, which is loaded with DOX. These hybrid systems with catalase-like properties significantly increased the effects of radiotherapy *in vitro* and *in vivo*, relieving hypoxia and synergistically suppressing tumor growth with low toxicity. These strategies illustrate the clinical potential for nanoparticle-based theranostic treatments.

7.3.3 Externally Activated Targeting

In addition to sensitivity to endogenous conditions, drug carriers can be designed with physicochemical properties responsive to external stimuli (such as magnetic fields, photothermal activation, or ultrasound treatment). These activators can change the properties or spatial structure of the carrier to release the drug payload, thus actively targeting the tumor site. Tumor targeting through external activation can be more accurate and controllable compared with carriers sensitive to the TME.

7.3.3.1 Magnetism-Based Tumor Targeting

Magnetic carriers are prepared by introducing magnetic materials into the carrier matrix. Under an external magnetic field, the drug carrier can accumulate at the

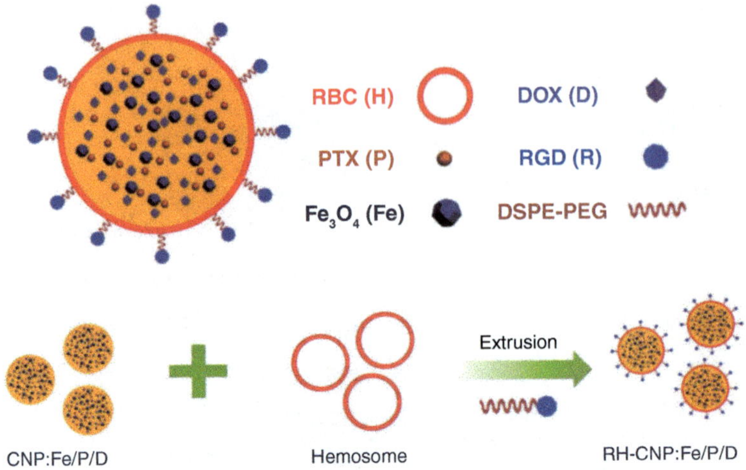

Figure 7.13 Schematic illustration of CNPs-loaded hemosomes for programmed codelivery against cancer. Source: Fu et al. [49]/ Reproduced with permission from Royal Society of Chemistry.

tumor site by circulating in blood. In addition, magnetic targeting can be combined with TME sensitivity and/or photothermal sensitivity, or other strategies to improve preferential targeting in drug delivery.

For example, we constructed a uniform, erythrocyte-membrane-camouflaged programmed codelivery system (RH-CNP:Fe/P/D) that incorporates hydrophilic–hydrophobic chemotherapeutic drug pair PTX/DOX and Fe_3O_4-loaded chitosan nanoparticles (CNPs) using a porous membrane emulsification process [49] (Figure 7.13). The red blood cell membrane enables the codelivery system to show stealth capability and an increased circulation time in the blood. The uniformly small size of the biomimetic NPs with a narrow size distribution ($PDI < 0.1$) was beneficial for EPR. In addition, RH-CNP:Fe/P/D has colloid stability, and both the particle size and zeta potential are well maintained in cell culture media, which can fulfill requirements *in vivo*. Owing to the homogeneous and efficient encapsulation of Fe_3O_4 nanocrystals, NPs were highly accumulated at the tumor site, mediated by magnetic targeting. Both uniform particle size and high drug loading ability enhanced the therapeutic effect. As a result, this combined codelivery system resulted in remarkable inhibition of tumor growth, increased survival rates, and decreased side effects of chemotherapy drugs.

Espinosa et al. reported development of the Janus magneto-plasmonic nanoparticles, made of gold nano-stars and Fe_3O_4 nanospheres, which improved drug delivery by magnetic targeting [50]. In addition to magnetic targeting, they combined magnetic hyperthermia and photothermia, resulting in synergistic cytotoxic effects in tumor cells. The precision of magnetic guidance of these particles and their potency under photothermal activation confers strong inhibition of tumor growth.

7.3.3.2 Photosensitive Tumor Targeting

The most typical application of light-sensitive materials in tumor-targeted therapy is to change the functional properties by changing the material structure. Functional groups that confer light sensitivity to carrier materials generally include azobenzene, coumarin, o-nitrobenzyl, and spiropyran [51]. Compared with the other photosensitive materials, nanoparticle drug carriers made with *o*-nitrobenzyl exhibit the greatest sensitivity and degradation under photostimulation [52]. Based on these characteristics, Yao et al. used *o*-nitrobenzyl in the synthesis of an ultraviolet (UV)-sensitive/reducible dual-responsive molecule for DOX-loaded liposomes. This tumor-targeting platform showed high excitability under UV or reducing agents, high biocompatibility, and allowed therapeutic guidance by magnetic resonance imaging (MRI) [53] (Figure 7.14). Specifically, UV stimulation triggered cleavage of an ester group in the liposome carrier, resulting in the release of DOX, and ultimately stronger therapeutic effects than DOX alone, with minimal detectable side effects associated with the treatment process.

It should be noted that near-infrared (NIR) and far-infrared light can penetrate skin and tissues more deeply than either UV or visible light, suggesting the suitability of NIR-sensitive materials for combination thermotherapies. In particular, Kuang et al. reported a multifunctional polyprodrug-/siRNA-loaded photon upconversion NPs to produce a NIR light-activated on-demand delivery system for drugs or gene therapies [54]. Upon activation, these nanoparticles function as "nanotransducers" that convert NIR light (980 nm) into local UV light, then visible light, which cleaves the photosensitive platinum (IV) backbone polymers (PPt), consequently releasing highly toxic platinum (II) and a siRNA targeting Plk1. In addition, the accumulation of these particles can be imaged by computed tomography (CT), MRI, or upconversion luminescence. This platform showed synergistic therapeutic effects *in vitro* and *in vivo* through NIR light-activated platinum-based chemotherapy and RNA interference-based gene therapy, thus indicating the potential for multifunctional, light-sensitive nanoparticle delivery

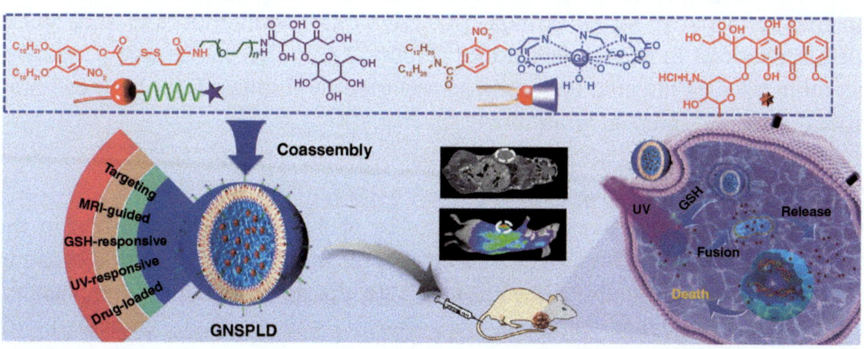

Figure 7.14 Schematic illustration of the MRI-guided targeting dual-responsive drug delivery system for liver cancer therapy. Source: Yao et al. [53]/ Reproduced with permission from Elsevier.

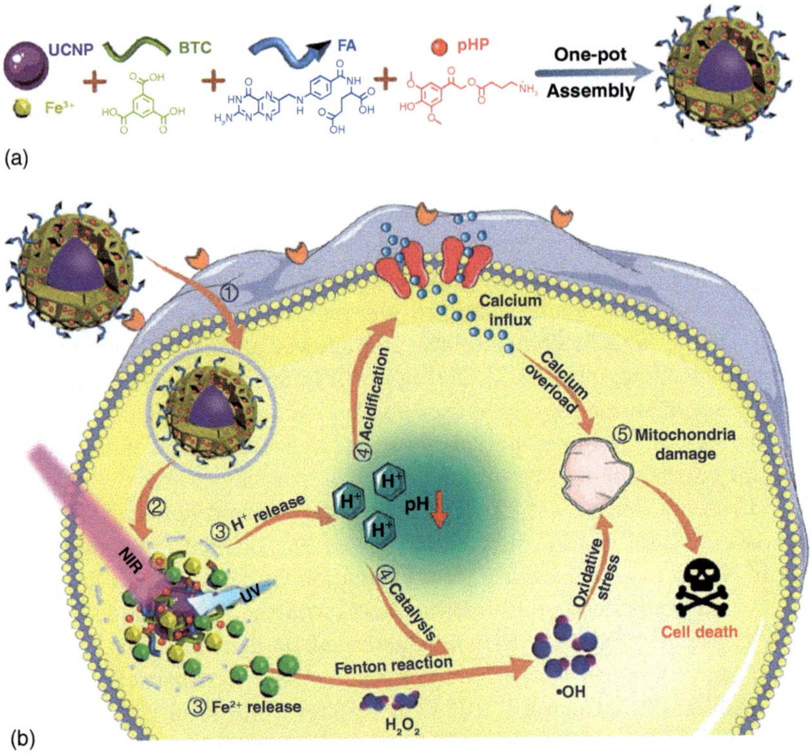

Figure 7.15 Schematic illustration of the construction of NIR-triggered nanoagent and the underlying anticancer mechanism (UCNP: upconversion nanoparticles). Source: Bao et al. [55]/ Reproduced with permission from Springer Nature.

platforms. In addition, the use of multifunction or multimodal components to target subcellular organelles could markedly enhance the antitumor potency of numerous drugs. For example, we developed a NIR-triggered nanoagent that can induce oxidative stress through photodynamic stimulation and subsequent calcium overload in mitochondria [55] (Figure 7.15). The therapeutic potency of this mitochondria-targeted system was supported by findings from both cell- and patient-derived tumor xenograft models *in vivo*.

7.3.3.3 Thermal-Responsive Targeting

Tumor tissues can be heated to temperatures of 40–44 °C, which results in the inhibition and death of tumor cells. Clinical application of hyperthermia began in the 1980s and was initially applied alone. With the development and application of antitumor drug delivery systems, hyperthermia was eventually combined with temperature-sensitive carriers. At normal body temperature, the carrier remains stable with the drug cargo during long-term circulation. However, once temperature increases to a threshold determined by carrier design, its stability decreases, and the drug payload is deposited at the heating site or leaks out, accumulating at the tumor site.

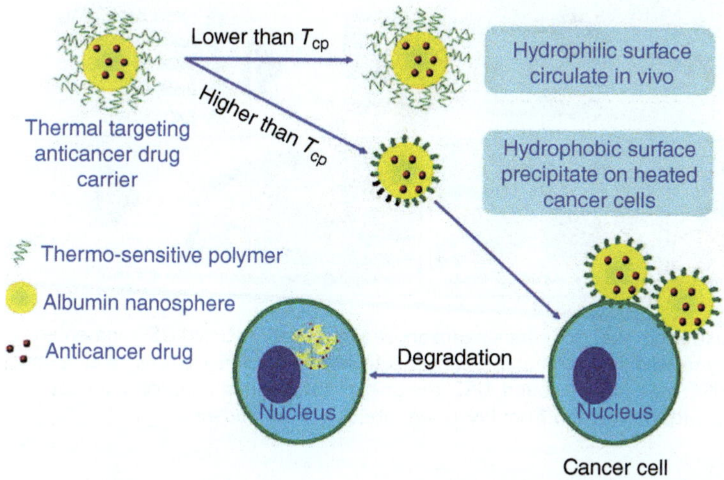

Figure 7.16 Target principal scheme of PNIPAM-based thermal targeting anticancer drug carrier (T_{cp}: cloud-point temperature). Source: Shen et al. [56]/ Reproduced with permission from Elsevier.

Poly(*N*-isopropylacrylamide) (PNIPAM) and poly(*N*,*N*-dimethylaminoethyl methacrylate) (PDMAEMA) are two thermal-sensitive polymers with low critical dissolution temperature used for this purpose. Using a free radical polymerization method, where acrylamide (AAm) was fed gradually, our group prepared PNIPAM-AAm polymers with a high transition temperature ($T_g \sim 42$–$43\,°C$) [56] (Figure 7.16). This carrier could specifically target HepG2 cells at 43 °C. Another example is that magnetic Fe_3O_4 nanoparticles were adopted as a photothermal core, while PNIPAM was used as thermal-sensitive shell, and DOX was loaded into the space between these two layers [57]. Upon NIR irradiation, the magnetic Fe_3O_4 core converts light energy into heat, resulting in shrinkage of the PNIPAM shell followed by drug release. Notably, tumor inhibition rates increased from 40.3% to 65.2% for either treatment alone, to 91.5%. This nanocarrier surprisingly incorporated light-, heat-, magnetism-, and chemotherapy-based strategies in an active attempt to develop a multimodal combined treatment for cancer.

7.3.3.4 Ultrasonic-Induced Targeting

In addition to the above external stimuli, such as light, heat, and magnetic fields, ultrasound can also be used to target and control the release of drugs to tumors. Ultrasound can reduce the pressure in tumors and temporarily enhance the permeability or leakiness of tumor microvasculature to increase the extravasation of drugs into the targeted tumor cells [58]. Basavarajappa and colleagues found that both the duration and distance of the ultrasound treatment impacted the effects of ultrasound on microvascular permeability. Aryal et al. used ultrasound to increase drug penetration through the blood–brain barrier (BBB) and blood–tumor barrier [59] and demonstrated that multiple focused ultrasound treatments could significantly improve the therapeutic effects of a DOX-loaded carrier on rat glioma.

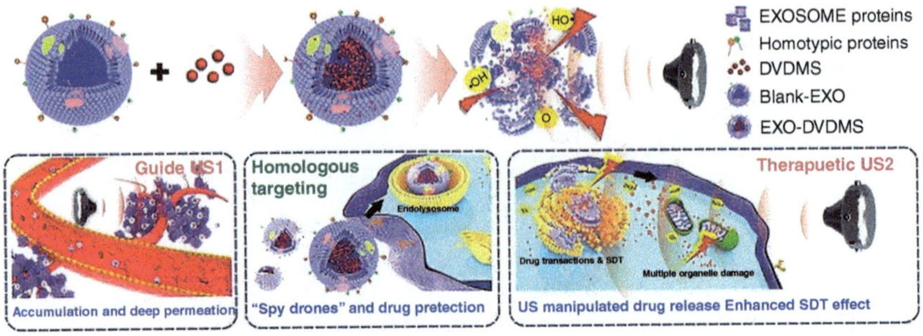

Figure 7.17 Illustration of endogenous nanosonosensitizers for focused US-augmented targeting delivery to recognize homotypic cancer cell, stimuli-responsive drug release, and enhanced SDT (US1: guided-ultrasound, US2: therapeutic-ultrasound). Source: Liu et al. [60]/ Reproduced with permission from Ivyspring International Publisher.

The strategy of focused ultrasound to direct nanosonosensitizers to specific tumor sites where they are activated to damage the surrounding tumor cells, or sonodynamic therapy (SDT), is considered a strong possible approach for cancer theranostics. For example, Liu et al. used sinoporphyrin sodium (DVDMS), a porphyrin sensitizer that has been studied for its purported therapeutic benefits and uses in imaging, to design a functionalized nanosonosensitizer (EXO-DVDMS) loaded into exosomes derived from homotypic tumor cells [60] (Figure 7.17). These highly stable exosomes showed striking specificity for localization to homotypic tumors and enabled controlled drug release in response to ultrasound. Activation of these particles induced production of reactive oxygen species (ROS), which in turn increased the effects of SDT.

7.4 Novel Biomimetic Delivery Strategies

Due to the high heterogeneity and complexity of tumors, the actual clinical efficacy of most anticancer formulations is substantially different from that expected based on preclinical investigation [61]. To overcome these problems attributable to tumor heterogeneity, more sophisticated and novel anticancer targeting strategies are needed. In recent years, some research groups have proposed anticancer strategies based on engineered biomimetic formulations. In these strategies, the structure and functions of proteins, cells, and even pathogens are integrated into the carrier design. Using the endogenous transport pathways involved in protein uptake, cell function, and bacterial infection, different agents, including drugs, antigens, and adjuvants, can be actively targeted to tumor tissues and even accurately delivered to subcellular structures. These strategies thus seek to overcome bottlenecks due to poor drug utilization or toxicity toward normal cells.

Using the fluidity of the cell membrane, the drug can be camouflaged and loaded by physical extrusion, electrostatic adsorption, and exosome secretion. The resulting cell membrane-modified NPs possess highly controlled biological functions.

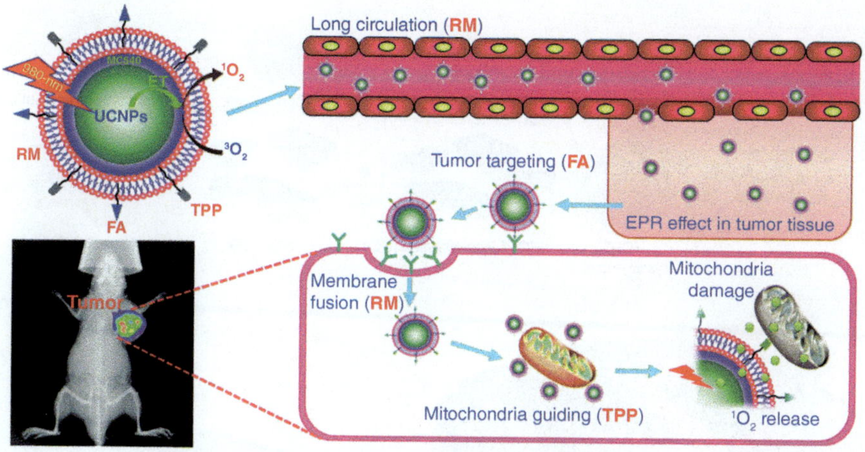

Figure 7.18 Schematic illustration of UCNP-based RM-coated dual-targeted biomimetic NPs (MC540: merocyanine 540, ET: energy transduction, FA: folic acid) Source: Ding et al. [62]/ Reproduced with permission from Royal Society of Chemistry.

The biomimic camouflage not only maintains the physicochemical properties of NPs but also protects NPs from being attacked by the immune system *in vivo* due to the proteins and polysaccharides on the cell membrane, allowing it to circulate for a long time in the blood and target tumors. The membrane emulsification technique could also be expanded and applied to the cell membrane coating and biomimetic carriers preparation, such as exosomes and other vesicles. Our group developed a red cell membrane (RM)-coated carrier to improve mitochondria targeting by functionalizing the upconversion NPs (UCNPs) with triphenyl phosphorus (TPP) [62] (Figure 7.18). In this platform, singlet oxygen ($^1O_2$) produced by the nanoparticles acts directly on the organelles responsible for powering cellular energy, efficiently killing cells.

To better mimic the tumor cells to realize active targeting, our group constructed one type of cancer cell membrane (CM)-coated HCPT nanocrystal using the soft template effect of serum albumin, which improved the solubility of chemotherapeutic drugs and facilitated homotypic targeted delivery using membrane proteins on the camouflaged nanocrystal [63] (Figure 7.19). We also used a cancer cell membrane to construct biomimetic antimonene nanoparticles (CmNPs) for high-performance photodynamic-photothermal theranostics [64]. Efficient enrichment of CmNPs at the tumor site, including improved stability and safety, was obtained via homologous targeting.

We also explored several leukocyte membrane-cloaking magnetic nanoclusters [65]. Macrophage phagocytosis was significantly inhibited due to the leukocyte membrane camouflage, which protected magnetosomes from being cleared rapidly *in vivo*. This system was better at programmed drug delivery, including a longer circulation time, better tumor accumulation, more tumor updates, MR imaging, and a positive intracellular fate. This produced potent therapeutic effects with few abnormalities.

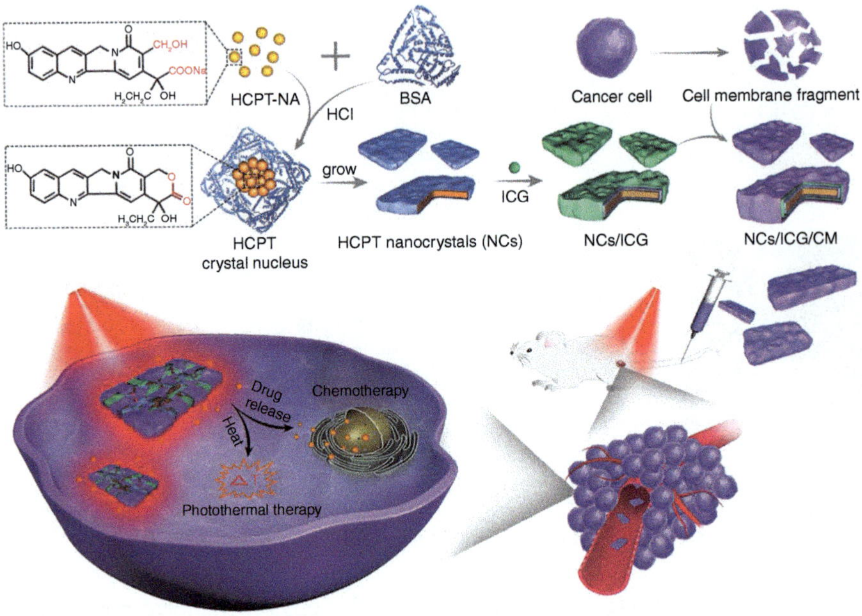

Figure 7.19 Schematic illustrating cancer cell membrane-coated HCPT nanocrystals sandwiched with photosensitizer for the orchestration of chemophotothermal combination therapy (BSA: bovine serum albumin, ICG: indocyanine green). Source: Zhang et al. [63]/ Reproduced with permission from John Wiley & Sons, Inc.

7.5 Summary

This chapter introduced various drug delivery systems for anticancer agents based on polymer nano/microspheres and the structure-activity relationship of the physical and chemical properties of antitumor carriers. It also summarized applications of nano/microspheres in tumor targeting and novel biomimetic delivery. By rationally designing size, charge, shape, and flexibility, nano/microspheres can enhance and improve the effectiveness of tumor treatments. For studies with such purpose, control of a single variable is of vital importance for ensuring quantitative research. The membrane emulsification technique developed by our group can help to achieve this goal. Using this technique, we were able to obtain particles with uniform size to ensure consistency in the structure–activity relationship between specific particle properties and effect on the target cell or physiological microenvironment, resulting in credible results in biological assessment. The uniform micro and nanoparticles prepared by our membrane emulsification technology will assist in the further development of studies on the structure-activity relationship. Furthermore, the membrane emulsification technique can be applied to the cell membrane coating and biomimetic carriers preparation, such as exosomes and other vesicles. Regarding anticancer drug delivery, integrating biological guidance, environmental response, and external stimulation (magnetic response) can all make the uniform carrier a biomimetic or intelligent drug delivery system and contribute to safe and accurate diagnoses and treatments.

Nevertheless, most of above results stopped in animal experiments. The main reasons are due to the complex delivery system construction and the use of materials not approved by the FDA, making it difficult to be approved for clinical trial. Therefore, using FDA-approved materials (e.g. PLGA) to design environment-sensitive drug delivery systems with specified size, morphology, deformability, etc. is an important strategy. In the near future, it is expected that the new strategies, platforms, and mechanisms will promote nano/microsphere delivery systems into clinical trials, and will bring microsphere products to the market.

References

1 Sung, H., Ferlay, J., Siegel, R.L. et al. (2021). Global cancer statistics 2020: GLOBOCAN estimates of incidence and mortality worldwide for 36 cancers in 185 countries. *CA: A Cancer Journal for Clinicians* 71 (3): 209–249.

2 Wei, W., Ma, G.H., Hu, G. et al. (2008). Preparation of hierarchical hollow $CaCO_3$ particles and the application as anticancer drug carrier. *Journal of the American Chemical Society* 130 (47): 15808–15810.

3 Rosenblum, D., Joshi, N., Tao, W. et al. (2018). Progress and challenges towards targeted delivery of cancer therapeutics. *Nature Communications* 9: 1410.

4 Mitragotri, S. and Lahann, J. (2009). Physical approaches to biomaterial design. *Nature Materials* 8 (1): 15–23.

5 Yang, Y.W., Nie, D., Liu, Y. et al. (2019). Advances in particle shape engineering for improved drug delivery. *Drug Discovery Today* 24 (2): 575–583.

6 Maeda, H., Wu, J., Sawa, T. et al. (2000). Tumor vascular permeability and the EPR effect in macromolecular therapeutics: a review. *Journal of Controlled Release* 65 (1–2): 271–284.

7 Kohane, D.S. (2007). Microparticles and nanoparticles for drug delivery. *Biotechnology and Bioengineering* 96 (2): 203–209.

8 Goldsmith, H.L. and Turitto, V.T. (1986). Rheological aspects of thrombosis and hemostasis - basic principles and applications - Icth-report - subcommittee on rheology of the international committee on thrombosis and hemostasis. *Thrombosis and Haemostasis* 55 (3): 415–435.

9 Kwiatkowska, K. and Sobota, A. (1999). Signaling pathways in phagocytosis. *BioEssays* 21 (5): 422–431.

10 Tabata, Y. and Ikada, Y. (1990). Phagocytosis of polymer microspheres by macrophages. *Advances in Polymer Science* 94: 107–141.

11 Rejman, J., Oberle, V., Zuhorn, I.S. et al. (2004). Size-dependent internalization of particles via the pathways of clathrin-and caveolae-mediated endocytosis. *Biochemical Journal* 377: 159–169.

12 May, R.C. and Machesky, L.M. (2001). Phagocytosis and the actin cytoskeleton. *Journal of Cell Science* 114 (6): 1061–1077.

13 Alexis, F., Pridgen, E., Molnar, L.K. et al. (2008). Factors affecting the clearance and biodistribution of polymeric nanoparticles. *Molecular Pharmaceutics* 5 (4): 505–515.

14 Ni, D.Z., Ding, H., Liu, S. et al. (2015). Superior intratumoral penetration of paclitaxel nanodots strengthens tumor restriction and metastasis prevention. *Small* 11 (21): 2518–2526.

15 Yue, Z.G., Wei, W., Lv, P.P. et al. (2011). Surface charge affects cellular uptake and intracellular trafficking of chitosan-based nanoparticles. *Biomacromolecules* 12 (7): 2440–2446.

16 Lv, P.P., Wei, W., Yue, H. et al. (2011). Porous quaternized chitosan nanoparticles containing paclitaxel nanocrystals improved therapeutic efficacy in non-small-cell lung cancer after oral administration. *Biomacromolecules* 12 (12): 4230–4239.

17 Wei, W., Lv, P.P., Chen, X.M. et al. (2013). Codelivery of mTERT siRNA and paclitaxel by chitosan-based nanoparticles promoted synergistic tumor suppression. *Biomaterials* 34 (15): 3912–3923.

18 Nel, A., Mädler, L., Velegol, D. et al. (2009). Understanding biophysicochemical interactions at the nano-bio interface. *Nature Materials* 8: 543–557.

19 Ma, Y.F., Lv, P.P., Yue, Z.G. et al. (2012). Chitosan-based intelligent charge-reversed system for high efficiency delivery of paclitaxel. *The Chinese Journal of Process Engineering* 12 (3): 460–465.

20 Walter, A. and Gutknecht, J. (1986). Permeability of small nonelectrolytes through lipid bilayer-membranes. *Journal of Membrane Biology* 90 (3): 207–217.

21 Li, Y., Chen, X., and Gu, N. (2008). Computational investigation of interaction between nanoparticles and membranes: hydrophobic/hydrophilic effect. *Journal of Physical Chemistry B* 112 (51): 16647–16653.

22 Yue, Z.G., You, Z.X., Yang, Q.Z. et al. (2013). Molecular structure matters: PEG-*b*-PLA nanoparticles with hydrophilicity and deformability demonstrate their advantages for high-performance delivery of anti-cancer drugs. *Journal of Materials Chemistry B* 1 (26): 3239–3247.

23 Champion, J.A. and Mitragotri, S. (2006). Role of target geometry in phagocytosis. *Proceedings of the National Academy of Sciences of the United States of America* 103 (13): 4930–4934.

24 Yue, Z.G., Wei, W., You, Z.X. et al. (2011). Iron oxide nanotubes for magnetically guided delivery and pH-activated release of insoluble anticancer drugs. *Advanced Functional Materials* 21 (18): 3446–3453.

25 Champion, J.A. and Mitragotri, S. (2009). Shape induced inhibition of phagocytosis of polymer particles. *Pharmaceutical Research* 26 (1): 244–249.

26 Gratton, S.E.A., Ropp, P.A., Pohlhaus, P.D. et al. (2008). The effect of particle design on cellular internalization pathways. *Proceedings of the National Academy of Sciences of the United States of America* 105 (33): 11613–11618.

27 Muro, S., Garnacho, C., Champion, J.A. et al. (2008). Control of endothelial targeting and intracellular delivery of therapeutic enzymes by modulating the size and shape of ICAM-1-targeted carriers. *Molecular Therapy* 16 (8): 1450–1458.

28 Chen, C.S., Mrksich, M., Huang, S. et al. (1997). Geometric control of cell life and death. *Science* 276 (5317): 1425–1428.

29 Yue, H., Wei, W., Yue, Z.G. et al. (2012). The role of the lateral dimension of graphene oxide in the regulation of cellular responses. *Biomaterials* 33: 4013–4021.

30 Chen, P.Y., Yue, H., Zhai, X.B. et al. (2019). Transport of a graphene nanosheet sandwiched inside cell membranes. *Science Advances* 5 (6): eaaw3192, 1–13.

31 Dalby, M.J., Riehle, M.O., Sutherland, D.S. et al. (2004). Use of nanotopography to study mechanotransduction in fibroblasts - methods and perspectives. *European Journal of Cell Biology* 83 (4): 159–169.

32 Verma, A., Uzun, O., Hu, Y.H. et al. (2008). Surface-structure-regulated cell-membrane penetration by monolayer-protected nanoparticles. *Nature Materials* 7: 588–595.

33 Discher, D.E. (2009). Shape effects of filaments versus spherical particles in flow and drug delivery. *Proceedings of the Asme Summer Bioengineering Conference 2008, Pts a and B,* Marco Island, Florida, USA (25–29 June 2008). USA: ASME.

34 Mitra, A.K., Agrahari, V., Mandal, A. et al. (2015). Novel delivery approaches for cancer therapeutics. *Journal of Controlled Release* 219: 248–268.

35 Wei, W., Yue, Z.G., Qu, J.B. et al. (2010). Galactosylated nanocrystallites of insoluble anticancer drug for liver-targeting therapy: an *in vitro* evaluation. *Nanomedicine* 5 (4): 589–596.

36 Shen, Z.Y., Wei, W., Tanaka, H. et al. (2011). A galactosamine-mediated drug delivery carrier for targeted liver cancer therapy. *Pharmacological Research* 64 (4): 410–419.

37 Lv, P.P., Ma, Y.F., Yu, R. et al. (2012). Targeted delivery of insoluble cargo (paclitaxel) by PEGylated chitosan nanoparticles grafted with Arg-Gly-Asp (RGD). *Molecular Pharmaceutics* 9 (6): 1736–1747.

38 Wang, C.L., Zhang, W., He, Y.J. et al. (2021). Ferritin-based targeted delivery of arsenic to diverse leukaemia types confers strong anti-leukaemia therapeutic effects. *Nature Nanotechnology* 16 (12): 1413–1423.

39 Ganta, S., Devalapally, H., Shahiwala, A. et al. (2008). A review of stimuli-responsive nanocarriers for drug and gene delivery. *Journal of Controlled Release* 126 (3): 187–204.

40 Qing, S., Lyu, C.L., Zhu, L. et al. (2020). Biomineralized bacterial outer membrane vesicles potentiate safe and efficient tumor microenvironment reprogramming for anticancer therapy. *Advanced Materials* 32 (47): 2002085.

41 Wang, C.L., Wang, X.J., Zhang, W. et al. (2022). Shielding ferritin with a biomineralized shell enables efficient modulation of tumor microenvironment and targeted delivery of diverse therapeutic agents. *Advacned Materials* 34 (5): 2107150.

42 Augoff, K., Hryniewicz-Jankowska, A., Tabola, R. et al. (2022). MMP9: a tough target for targeted therapy for cancer. *Cancers* 14 (7): 1847.

43 Hatakeyama, H., Akita, H., Kogure, K. et al. (2007). Development of a novel systemic gene delivery system for cancer therapy with a tumor-specific cleavable PEG-lipid. *Gene Therapy* 14 (1): 68–77.

44 Iaccarino, G., Profeta, M., Vecchione, R. et al. (2019). Matrix metalloproteinase-cleavable nanocapsules for tumor-activated drug release. *Acta Biomaterialia* 89: 265–278.

45 Yue, Z.G., Lv, P.P., Yue, H. et al. (2013). Inducible graphene oxide probe for high-specific tumor diagnosis. *Chemical Communications* 49 (37): 3902–3904.

46 Harris, A.L. (2002). Hypoxia - a key regulatory factor in tumour growth. *Nature Reviews Cancer* 2 (1): 38–47.

47 Lyu, C.L., Lu, G.H., Bao, W.E. et al. (2020). Engineering magnetosomes with chimeric membrane and hyaluronidase for efficient delivery of HIF-1 siRNA into deep hypoxic tumors. *Chemical Engineering Journal* 398: 125453.

48 He, Z.M., Huang, X.L., Wang, C. et al. (2019). A catalase-like metal-organic framework nanohybrid for O_2-evolving synergistic chemoradiotherapy. *Angewandte Chemie International Edition* 58 (26): 8752–8756.

49 Fu, Q., Lv, P.P., Chen, Z.K. et al. (2015). Programmed co-delivery of paclitaxel and doxorubicin boosted by camouflaging with erythrocyte membrane. *Nanoscale* 7 (9): 4020–4030.

50 Espinosa, A., Reguera, J., Curcio, A. et al. (2020). Janus magnetic-plasmonic nanoparticles for magnetically guided and thermally activated cancer therapy. *Small* 16 (11): 1904960.

51 Barhoumi, A., Liu, Q., and Kohane, D.S. (2015). Ultraviolet light-mediated drug delivery: principles, applications, and challenges. *Journal of Controlled Release* 219: 31–42.

52 El Founi, M., Soliman, S.M.A., Vanderesse, R. et al. (2018). Light-sensitive dextran-covered PNBA nanoparticles as triggered drug delivery systems: formulation, characteristics and cytotoxicity. *Journal of Colloid and Interface Science* 514: 289–298.

53 Yao, W.H., Liu, C.Y., Wang, N. et al. (2021). An MRI-guided targeting dual-responsive drug delivery system for liver cancer therapy. *Journal of Colloid and Interface Science* 603: 783–798.

54 Kuang, G.Z., Lu, H.T., He, S.S. et al. (2021). Near-infrared light-triggered polyprodrug/siRNA loaded upconversion nanoparticles for multi-modality imaging and synergistic cancer therapy. *Advanced Healthcare Materials* 10 (20): 2100938.

55 Bao, W.E., Liu, M., Meng, J.Q. et al. (2021). MOFs-based nanoagent enables dual mitochondrial damage in synergistic antitumor therapy via oxidative stress and calcium overload. *Nature Communications* 12 (1): 6399.

56 Shen, Z., Wei, W., Zhao, Y.J. et al. (2008). Thermosensitive polymer-conjugated albumin nanospheres as thermal targeting anti-cancer drug carrier. *European Journal of Pharmaceutical Sciences* 35 (4): 271–282.

57 Shen, S., Ding, B., Zhang, S.C. et al. (2017). Near-infrared light-responsive nanoparticles with thermosensitivse yolk-shell structure for multimodal imaging and chemo-photothermal therapy of tumor. *Nanomedicine: Nanotechnology, Biology and Medicine* 13 (5): 1607–1616.

58 Basavarajappa, L., Rijal, G., and Hoyt, K. (2021). Multifocused ultrasound therapy for controlled microvascular permeabilization and improved drug delivery.

IEEE Transactions on Ultrasonics Ferroelectrics and Frequency Control 68 (4): 961–968.

59 Aryal, M., Vykhodtseva, N., Zhang, Y.Z. et al. (2013). Multiple treatments with liposomal doxorubicin and ultrasound-induced disruption of blood-tumor and blood-brain barriers improve outcomes in a rat glioma model. *Journal of Controlled Release* 169 (1–2): 103–111.

60 Liu, Y.C., Bai, L.M., Guo, K.L. et al. (2019). Focused ultrasound-augmented targeting delivery of nanosonosensitizers from homogenous exosomes for enhanced sonodynamic cancer therapy. *Theranostics* 9 (18): 5261–5281.

61 Shi, J.J., Kantoff, P.W., Wooster, R. et al. (2017). Cancer nanomedicine: progress, challenges and opportunities. *Nature Reviews Cancer* 17 (1): 20–37.

62 Ding, H., Lv, Y.L., Ni, D.Z. et al. (2015). Erythrocyte membrane-coated NIR-triggered biomimetic nanovectors with programmed delivery for photodynamic therapy of cancer. *Nanoscale* 7 (21): 9806–9815.

63 Zhang, L.J., Zhang, X., Lu, G.H. et al. (2019). Cell membrane camouflaged hydrophobic drug nanoflake sandwiched with photosensitizer for orchestration of chemo-photothermal combination therapy. *Small* 15 (20): 1805544.

64 Lu, G.H., Lv, C.L., Bao, W.E. et al. (2019). Antimonene with two-orders-of-magnitude improved stability for high-performance cancer theranostics. *Chemical Science* 10 (18): 4847–4853.

65 Zhang, F., Li, F., Lu, G.H. et al. (2019). Engineering magnetosomes for ferroptosis/immunomodulation synergism in cancer. *ACS Nano* 13 (5): 5662–5673.

8

Applications of Uniform Particles in Vaccine Formulations

8.1 Introduction

Vaccination is one of the most successful medical interventions in human history [1]. Nowadays, subunit vaccines (recombinant proteins, peptides, and nuclear acids), which only extract the key antigenic components of the pathogens, are becoming the mainstream in the modern vaccine industry [2]. Compared with the attenuated or inactivated pathogens, these subunits harbored higher safety profiles, smaller molecular weight, and simpler structures for rapid development during the pandemics, but, at the expense of immunogenicity due to the loss of pathogenic components [3]. To compensate for this, there is an increasing demand for the efficient adjuvant to enhance the immune responses, such as aluminum salts (alum), oil-in-water emulsions (MF59), and immunostimulatory molecule, MPLA (AS04) [4]. As shown in Figure 8.1, vaccine formulations were evolved from the fluidic mixture of antigens and adjuvants (alum) to the particulate vaccines (AS01). Additionally, to better provoke the responses, it is becoming increasingly important to strengthen the delivery efficacy of the antigenic and immunogenic components [5]. Particles, such as liposomes, polymeric particles, and protein aggregates, were engineered with multiscale structures and tunable physicochemical properties to load and deliver the vaccine components [6]. As viruses or bacteria are the perfect delivery vehicles by nature selection, the infection process may offer insightful implications for the particle design [7]. Recent literature demonstrated the immense potential of synthetic particles to enhance the lymph-node (LN) accumulation, antigen-presenting-cell internalization, and the intracellular transfer of the delivered antigens to dictate the immune potency and tendency [8, 9]. Additionally, by mimicking the pathogenic traits, such as lipid-covered surfaces, dense array of antigens, and virus-like morphologies, these particles can trigger the inherent danger signals and the immune recognitions and induce the enhanced antibody secretions and T cell responses [10].

Despite the progress, vaccines still frequently suffered from the limitations, such as unpredictable efficacy, deficient immunogenicity, and limited immune maintenance [11]. To address this, understanding the designing principles of the particulate vaccines are the main access to the efficient vaccine formulations to enhance immunogenicity and delivery efficacy. Under these circumstances, uniform

Novel Membrane Emulsification: Principles, Preparation, Processes, and Bioapplications,
First Edition. Guanghui Ma.

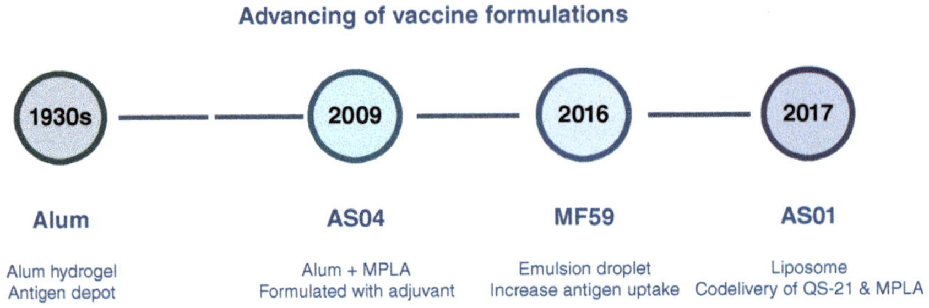

Figure 8.1 Evolution of vaccine adjuvant and delivery system.

particles offer almost homogeneous structures and well-defined properties, which can serve as the tailorable platforms to study the bioeffects of vaccine adjuvant and delivery systems. Additionally, the uniformity of the particulate adjuvant harbors major implications for the repeatability of the bioeffects, which give rise to the ready-to-use formulation strategies for the emerging pandemics with the enhanced immune responses and batch-to-batch stability [12].

To better illustrate the immense potential and the underlying mechanisms, this chapter discussed the applications of uniform particles in vaccine formulation (adjuvant and delivery systems). We centered on the assembly of various antigenic components for higher loading efficiency, the perspectives to tailor the structures and physicochemical properties of the clinical-approved or translatable materials, as well as the strategies to strengthen the multiscale delivery of vaccines.

8.2 Adjuvant and Delivery System: Assembling the Vaccine Components

Vaccines can stimulate antigen-specific antibody secretion and cellular immune responses to circumvent the pandemic infections or exert the cytotoxic effect against the tumor cells or chronic diseases. In these scenarios, modern vaccines are expected: (i) integrating with target molecules, namely aptamer, antibody, or peptide, to accumulate within the LN, tumor, or the specific cells, such as professional antigen-presenting cells (APCs), B cell, or T cell; (ii) enhancing cellular uptake to boost the intracellular concentration of antigens for the increased the immune responses; and (iii) formulating with molecular adjuvant, such as Toll-like receptor (TLR) agonists, cGAS-STING activators, or other pathogen-associated molecular pattern (PAMP), to stimulate the innate immune responses for the enhanced vaccinations [13]. Thus, multicomponent vaccines were developed to codeliver antigens, molecular adjuvants, and target molecules. Due to the free diffusion and easy degradation, fluidic mixtures can hardly exert potent immune activations [6]. Uniform particles, by contrast, can offer hierarchical structures and tailorable properties. Here, we poise the idea of "vaccine chassis" based on the uniform particles, which may serve as a "plug and play" platform to modularize the assembly of target molecules, adjuvants, or antigens, offering standardized tools

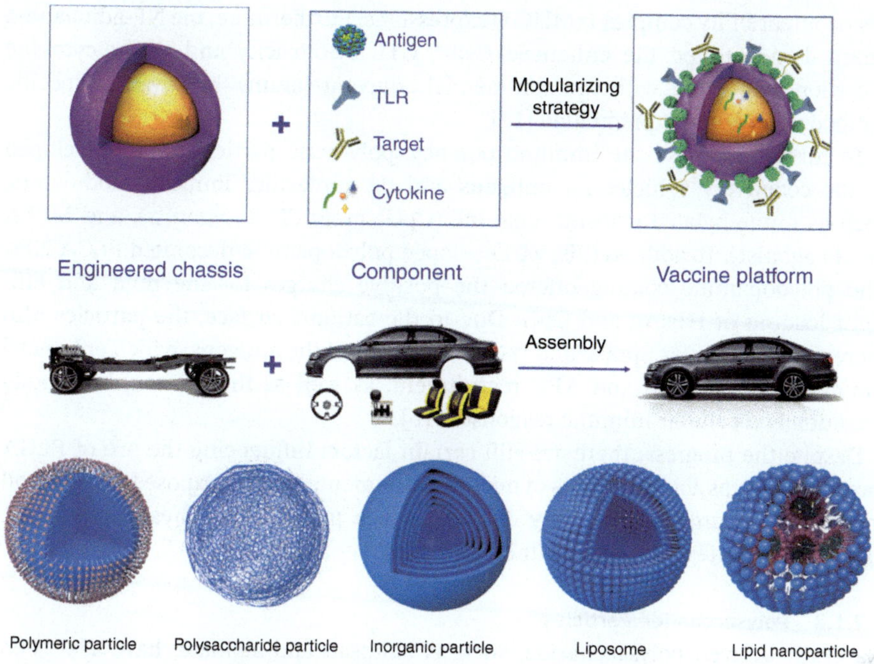

Figure 8.2 "Plug and play" platforms for vaccine assembly and delivery.

to unravel the underlying mechanisms for rational design of advanced vaccines (Figure 8.2). Additionally, in response to the emerging pandemic, the synthetic vaccines can be strategically developed in a streamlined way to maximize the delivery efficiency.

In this section, from a translational medicine perspective, we focused on particulate platforms based on the clinically approved or translatable materials. In addition, this section also discussed modularizing strategies to control the surface charges, hydrophobicity, and chemical properties of particles for rational and efficient assembly of vaccines.

8.2.1 Particulate Vaccine Platforms

8.2.1.1 Polymeric Particles

Poly(lactide acid) (PLA) and poly(DL-lactide-*co*-glycolide acid) (PLGA) can be metabolized to their constituent monomers in aqueous media, offering the excellent biodegradation and safety profile, which were clinically approved as injectable polymeric particle. The controllable sizes and properties favored for the encapsulation or adsorption of antigens, chemical linking with the targeting molecules, and the efficient incorporations of the molecular adjuvants, especially the hydrophobic ones [14]. Our group, as one of the pioneers in this area, prepared a series of PLA-based nanoparticles (NPs) and employed as adjuvant for hepatitis B surface antigen (HBsAg). We found that NPs facilitated cellular uptake and cytosolic delivery of the antigens, thus boosted the costimulators (CD80/86), as well as major histocompatibility complex I (MHC I) expressions. Furthermore, the NP-adjuvanted group demonstrated the enhanced $CD8^+$ CTL cytotoxicity and IFN-γ cytokine secretion, compared with the commercial adjuvant (alum). Meanwhile, specific antibody titer is also upregulated [15].

To further increase the immune potency, polymeric particles were developed as the codelivery vehicles for antigens and the molecular immune-modulators, such as unmethylated cytosine–guanine (CpG) motif (TLR9 agonist), and MPLA (TLR4 agonist). To address this, we developed polydopamine-decorated PLGA NPs. The polydopamine coating offered the positive charges for the mild and efficient loading of HBsAg and CpG. Due to the cationic surface, the particles also increased the cellular uptake and lysosome escape of the antigens, which enhanced the cytokine secretion and APC recruitment, as well as the increased antibody secretion and cellular immune responses [16].

Despite the progress, there are still certain factors influencing the use of PLGA particles, such as acidic features of microenvironment, proteins exposed to water–oil interface, and protein instability during particle preparation. These factors may affect protein integrity and immunogenicity [17].

8.2.1.2 Polysaccharide Particles

Naturally derived polysaccharide, such as chitosan and alginate, have also been employed in the clinical trials. As for alginate polysaccharide, it is a kind of negatively charged polysaccharides, which is extracted from marine brown algae. The water-soluble linear polysaccharide is made up of irregular blocks of β-D-mannuronic acid (M) and 1–4 linked α-L-guluronic residues (G). Investigation of alginate-based particles has several advantages such as biocompatibility, mucoadhesive properties, nontoxicity, hydrophilicity, bioavailability, and low-cost [18]. In case of chitosan, it is derived by partial deacetylation of chitin obtained from the shells of crustaceans or the mycelium of fungi, which is a natural cationic polysaccharide composed of glucosamine (*(1–4)-linked 2-amino-2-deoxy-D-glucose) and *N*-acetylglucosamine (2-acetamido-2-deoxy-D-glucose) [19]. Chitosan-based particles were developed with well-defined properties, including biocompatibility and the capacity to stimulate robust Th1 responses [20].

In a previous work, we explored the immunogenic effect of the free amine group ($-NH_2$) on chitosan particles. Typically, chitosan particles were typically solidified with glutaraldehyde, which linked the surface amine and hampered the binding with the complement component. To address this, we prepared the chitosan/alginate hybrid particles with layer-by-layer process. Chitosan particles were first solidified with glutaraldehyde, subsequently coated with alginate sodium with electrostatic interactions, and afterward adsorbed another layer of chitosan, which left the free amine group on the particle surface ($CS-NH_2$). Compared with the traditional chitosan particles, $CS-NH_2$ evidently triggered the complement activations, which thus induced an up-regulated antigen-specific IgG titers, and the increased production of IL-4 and IFN-γ among the splenocytes [21].

In addition, we developed uniform-sized chitosan hydrogel microparticle (Gel MP) by rapid MET and employed it as vaccine adjuvant. First, a mixture of quaternized chitosan and sodium glycerophosphate triggered the ionic cross-linking of the particles, which were granted with the increased thermal-gelation and pH-sensitive properties. By exploiting this, Gel MP achieved high lysosomal escape of the antigens (close to 80%), allowing for the cytoplasm presentation of antigens, and the increased MHC I engagement. During the acidic microenvironment, Gel MP might swell and disintegrate, resulting in osmotic imbalance and lysosome rupture. This subsequently led to the lysosomal escape of the antigens. As a result, vaccination with Gel MP evidently increased the antigen-specific antibodies and cytokine production [22].

Furthermore, chitosan-based particles have immense potential as mucosal adjuvants. With positively charged surface and gel-like texture, they can demonstrate mucoadhesive properties, which may prolong the antigen retention time. Simultaneously, chitosan particles can also open the tight junctions between the epithelial cells so antigen can be transported more quickly to the site of immune initiation and execution [23]. Parra et al. developed a norovirus virus-like particle (VLP) vaccine, which was formulated with chitosan (ChiSys™), norovirus VLP antigen, and MPLA and spray-dried as microparticle (NV-VLP). In a Phase I clinical trial, two randomized, double-blind, controlled studies of the NV-VLP vaccine were conducted. A group of healthy volunteers (18–49 years old) was randomly assigned to receive two intranasal administrations of NV-VLP or to act as controls (over an interval of 21 days). After vaccination, the symptoms were recorded for 7 days, and the safety evaluation was followed up for 180 days. Serum and peripheral blood mononuclear cells were also sampled. According to the results, nasal stuffiness and discharge, and sneezing were the most common symptoms. There were no vaccine-related serious adverse effects (SAEs). 100 μg dose induced the greatest immune response, with IgG and IgA titers increasing by 4.8 and 9.1-fold, respectively, compared with the placebo group. All subjects, who received the vaccine dose of 50 or 100 μg, produced IgA antibody-secreting cells (ASCs), with surface molecules that are associated with homing to mucosal and peripheral lymphoid tissues ($CD19^+$, $CD27^+$, $CD62L^-$, integrin $\alpha4/\beta7^+$, $CD19^+$, $CD27^+$, $CD62L^+$, and integrin $\alpha4/\beta7^+$). The results indicated that intranasal administration of NV-VLP may evoke IgA secretion by intestinal mucosal tissues (the entry site of norovirus) [24].

Although chitosan and alginate have been investigated for many years as a basic material for vaccine adjuvants, it is not yet present in any marketed vaccine formulation. The main obstacles may be a lack of understanding of its metabolism, difficulty in quality control, and concerns regarding safety issues.

8.2.1.3 Inorganic Particles

The immunomodulatory effects of inorganic particles, such as aluminum compounds, calcium phosphates, mesoporous silica, and zinc compounds, are receiving increasing attention [25]. Compared with the polymeric particles, it is easier to regulate the morphologies of the inorganic particles [26]. Combining inorganic particles with antigens can prompt specific immune responses and encourage cross-presentation of antigens, which is critical for generating $CD8^+$ T cells [27]. Due to their inherent immuno-regulating properties, the composition of inorganic nanoadjuvants is essential in achieving the desired immune responses. Inorganic NPs have inherent immune-regulating properties that make their composition an important factor in achieving the desired immune responses [28].

Aluminum-salt-based formulation was the first and only vaccine adjuvant in most countries. As a result, the traditional alum adjuvant has a strong effect on the humoral immune response, but it lacks the ability to stimulate the cellular immune response [29]. It is necessary to transform the traditional aluminum adjuvant into new formulation. c-AlOOH NPs and aluminum oxide (α-Al_2O_3) NPs have emerged as new aluminum-adjuvant platforms for the advanced immunotherapy, which showed significantly improved antigen cross-presentation and serum antibody titers in the melanoma tumor model (B16F10-OVA) [30].

In addition, CaXs biomaterials have been widely used in the biomedical field due to their excellent biocompatibility and biodegradability [31]. Calcium phosphates, calcium carbonates, calcium silicate, and calcium fluoride are all CaX-based materials [32]. As opposed to other inorganic biomaterials like silica-, carbon-, and gold-based biomaterials, CaXs biomaterials can dissolve into nontoxic ions and participate in the normal metabolism of organisms. As such, they may provide safer clinical solutions for disease theranostic [33].

8.2.1.4 Liposome

Similar to bilayer membranes, liposomes can interact readily with cellular membranes. Liposomes have been deemed a good candidate for vaccine delivery systems due to their safety, biocompatibility, size controllability, capability for easy functionalization, and effective delivery of antigens. In recent years, they have been modified and tailored to advance the delivery efficiency of the vaccines [34].

The immunostimulatory properties of liposomes enhanced when they were loaded with immunostimulatory molecules, such as agonists of TLRs, oligonucleotides, and bioactive polysaccharides [35]. Furthermore, combination with immunosuppressive systems in tumor microenvironment enhanced antitumor immunity. Zhou et al. combined cationic liposomes with CpG (CpG DNA lipoplex) to prevent pulmonary metastasis following intranasal administration. Compared to naked CpG, CpG DNA lipoplex significantly increased the inhibition of the tumor cell proliferation. Providing mice with CpG DNA lipoplex prior to intranasal administration evidently induced tumor regression and prolonged survival time [36].

8.2.1.5 Lipid Nanoparticle

Lipid nanoparticles (LNP) are regarded as the best-in-class delivery systems for mRNA vaccines [37]. As the complex of nuclear acid and lipids, LNP is composed by the ionizable lipids, which offered the pH-responsible charge reversal to encapsulate and release the mRNA antigens; PEGylated lipids, such as DMG-PEG, DSPE-PEG; supporting lipids (namely, DPPC, and DOPE); and cholesterol to modulate the stability of the LNPs [38]. The well-defined formulation can offer high encapsulation efficiency, increase the cellular uptake via the lipid/membrane interactions, and cause the burst release of the loaded mRNA by the charge-switch within the acidic lysosomes [39]. For instance, Prof. Qin and colleagues developed a lipid nanoparticle-encapsulated mRNA (mRNA-LNP) encoding the receptor binding domain (RBD) of SARS-CoV-2 (called ARCoV). Two doses of ARCoV immunization in mice conferred complete protection against the challenge of a SARS-CoV-2 mouse-adapted strain [40].

Additionally, researchers have focused on optimizing ionizable lipids to control tissue targeting, encapsulation, and delivery efficiency. Cornebise et al. found that the hydrogen bonding between hydroxyl headgroup and the mRNA plays a key role in improving LNP *in vivo* expression. Based on this finding they identified a set of ionizable lipids with squaramide as headgroup. These ionizable lipids can promote additional interactions with mRNA beyond Coulombic attraction, demonstrating higher mRNA encapsulation efficiency, and ultimately enabling the enhanced *in vivo* performance [41].

Additionally, high-end techniques, such as deep learning, have also been employed to screen the ideal lipids. Anderson et al. designed an isocyanide-mediated one-step three-component reaction to simultaneously couple primary or secondary amines, ketones and isocyanides or isocyanide derivatives. Using this system, they synthesized and evaluated the vaccine potential of over one thousand lipid formulations. Their study provided evidence of lipid adjuvant-assisted mRNA vaccination and showed that lipids with cyclic amino head groups could activate the immune system through an mRNA-independent STING pathway [42].

8.2.2 Modularizing Strategies for Vaccine Delivery System

As aforementioned, particulate platforms can offer tailorable morphology and properties to load various vaccine components. To some extent, it is the modularizing strategy that may dictate the immune recognition, release kinetics, and delivery efficiency of the vaccines [43]. For instance, MPLA can activate the TLR4 receptor on the cellular membrane. If MPLA is encapsulated in the hydrophobic core, instead of linking on the shell of the polymeric particles, it may hinder the surface recognition of the TLR4 signaling, impeding the MPLA-mediated immune activations. To address this, the up-to-date modularizing strategies were discussed in this section (Figure 8.3).

8.2.2.1 Encapsulation

The encapsulation within the particles favors the controlled release of the vaccine components, as well as the stability against varied pH or enzyme conditions or the tissue clearance [44]. Furthermore, high antigen concentration within the injection site can also form the antigen depot, leading to the recruitment of APCs and the increased inflammatory cytokine levels for the enhanced innate immune activations [45]. But, for the hydrophilic components, such as antigens or antibodies, encapsulated by PLA/PLGA hydrophobic materials, the conventional encapsulation involves high concentration of organic solvent or high shear stress. During the emulsification and subsequent solidification of the particles, it may hamper the natural conformation and bioactivities of the encapsulated component [46].

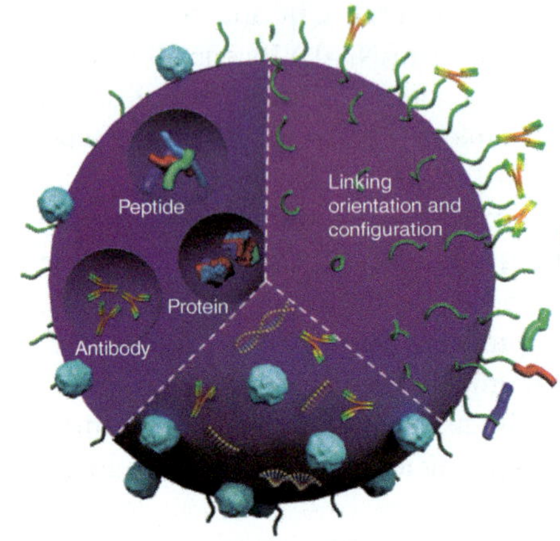

Figure 8.3 Modularizing strategies for the assembly of particulate vaccine.

To address this, membrane emulsification techniques were developed to overcome the high shear force in the conventional encapsulation methods, such as spray drying, mechanical stirring, homogenization, and ultrasonication. Through passing through the porous membrane, PLA/PLGA materials-based or hydrophilic chitosan-based particles were formed in a facile and energy-efficient way, with high loading efficiency of the encapsulated antigens [47].

8.2.2.2 Adsorption

Particulate delivery systems are enhanced by a variety of complex, synergistic, and highly functional vaccine-based multicomponent assemblies [48]. Physical adsorption of moieties onto particle surfaces does not require complex chemical reactions [49]. The adsorption process relies on the weak interaction between distinct individual building blocks, such as hydrophobic interactions, electrostatic interactions, hydrogen-bonding, and π–π stacking [50]. This may lead to the formation of a complex that then assembles into the required architecture. The adsorption strategies not only confer the landing spot for higher loading efficiency of the antigenic components but also increase the immune recognitions of the target molecules and boost the immune responses via preventing the clearance of the vaccines. For example, neoantigen-based vaccines have gained clinical insights to induce T cells that target heterogeneous clones of tumors. While neoantigen peptides are rapidly cleared and low immunogenic, this limits their ability to initiate strong T cell responses by APCs [51]. To overcome these challenges, macro and nanoengineered approaches were developed, but many of them require chemical modification or physical emulsification of the peptides, potentially changing their presentation properties. Additionally, as neoantigen vaccines typically require many peptides, changing each one individually is cumbersome for clinical translation, and is likely to result in high batch-to-batch variability. David Mooney's group proposed a simple strategy for enhancing antigen immunogenicity using polyethylenimine

(PEI) and a mesoporous silica microrod (MSR) vaccine. MSRs (MSR-PEI) were prepared by simply mixing MSR with PEI solution for 15 minutes. An antigen pool was then directly adsorbed onto the MSR-PEI particles. As a result, the MSR vaccine can be injected using standard needles. They were demonstrated to concentrate and activate large numbers of host APCs and induce more potent humoral responses and prophylactic antitumor response than conventional vaccine formulations [52].

To start with, electrostatic interaction is widely used for vaccine loading. The polyionic-composite NPs generally require a combination of positive and negative molecules in order to achieve a first-order assembled structure, and then further aggregate to reach a nanodimensional state. A major advantage is that the preparation process is simple and the multistage assembly is parallel. Guan et al. developed a nanovaccine that induces antigen-specific antitumor immune responses. Electrostatic binding was employed to codeliver ovalbumin (OVA) and unmethylated CpG using polycationic PEI. Positively charged PEI enhanced the uptake of PEI/CpG/OVA nanovaccine in dendritic cells (DCs) and facilitated the endosomal escape of the nanovaccine for antigen delivery to the cytoplasm [53]. Although efficient, cationic polymers, such as PEI, may also cause side effects *in vivo*, and induce complex manufacture and quality control procedures during the clinical translation. To address this, our group developed a novel nanovaccine without the addition of extra polymeric components. Through the addition of glutathione, the breakage of the disulfide bonds within the antigens OVA was generated to reverse the molecular charge from negative to positive. Afterward, the positive-charge antigen and negatively charged CpG were assembled and formed NPs via the electrostatic interactions. The resulting particles were with the smaller sizes (less than 80 nm), negative charges, and good stability, which facilitated the LN accumulation after subcutaneous injection. As a result, the draining LN was significantly activated with increased cellular immune responses [54].

8.2.2.3 Conjugation

Chemical conjugation is effective in the assembly of the moieties that cannot be introduced via adsorption [55]. Additionally, the chemical bond is broken once encountering the variations in the pH, light, thermal, or redox condition, which shows the immense potential in the control release of the vaccine [56]. By optimizing the linking site, the well-defined conjugation can regulate the orientation and configurations of the moieties, such as targeting aptamer, antibody, or antigen.

By utilizing the reactive groups of naturally occurring reactive groups exposed on their outer or inner surfaces, functional moieties can be chemically conjugated. The most used cross-linking agents in bioconjugation of vaccine chassis include maleimides for cysteine thiols, *N*-hydroxy-succinimides (NHS) for lysine amines, and carbodiimides for glutamate or aspartate carboxylates. These agents have been extensively used in bioconjugate vaccines, antibodies, fluorescent dyes, folic acid, cell-targeting and cell-penetrating peptides, and small-molecule drugs [57]. There may be limitations to this strategy. Some conjugation processes require highly abrasive chemical reagents (e.g. organic solvents) and high temperatures, which may compromise the integrity of the particles.

Click chemistry has recently been shown to be a viable functionalization technique. As one of the most useful synthetic tools with possibilities for a broad range of applications, click chemistry describes a family of modular, efficient, versatile, and reliable reactions [58]. Despite being the most widely adopted click reaction in the family, copper(I)-catalyzed alkyne-azide cycloaddition is cytotoxic, and therefore, not suitable for some applications, including bioconjugation. In order to develop copper-free click reactions, scientists have been exploring strain-promoted alkyne-azide cycloadditions, which are the most popular example so far. But, the safety profile of the labeling reagent, such as dibenzocyclooctyne (DBCO), is still uncertain, hampering the clinical applications of this technique [59].

8.3 Physicochemical Traits for the Enhanced Vaccination

It is the multivalent structures and physicochemical properties of particulate vaccines (e.g. particulate, small size, softness, dynamic restructuring, and highly repetitive surface structures) that determine the level of immune response. Thus, there is a broad interest in the structural effect of the particles in vaccine delivery. The uniformity of particles may facilitate the repeatability of the bioeffects, which may shed light on the rational design of particulate vaccines, which can aid the enhanced vaccine distribution, APC internalization and presentation, and the enhanced immune response. Our focus is on self-adjuvant vaccine delivery systems using particles instead of pathogen- or cell-based formulation, although this is also an important part of biomimetic vaccines and we refer the interested reader to other excellent reviews [60–64]. For biomaterial engineering and vaccine delivery, the structure-immunogenicity effect was elaborated in terms of size, hydrophobicity, surface pattern, antigen display, charge, shape, and deformability.

8.3.1 Size

When antigen-loaded particles are encountered by APCs, the particle size is the primary factor that determines the fate of antigen. With the aid of membrane emulsification technique, particles can be uniformly prepared, which may guarantee the reproducibility and bioeffects of the particulate adjuvants. In addition, the uniform size may also provide increased implications to illustrate the underlying mechanisms of the particles' adjuvanticity.

An optimal goal of antigen delivery is to increase intracellular concentration. Previously, we found that APCs tended to uptake smaller particles [15, 65]. Nevertheless, smaller sizes do not assure a robust immune response. In addition, small particles, despite being prone to endocytosis, tend to have low antigen loads, which may adversely affect the concentration of antigens within the APCs. Additionally, particles with larger particle sizes would provide stable landing sites for microtubules and initiate predominantly receptor-mediated phagocytosis, which contributes to enhanced cytokine secretion, migration ability, and immune activation [66–68]. Based on the microporous membrane emulsification technology, we prepared multisized chitosan particles with narrow size distributions. Afterward, the intracellular fate was proved. Due to their larger surface areas, nanoparticles were more suitable for antigen adsorption and transport than microparticles, as the cells were more readily able to take up smaller particles. Microparticles (~1.9 um) on the other hand, had a larger volume and were more suitable for encapsulating antigens. The mechanism of endocytosis and intracellular transport demonstrated that NPs were taken up by macrophage-mediated endocytosis and clathrin pathway. Microtubule movements, phagocytosis dependent on actin, and cytoskeletal rearrangements are other pathways that macrophages use to take up particles, especially microparticles [69].

Additionally, particle size processes a significant effect on the LN accumulation of delivered vaccines. The interstitial fluid drainage from the injection sites delivered antigen to LNs efficiently by delivering nanoscale particles, with size around 10–100 nm [70, 71]. Using poly(γ-glutamic acid)-based vaccine nanoparticles (SVNPs), Kim et al. recently created a series of aminated NPs with sizes between 20 and 70 nm. The model antigen OVA and the TLR 3 agonist (poly(I : C)) were electro-adsorbable with the positive charges. These particles rapidly accumulated in the LNs after subcutaneous administration, causing them to retain antigens for 168 hours. As a result, SVNPs induced higher levels of tumor necrosis factor-α (TNF-α) and IFN-γ, significantly reducing tumor growth and improving overall survival of the tumor-bearing mice [72].

However, nanosized NPs tend to drain into the LN rapidly, while also being removed quickly by the body [73]. Microparticles, on the other hand, tend to create an antigen depot at the injection site, causing APC engagement and recruitment. As a result of clathrin- and caveolin-mediated endocytosis or free diffusion, APCs interact with smaller particles for cellular internalization [74]. APCs typically secreted higher amounts of cytokines during phagocytosis, as larger particles trigger receptor-mediated phagocytosis [75, 76]. Moreover, microparticles were less likely to be exocytosed, despite consuming more energy during the phagocytosis, contributing to some degree to the internalization of the delivered antigens. Toward this end, future efforts were expected to develop a size-tunable particulate vaccine [76–78].

8.3.2 Charge

Due to the negative charges of the cellular membrane, positive-charged particles were more likely to transduce across the membrane and gain access to the cellular interior [79]. To achieve trans-membrane delivery efficiently, positive-charged particles could exploit the electrostatic adsorption between the cationic surface and the anionic cell membranes [80]. The engineered particles may also have implications on the antigen processing within the APCs [81, 82]. In a study, Han et al. engineered PLGA/protamine NPs with different charges. Positively charged particles had a larger area of contact with lysosomes than negative-charged ones, and their expression of costimulatory molecules, such as CD80, CD86, and CD83, were significantly increased indicating enhanced cross-presentation of antigen [83].

Within the lysosome, antigen-conjugating cationic particles were protonated after being captured. In response to such protonation effects, a rapid influx of hydrogen ions and chloride ions will occur from the cytoplasm in a short time, which led to the swelling and rupture of lysosomes. Antigens were released from the lysosome and subsequently interacted with the endoplasmic reticulum and intracellular proteasome in the cytoplasm, binding to MHC I and triggering cellular immunity. Xu et al. developed PEI-OVA NPs in a recent study. By chemical modifications with fluoroalkyl chains, the delivery systems became hydrophobic and lipophilic, allowing NPs to diffuse through lipid membranes and escape from endosomes. As another result of the positive charge of PEI, the proton-sponge effect induced cytosolic delivery of antigen, evidently increasing the prophylactic and therapeutic effects of the vaccines [84]. Using the zwitterionic lipid distearoyl phosphoethanolamine–polycarboxybetaine (DSPE-PCB), Qiao et al. prepared a mannosylated zwitterionic-based cationic liposome (man-ZCL) for the delivery of a DNA vaccine. The modified surface of the particles allowed them to target DC cells and increase the internalization of antigens. Later, after entering the lysosome, the man-ZCl molecules were rapidly protonated with positive charges, allowing them to destabilize the negative-charged lysosomes and release DNA antigens into the cytoplasm to facilitate cross-presentation. Upon intramuscular injection, Man-ZCL formed an antigen depot at the injection sites and enhanced the presence of DNA antigen in the drainage LN, which produced an immune response against HIV *in vivo* [85].

Furthermore, particles with positive charges adsorbed larger quantities of antigens, resulting in the increased cellular recruitment [86, 87]. For instance, we prepared a polymer/lipid hybrid nanoparticle (PLNP), based on PLGA and the cationic lipid 3*-[*N*-(*N*, *N*-dimethylaminoethane)-carbamoyl] cholesterol hydrochloride (DC-Chol, a cationic lipid). The hydrophobic core of PLNPs encapsulated a water-insoluble drug, all-trans-retinoic acid (atRA), which was proved to provoke the expression of gut-homing receptors of the parenteral lymphocytes for potent mucosal immune activations. Moreover, the effect can be further boosted by the coadministration of the immunogenic components, such as antigens or adjuvants. In this case, through the cationic surface, the particles also enabled the adsorption of the negatively charged antigen and CpG (TLR 9 agonist), and thus form the codelivery system of atRA, antigen, and CpG. After intramuscular injection, PLNP, contrasting with Alum-adjuvanted group (OVA-Alum-atRA), induced comparable antigen depot at the injection site (Figure 8.4a), but the increased IL-6 and IL-15 cytokine levels, which generated an immunogenic environment for atRA-induced gut-tropism of the immunocytes, suggesting the robust mucosal immune activations (Figure 8.4b,c). Furthermore, antigen-specific $CD8^+$ T cell was also evidently boosted among the splenocytes, indicating the potent systemic immune response (Figure 8.4d) [78].

In conclusion, these findings support the idea that surface charges are important in regulating cellular uptake and tissue distribution. While positively charged particles are capable of capturing cells, their immunogenicity is inhibited evidently by nonspecific binding. Therefore, the charges need to be rationally engineered in

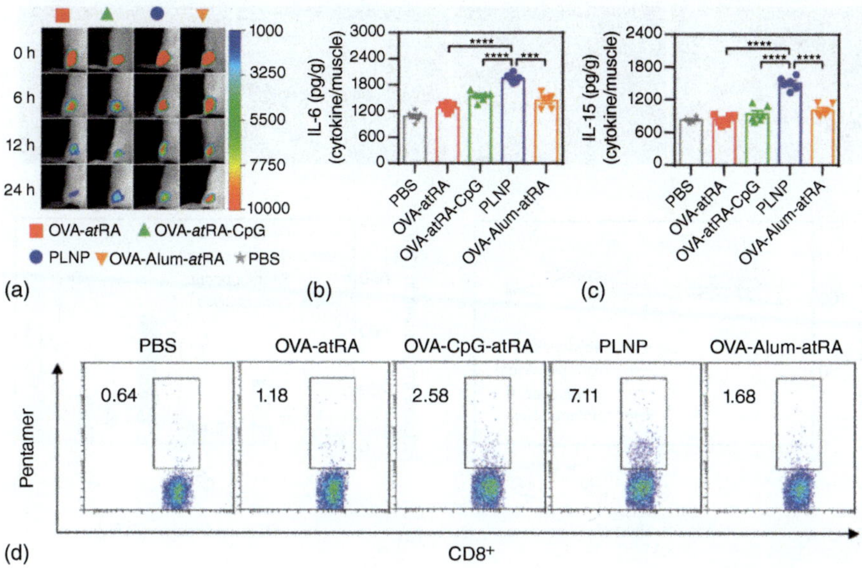

Figure 8.4 Cationic polymer/lipid nanoparticles (PLNPs) formed an antigen depot for proactive antigen delivery. (a) Antigen depot at the injection site. (b) IL-6 and (c) IL-15 concentrations at the injection site. (d) Antigen-specific $CD8^+$ T cell levels among the mouse spleen cells (OVA: ovalbumin; atRA: all-trans retinoic acid; CpG: CpG oligodeoxynucleotides; PBS: phosphate-buffered saline). Source: Du et al. [88]/American Chemical Society.

different situations. Under these circumstances, future efforts in the charge-flipping and stimulus-responsive materials may provide optimal strategies to harness the immunogenic properties of particles.

8.3.3 Shape

The shapes of microscopic biological particles also affect the immunological behavior because of tissue distribution, cellular uptake, intracellular signaling pathways, and excretion. In a recent study, researchers investigated the influence of the size and shape of gold NPs on the anatomical distribution of the NPs. Because of the kinetics of accumulation and clearance, varied particle shapes may be distributed in the different tissues. It was found that there was no difference between spherical and star-shaped gold NPs in the liver. But it was only star-like GNPs that accumulate in the lung [89].

It is thought that the varied internalization pathway may be attributed by the contact angle during the endocytosis. Based on the hydrothermally treated bacteria backbones (demi-bacteria, DB), we fabricated spherical, chain-like, rod-like, and *Y*-shaped particles derived from *Pediococcus acidilactici* (Figure 8.5a). When treated with DCs, rod-shaped *Lactobacillus casei* had a reduced contact angle and achieved an increased endocytosis (Figure 8.5b,c), in addition offering a hollow/porous structure that would accommodate antigens and adjuvants. In this way, this nanoarchitecture allowed the integration of multiple immunostimulatory elements

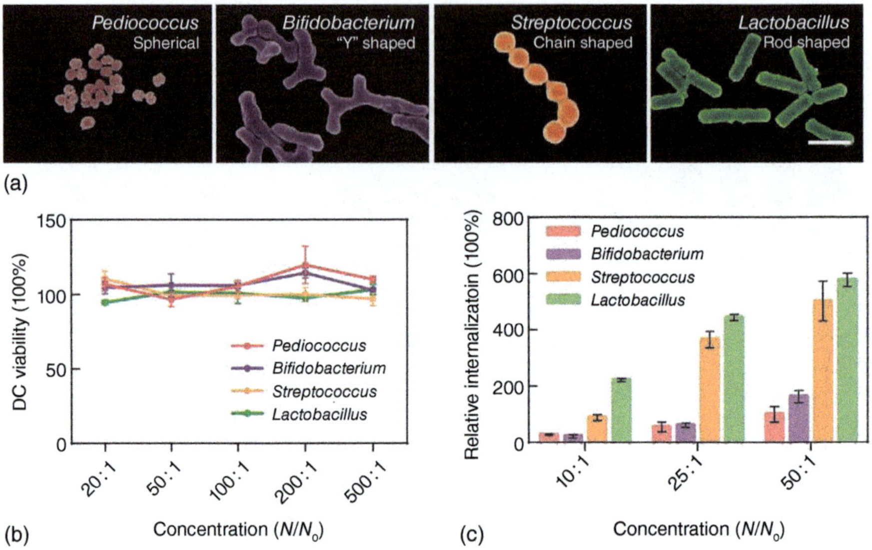

Figure 8.5 DBs of various shapes showed different endocytosis efficiencies. (a) Schematic diagram of hydrothermal process to prepare DB with different shapes. (b) Illustration on the contact angle (Ω) during the cellular encountering. (c) Dose-dependent endocytosis with DC of DB, indicating that rod-shaped DB inclined to be internalized due to the smaller contact angle. Source: Ni et al. [90]/John Wiley & Sons/CC BY 4.0.

in a pattern that closely resembles that of bacterial pathogens. Through pathogen mimicry, antigens were greatly enhanced in uptake and cross-presentation, leading to much stronger immune activation for cancer vaccines [90].

The difference between spherical and nonspherical particles' immunogenicity still remains a controversial topic. In one study, nanofibers and nanospheres were prepared via the self-assembly of amphiphile peptides. Cytokine production and cell surface markers showed that nanofibers induced a higher level of IFN-γ, TNF-α, and CD86 expression, while nanospheres induced a slightly higher level of Th2-associated immune response [91]. In another study, carboxylated polystyrene particles were shaped into distinct shapes. Researchers found that cylindrical (193 nm) particles elicited a Th1-biased immune response, whereas rod-shaped (1530 nm) particles elicited a Th2-biased response [92].

The underlying mechanisms by which shapes influence immune cell function are still relatively obscure due to a lack of single-factor studies. There is often an association between size and shape effects on phagocytosis. As an example, oblate ellipsoidal particles have higher internalization efficiency than spheres of small sizes, but this is no longer the case when the particle size increases [93]. In addition, the shape effect is related to how the particles and APCs contact each other. The radical contact in the case of rod-like particles may consume less energy for cellular uptake compared with the cellular interactions in the axial direction. To develop rational vaccine formulations, it is, therefore, necessary to develop novel methods to precisely control particulate shapes within a single particle platform.

8.3.4 Hydrophobicity

Since the cellular membranes consist of hydrophobic bilipids, APCs interact better with hydrophobic particles than with soluble proteins. According to a recent research, hydrophobic microparticles greatly enhanced the antigen internalization, MHC class II expression, and CD86 expression on the DCs. A force measurement of adhesion between live cell membranes and particles was conducted to investigate further the physical interaction. Through atomic force microscopy, detachment forces can be measured by probing cells with particle-modified tips. As a result of the increased surface hydrophobicity, the particles and cell membranes were able to interact more physically, resulting in greater microparticle uptake and therefore higher antigen concentrations in the cells [21]. Also, molecular dynamics analysis determined that hydrophobic particles tend to bind with the raft domain on membranes, which enabled them to endocytose more easily and deliver vaccines to cell sites [21, 94].

Additionally, the particles' intrinsic hydrophobicity may have aided the lysosomal escape of antigen by increasing their affinity toward endo/lysosomes. Within the cytoplasm, some antigens were processed by the proteasome and presented via MHC I, inducing immune responses [95, 96]. In one study, self-assembled NPs were synthesized by combining Group A Streptococcus (GAS) peptide antigen with different types of hydrophobic amino acids. The conformation and hydrophobicity of the particles were altered by changing the amino acid type and molecular weight. According to circular dichroism spectra analysis, the peptide, which contained 15 leucine residues, formed a typical helical conformation, which in turn mimicked the native conformation of GAS M-proteins and thus promoted the immune recognition. Also, the hydrophobicity of peptides is positively correlated with antibody response. The NPs with a higher hydrophobicity induced the increased DC maturation and higher antibody titers without inducing undesirable inflammatory side effects in C57BL/6 mice [97].

Additionally, particle-stabilized emulsions showed good hydrophobic effects. As shown in Figure 8.6a, we developed an alum-stabilized Pickering emulsion (PAPE) by packing alum on the squalene/water interphase. By exposing the alum-supported oil/water interphase between the inter-alum gaps, the emulsions were proved to be more hydrophobic, compared to the surfactant-stabilized droplets, whose hydrophilic domains were densely covered, and increased the affinity of the cellular membrane (Figure 8.6b). Consequently, PAPE induced potent cellular internalization. Traced by lumogallion, evident alum (red) was accumulated within the cells, which may dictate the intracellular transfer of antigens for higher immune activations. On the other hand, the commercial alum failed to elicit similar results, due to the hydrophilic surface and the plate-like configuration (Figure 8.6c). When comparing PAPE group with alum-treated group (Alum) and the fluidic antigen RBD, higher levels of antigen-specific antibodies and the elevated interferon-γ-producing T cells were observed (Figure 8.6d), indicating the enhanced vaccinations [98].

The innate immune system senses and detects pathogens by recognizing the highly conserved danger signals. In these danger signals, it is found that

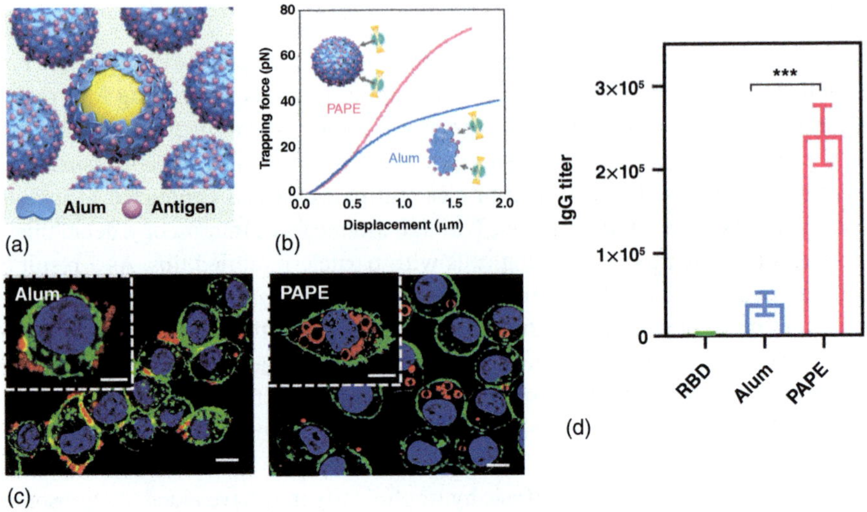

Figure 8.6 (a) Alum-stabilized Pickering emulsion allowed for the improved hydrophobicity; (b) higher cellular affinity via nanotweezers; and (c) increased cellular uptake compared with alum-treated cells. Cell membrane, nuclear, and alum were labeled in phalloidin (green), DAPI (blue), and lumogallion (red), respectively; (d) stimulate the antigen-specific IgG titer by six times contrasting to alum-adjuvanted group. Source: Peng et al. [98]/John Wiley & Sons.

hydrophobic aliphatic groups and aromatic groups play a crucial role in activating the immune system. Interestingly, the hydrophobic substances hardly reveal themselves to the external environment under healthy conditions. Moyano et al. showed an increase in the hydrophobicity essentially boosted the immune potency. By interacting with membranes and specific receptors on the surface of necrotic cells, these signals became exposed and triggered an enhanced innate immune response. Due to these reasons, increased hydrophobicity of antigen carriers contributed to the activation of the immune system.

Collectively, tuning on the surface hydrophobicity allows for the enhanced immunogenicity and the strong immune response, eliciting the increased cellular uptake and lysosome escape of the antigens.

8.3.5 Softness

Bioinspired design can be used to develop robust immune recognition and potentiation. By engineering the physiochemical nature of synthetic particles, synthetic particles were emerged to mimic the pathogen sizes, surface structure, hydrophobicity, and antigen display [99]. But a critical factor has been overlooked: the pathogens' softness. Soft materials, such as pathogens, can deform on APC surfaces to enlarge the area of contact. As a result, the surface ligands will flow to the contact zone to trigger the immune recognition, which can hardly be replicated by the rigid particles.

To address this, we developed a two-dimensional graphene oxide (GO) to demonstrate the soft contact. Firstly, GO sheets with the sizes of 350 nm and 2 μm

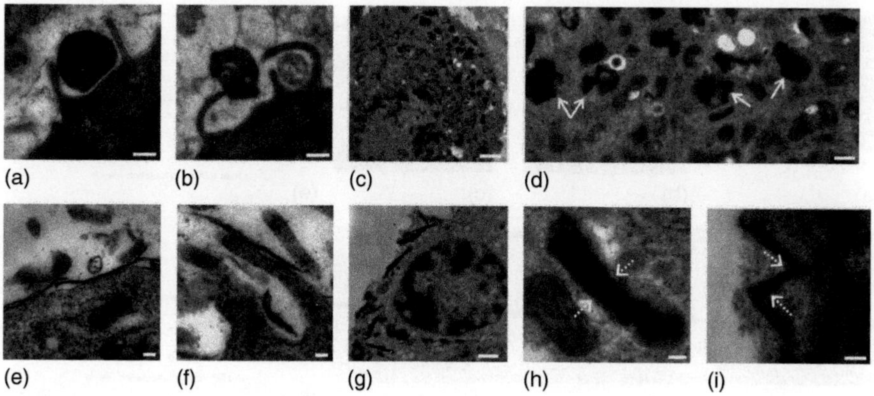

Figure 8.7 TEM images of the internalization of 350 nm GO (above, a–d) and 2 μm GO (bottom, e–i) in macrophage. (a, b) and (e, f) indicated the interactions with the cellular membrane at the initial uptake of 350 nm GO and 2 μm GO, respectively. (c, d, g, h, and i) demonstrated the intracellular transfer of the 350 nm GO (solid arrows) and 2 μm GO (dashed arrows). Scale bars: (c) and (g), 1 μm; others, 200 nm. Source: Yue et al. [100]/Elsevier.

were prepared by sonification and separated by varied sedimentation rates. Both sizes induced comparable cellular uptake among macrophages. But, 2 μm GO experienced deformation during the cellular encountering, which first attached and then deformed on the cellular membrane (Figure 8.7). Within the cytoplasm, the microsized GO can also be folded evidently, indicating the increased softness contrasting with the nanosized GO. As such, in comparison with 350 nm GO, 2 μm elicited significantly boosted cytokine secretion and much stronger inflammation responses, demonstrating the immense potential of the two-dimensional soft material as enhanced vaccine adjuvant [100].

As aforementioned, membrane dynamic curvature and lateral diffusion of endocytosis targets have long been neglected. To address this, we developed a particle-stabilized emulsion (Pickering emulsion), to recapitalize the oily core for the dynamic interactions and introduce the enhanced bioeffects by optimizing the interracial particles (Figure 8.8a). Firstly, Pickering emulsion was typically employed in petroleum separation and cosmetics. To increase the biocompatibility and adjuvanticity, Pickering emulsion was redesigned with PLGA NPs as the stabilizers, instead of surfactant, and was prepared using the clinical-approved squalene. The prepared droplet was named as "PPAS," and functioned as an elastic and biodegradable adjuvant. The particle-adsorbed o/w interface boosted the surface area, which increased the antigen loading efficiency (Figure 8.8b, c). PPAS demonstrated the increased pliability and lateral mobility, compared with surfactant-stabilized emulsion or solid micro/nanoparticles. As shown in Figure 8.8d, the soft droplets were first deposited and deformed on the cell membrane with the cellular wrapping, as to increase the contact area. The antigens on the droplet surface were then transported to the contact zone, triggering the dynamic interactions with the cells, and boosting the antigen uptake afterward. In this way, the force-dependent deformation and dynamic fluidity of PPAS stimulated the multivalent three-dimensional interactions

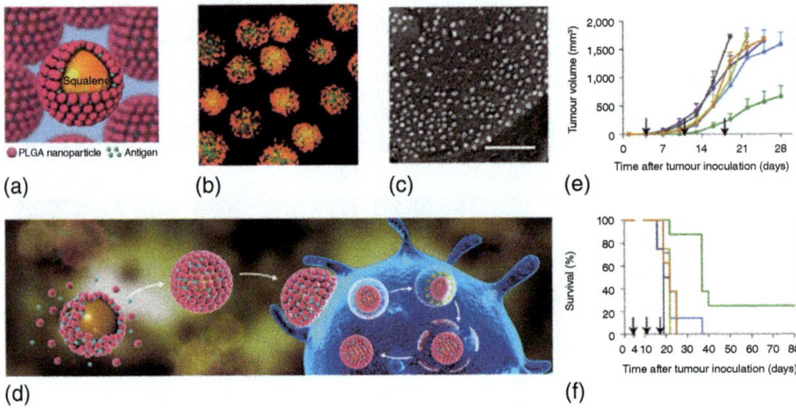

Figure 8.8 Exploiting the softness of Pickering emulsion for the enhanced vaccine efficacy. (a) Schematic representation; (b) confocal image (scale bar = 5 μm, antigen = green, particle = red); and (c) Cryo-SEM images (Scale bar = 500 nm) of the soft droplet (PPAS). (d) Soft contact for the elevated immune recognition and phagocytosis. (e, f) Tumor growth curves (e) and survival rate (f) of the vaccinated mice, indicating the significant tumor regression and a higher antimelanoma efficacy. Source: Xia et al. [101]/Springer Nature.

with APCs. PPAS induced significantly increased humoral and cellular immune responses *in vivo*. In comparison with commercial adjuvants (alum, MF59, and AS04), In addition, Pickering emulsions exerted potent immune protections against influenza virus challenge. In both E.G7/OVA and B16/MUC1 antitumor therapies, PPAS caused evident tumor-regression and significantly increased the survival rates of the tumor-bearing mice (Figure 8.8e,f). The soft PPAS may provide an alternative strategy to exploit the force-dependent deformation and lateral mobility of particulate adjuvants for robust vaccinations [101].

When it comes to liposomes, the soft particles may enter the cell either through cell fusion or endocytosis. Additionally, the stiffer liposomes might only enter cells via endocytosis. In this case, the membrane bending, the surface tension, and the coated pits required higher membrane deformation energy. Therefore, soft particles were more likely to be internalized by the cells than their stiffer counterparts [102]. Despite this, the principles may not be applicable in all cases, especially since some studies revealed that macrophages prefer to take up stiffer particles [102, 103]. A rational design was required in terms of the cell types, particle formulations, and uptake modes in order to control softness.

8.4 Connecting the Dots: Strengthening on the Multi-Scale Delivery of Vaccines

Efficient vaccine delivery is expected to break through the multiscale barriers: (i) tissue level: distributed to specific tissues or organs to activate the immune system; (ii) cellular level: immune recognition and cellular uptake for potent antigen delivery; and (iii) subcellular level: antigen delivery to the cytosol to prevent

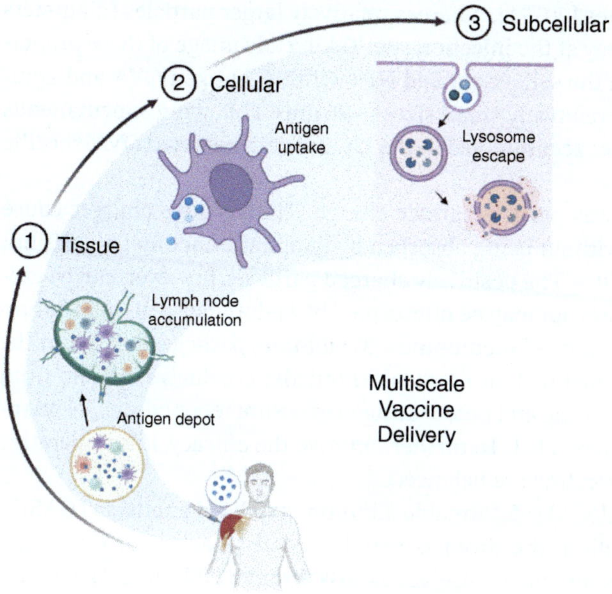

Figure 8.9 Particulate vaccines to enhance the distribution, internalization, and presentation of vaccines.

enzyme-mediated degradation, and combine with the MHC molecules for onset of the immune response (Figure 8.9). As a result, promising vaccine formulations are expected to strengthen the multiscale delivery of vaccines [104, 105]. A combination of controllable sizes, softness, and surface properties may favor LN accumulations of the antigenic charges. Afterward, the particles are able to increase the effectiveness of cellular internalization by mimicking the sizes, shapes, charges, softness, and compositions of pathogens. They may also be capable of stimulating innate immune responses and activating immune responses through regulating intracellular transfer. In this section, the latest development in particle platforms for vaccine delivery systems and adjuvants were discussed in relation to how vaccines were presented, distributed, internalized, and presented.

8.4.1 Distribution

Direct transfer of concentrated cytokines and immunocytes to LNs may drive the differentiation and proliferation of T cells and B cells to activate the adaptive immune responses with the antigens and costimulatory signals [106–109]. In addition, accumulating research suggests that prolonged antigen persistence within the LNs would benefit long-term memory. In this sense, LN delivery may contribute to the effectiveness of vaccinations.

In order to accomplish this, the particle properties were intensively optimized. A recent study found that NP-based carriers with sizes of 10–100 nm could passively deliver vaccines to the LNs via the lymphatic drainage of interstitial fluid from the peripheral injection sites [70, 71], whereas relatively larger particles (diameters above 100 nm) tended to stay at the injection site. Taking advantage of these properties, our team constructed the self-assembled NPs with the antigen OVA and adjuvant (CpG) to achieve the relatively small sizes (~80 nm). Following subcutaneous administration, the particles accumulated in the LN and demonstrated OVA-specific adaptive immunity [54].

Another critical factor was particle surface charge. The negative charges cause particles to drain to LNs without being absorbed by lymphatic endothelial cells but inhibited their uptake by DCs. The positively charged particles, however, can be efficiently uptake by DCs in LNs but may be intercepted by endothelial cells at injection sites due to the negatively charged membranes. In addition, particles that hitchhike on the interstitial flow are inclined for LN transfer but also conducive to drain from them. In this regard, long-lasting and potent antigen exposure relies on the LN retention of the delivered vaccines [110]. To further improve the efficacy, it is imperative that cell uptake and LN targeting are balanced.

To address this, we developed a deformable albumin-stabilized emulsion (DASE). After intramuscular injection, the droplets could become attached to interstitial flow and approach cells, adjusting themselves to pass through endothelial gaps, and passively draining to the surrounding lymphatic vessels (Figure 8.10a). As depicted in Figure 8.10, we observed that the DASE passed through the intercellular gaps by labeling antigens, the cellular skeleton, and lymphatic endothelial cells with Cy5, FITC-phalloidin, and Fluro@555-VEGFR (an indicator for LN cells). With the sizes around 330 nm, which was three times of the intercellular gaps, the droplets were observed to evidently accumulate within the LNs (Figure 8.10b). But, the solid albumin particles (SAP) with similar sizes failed to deform and squeeze themselves through the cellular gaps, thus limited antigens were presented in the muscle sections. And, only a small number of solid-particle-loaded antigens accumulated within the draining LN (Figure 8.10c). As a result, the deformable droplets evidently increased the antigen uptake, CD86 expression, and cross-presentation of the LN-residing DCs, leading to the enhanced $CD8^+$ T cell activations (Figure 8.10d) [111].

Together, the size and charge of NPs all contribute to their ability to target the lymphoid organs. Other than that, in order to prolong the LNs retention time, we optimized the delivery modes of nanocarrier vaccines. To study the deliver kinetics, we designed immunostimulatory CpG-modified tumor-derived nanovesicles (CNVs) in three delivery modes, including mono-pulse CNVs, staggered pulse CNVs, and gel-confined CNVs for antitumor vaccines. As a result of their nanometer size, mono-pulse and staggered-pulse CNVs could passively drain to LNs, whereas gel-confinement CNVs were actively delivered to LNs via DC migration (Figure 8.11). All the formulations were subcutaneous injected in the backs of mice. The results of frozen-section studies showed that CNVs were rapidly drained to LNs after mono-pulse administration, then drained out via efferent lymphatic vessels, and were mainly found in the lymphatic vessels. As a result of staggered-pulse administration, CNVs tended to remain longer in the LNs, then spread to the subcapsular sinus region, and finally spread outside the T cell areas.

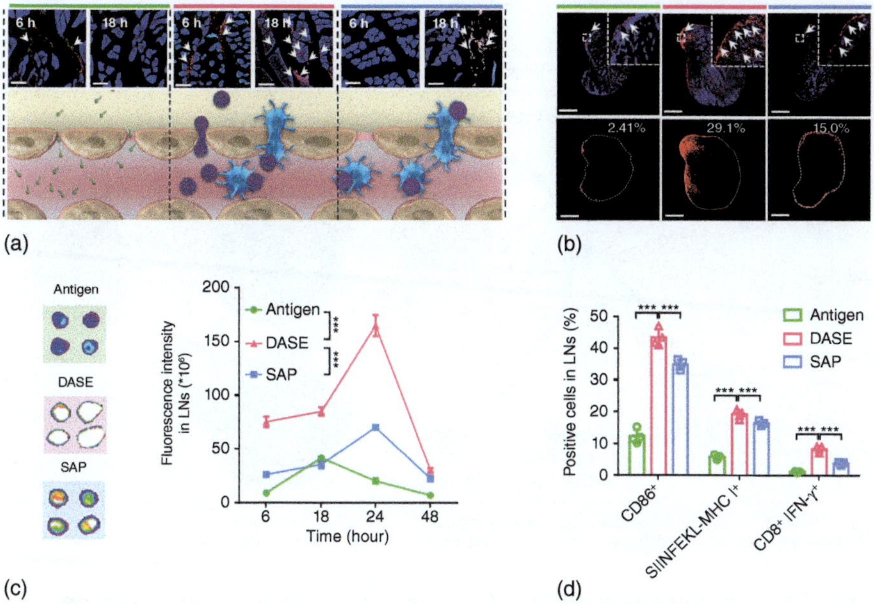

Figure 8.10 Deformable Pickering emulsion to potential the LN accumulation of vaccines (a) Muscle sections and schematic illustrations of the intercellular and intracellular transfer of Cy5-labeled antigens (red) between muscle cells (blue) or within lymphatic vessels (green). From left to right are fluidic antigen (antigen), deformable droplet DASE, and SAP, respectively. The white arrow indicated the presence of antigens. (b) LN accumulations of vaccines. The nucleus and antigens are labeled by DAPI (blue) and Cy5 (red), respectively. Scale bar = 400 μm. From left to right are fluidic antigen (antigen), deformable droplet DASE, and SAP, respectively. (c) The presence of antigens and the corresponding quantitative fluorescence intensity in LN drainage over time. The left picture shows the representative fluorescent images of LNs 24 hours after administration. (d) Proportions of $CD86^+$ DCs, SIINFEKL-MHC I-presenting DCs, and $CD8^+$ IFN-γ^+ T cells within the LNs 24 hours after administration. Source: Song et al. [111]/John Wiley & Sons.

With regard to the gel-confined administration, it primarily involved LN homing through APCs. Despite the slow transport rate, CNVs gradually increased in the LNs over time. The gel-confined approach appeared to have a superior tumor inhibition effect both in prophylactic and therapeutic tumor models compared to the other two groups [112]. In conditions like these, it is important to strike a balance between targeting LN and retaining the depot effect. Consequently, there is still controversy surrounding the need for LN targeting, and further studies are required.

It is worth-noting that we are not implying that LN targeting is the sole adjuvant strategy. Uniform particles with large sizes, but well-controlled release behavior may serve as the antigen repertoire for the increased immune potency and duration. In our recent work, a tumor vaccine strategy was developed to prepare uniform-sized self-healing microcapsules based on PLA (FDA approval) (Figure 8.12). This self-healing characteristic facilitated easy and effective antigen microencapsulation. Following vaccination, the degradation of PLA produced a lactic-acid-enriched microenvironment, which enhanced the activations of the

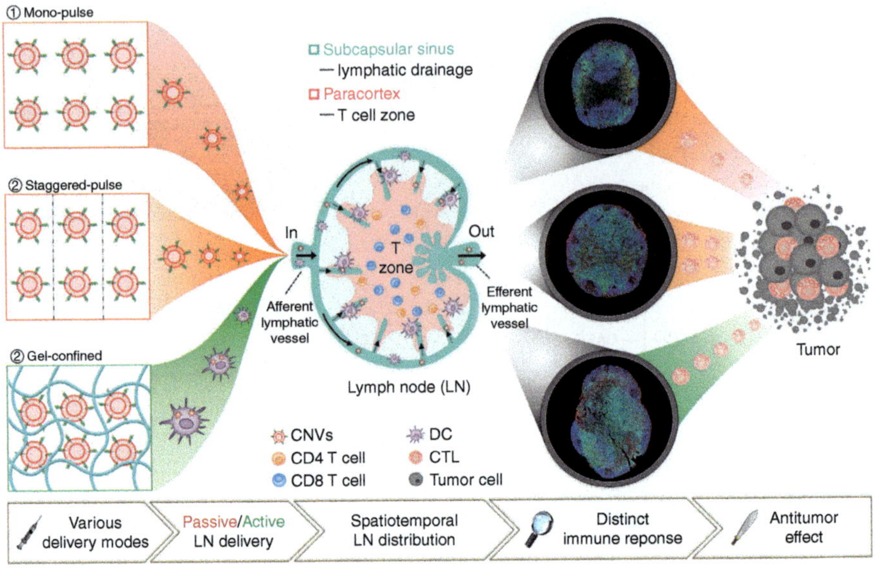

Figure 8.11 Spatiotemporal delivery of nanovaccines for tumor immunotherapy. we engineered CNVs, tested single injection and multiple injection (3×) delivery modes (mono-pulsed and staggered-pulsed, respectively) for passive drainage into LNs, and developed a hydrogel-based antigen depot delivery mode (gel-confined) for active LN delivery by APCs. Three different CNV delivery modes exhibited different spatiotemporal distributions inside LNs, which subsequently affected the use of CNVs and the interaction between DC and T cells in the T cell zone, resulting in distinct antitumor immunity. Source: Wang et al. [112]/John Wiley & Sons/CC BY 4.0.

recruited APCs. Moreover, surrounding the microparticles, the released acids, cell behavior, and antigen-released kinetics worked synergistically. In this case, antigen utilization could be effectively increased, improving APC activation and antigen presentation [114]. Furthermore, our group employed the self-healing microcapsules in therapeutic leukemia vaccines with the immune checkpoint inhibitor antiprogrammed-cell-death-protein-1 (anti-PD-1) and demonstrated better therapeutic outcomes [115].

8.4.2 Internalization

Following the administration, APCs are recruited to the site of invasion, they phagocytose the pathogens and secrete cytokines and/or chemokines to recruit more APCs, causing an immune response cascade. As soon as the draining LNs are filled with APCs, they begin processing the exogenous antigens for the adaptive immune response. As antigenic peptides bind to MHC-I or MHC-II, they travel to the cell surface and are presented to the naive T cells, leading to the differentiation of these cells. Following that, naive $CD4^+$ T and $CD8^+$ T cells are activated and amplified to acquire immune-helper functions and cytotoxic capabilities. APCs thus play an important role in triggering adaptive immunity. Accordingly, it is

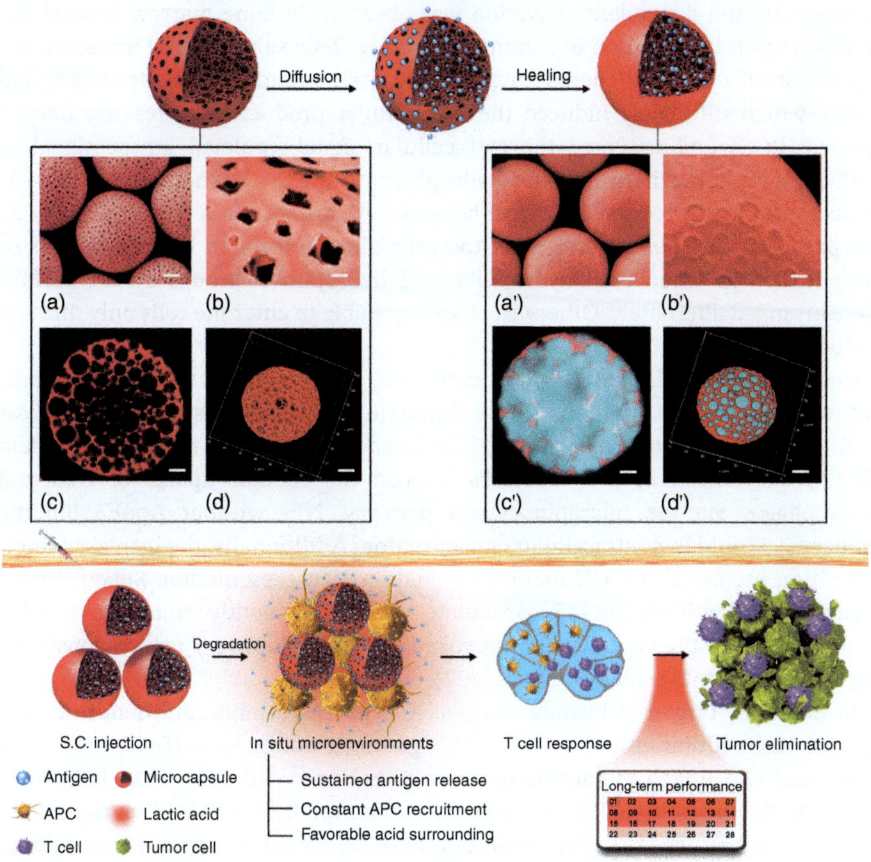

Figure 8.12 Strategy of using self-healing microcapsules to modulate immunization microenvironments for cancer vaccination. The antigen was loaded within the microparticle via a diffusion manner. Image (a–d) and (a'–d') demonstrated the morphology of the microparticles before and after antigen loading, respectively. Scanning electron microscopy (SEM) images of the microparticles in a magnified view (a & a', Scale bars, 10 μm) and (a) local feature (b & b', Scale bars, 1 μm). Confocal images of the microparticles in two-dimensional (2D) cut view (c and c′, Scale bars, 5 μm), and 3D reconstruction (d and d′, Scale bars, 10 μm). After administration, the microparticles generated the immunization microenvironments, contributing to sustained antigen release, increased antigen-presenting cell recruitment, and higher acidic concentration to induce innate immune activations. Consequently, the microparticle-based vaccine stimulates potent T-cell responses and tumor regression. Source: Liu et al. [113]/American Chemical Society.

necessary to enhance vaccine efficacy by promoting APC phagocytosis as well as LN metastasis [89, 116, 117].

As particulate vaccines interact with APCs, particle size, charge, shape, hydrophobicity, and softness play significant roles in endocytosis [118]. The details concerning the relationship between particle properties and endocytosis were discussed in Section 8.3. It has been found that, in comparison to negatively or neutrally charged particles, cells tend to take up more positively charged particles.

A concentration-dependent activation was observed in bone marrow-derived DC (BMDC) upon the addition of cationic liposomes. This subsequently increased the expression of cytokines/chemokines and the costimulatory molecules CD80 and CD86, which ultimately induced the intracellular production of reactive oxygen species (ROS) and triggered the extracellular signal-regulated kinase signaling pathway [119]. Furthermore, the hydrophilicity and hydrophobicity of particle surfaces can influence the biological behavior of particles. In a study by Xia et al. the particles were observed to enter the cells directly through the membranes or via endocytosis when both hydrophilic and hydrophobic groups on the surface were arranged alternately. Otherwise, they were able to enter the cells only through endocytosis [120].

Interestingly, the different uptake kinetics may be explained by the differed uptake pathways. When DCs internalize extracellular fluid-phase solutes, macropinocytosis is dominant rather than receptor-mediated endocytosis [121]. The apolipoprotein E3 (ApoE3)-loaded lipid hybrid NPs boosted the cellular uptake of DCs and macrophages via the micropinocytosis pathway. NPs without ApoE3 binding induced 2.37-fold less intracellular concentration. Additionally, the increased internalization increased the DC maturation and antigen presentation. Subcutaneous injection of ApoE3-decorated particulate vaccines evidently enhanced the LN accumulation of the vaccines and provoked stronger adaptive T cell responses for higher antitumor therapeutic effects [122].

In addition, targeting ligands, such as C-type agglutinins and dendritic cell-specific intercellular adhesion molecule 3-grabbing nonintegrin (DC-SIGN), were developed and linked to vaccine particles to enhance cellular uptake [123–125]. Yang et al. encapsulated TLR7 agonist (imiquimod, R837) within the PLGA NPs, and camouflaged the NPs with mannose-decorated cancer cell membranes. A confocal fluorescence image showed that mannose modification increased macrophage and BMDC uptake rates significantly. Moreover, the surface markers (CD80 and CD86) of DC cells were evidently induced as well as cytokine secretion. In addition, mannose-modified NPs can bind with DC-SIGN on the cellular surface, significantly causing tumor regression in B16-OVA melanoma tumor models [124].

8.4.3 Presentation

Presenting antigens to T cells for antitumor effect is another important feature after endocytosis and activation of APCs. To facilitate humoral and cellular immune responses, antigens are generally processed by the lysosomal pathway and presented to $CD4^+$ T cells via MHC-II complexes. The eradication of pathogens and the commencement of tumor lysis also require the engagement of $CD8^+$ T cells. As such, some tumor antigens need to escape from lysosomes and combine with the proteasome in the cytoplasm, where they are further decomposed into immunogenic polypeptide fragments, carrying them to the endoplasmic reticulum. The complex presents antigens to $CD8^+$ T cells and elicits CTL responses once incorporated with the MHC-I molecules (known as "cross-presentation") [126].

Our group devised a variety of lysosome escape strategies to cross-present antigens using rational-designed particles under these circumstances. First off, the acidic environment in the endosomes (pH = 5–6.5) may favor protonation of amino groups. Protonated NPs with a large specific surface area consume a large amount of hydrogen ions within the lysosomes, triggering the rapid influx of hydrogen ions and chloride ions, causing lysosomes to swell or rupture in preparation for antigen release. By using this phenomenon, we prepared positively charged chitosan nanospheres for cytosolic delivery, which were intensely internalized by the APCs after intramuscular injection. After the particles were exposed to rapid acidification, the codelivered antigens were released into the cytoplasm. As a consequence, more antigens were presented to T cells via MHC-I molecules [79]. Moreover, Lee et al. demonstrated that particles' curvatures affected their distribution within the endosome, resulting in immune activation. In order to boost the immune response, CpG was loaded into NPs with mixed curvatures (spiky NPs) or constant curvatures (spherical NPs). The transmission electron microscopy (TEM) results indicated that spiky NPs were more likely to be distributed around the endosomal membrane. Moreover, the endosome diameter increased as maturity increased, thereby enhancing the interaction between endosome and TLR-9 receptors and augmenting immune system activation [127].

Concurrently, another promising approach is to create internal air pressure by using the acidic environment. To demonstrate this, we coencapsulated antigens and pH-responsive agents (ammonium bicarbonate) inside PLGA hollow NPs (pH-responsive PLGA NPs). Following internalization, the endogenous low pH led to the decomposition of ammonium bicarbonate, which caused sufficient carbon dioxide (CO_2) bubbles to be produced within the lysosomes. A subsequent release of antigens to the cytoplasm is shown in Figure 8.13. Therefore, $CD8^+$ T cells were significantly generated and secreted more than 200% more IFN-γ compared to normal PLGA NPs [113].

Using the above results as inspiration, we developed the second strategy for generating gas using antigen-doped $CaCO_3$ nanoparticles (OVA@NP). By coprecipitating antigens with $CaCO_3$, high antigen-loading composite NPs were obtained. As a result, a vaterite form and hierarchical structure of the composite particles significantly facilitated decomposition, resulting in the dramatic release of CO_2 and subsequently damage to lysosome membrane. At the same time, antigens were transported into the cytosol, allowing for cross-presentation (Figure 8.14). In addition to the lysosome escape, the physical stress that is offered by the generated CO_2, induced autophagy through the LC3/Beclin1 pathway. By this way, the lysosome-resident antigens and cytosolic proteasomes would be enwrapped by the autophagosomes, which further enhanced the efficacy of cross-presentation. In turn, gas-responsive $CaCO_3$ NPs enhanced the antitumor responses, including $CD8^+$ T cell proliferation and cytotoxic tumor reduction, thereby indicating a new paradigm for safe and effective therapeutic cancer vaccines [128].

Based on these findings, it seems that the physicochemical properties and the loading mode of antigens have a direct effect on antigen-presenting mechanisms, dominating the subsequent adaptive immune responses.

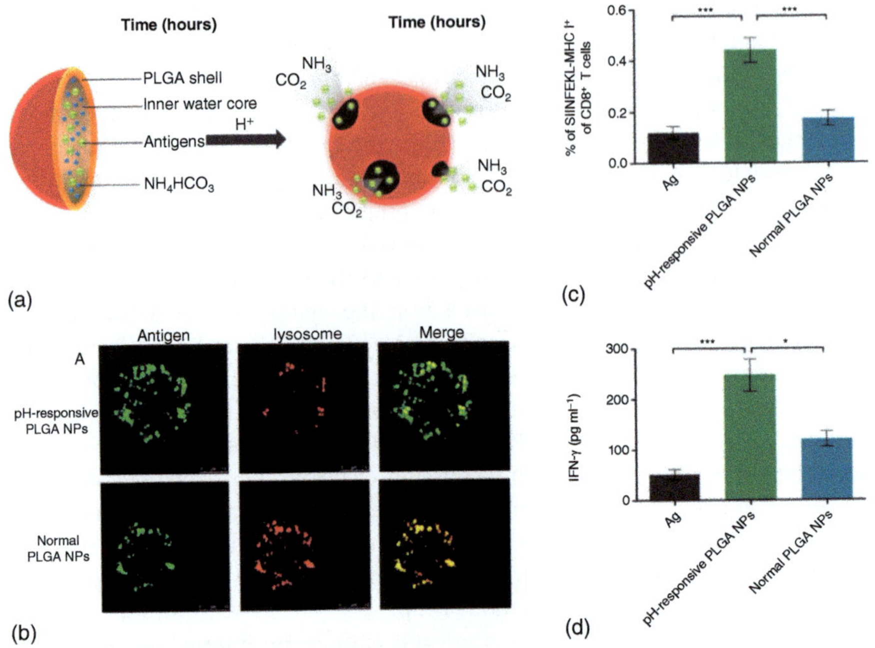

Figure 8.13 Gas generating strategy for cytosolic delivery (a) Schematic diagram. (b) Colocalization of the lysosomes (red) and antigens (green) via confocal images. (c) Antigen-specific CTLs among the splenocytes. (d) IFN-γ secretion from splenocytes, indicating the engagement of Th-1 immune response. Data are expressed as means ± SEM ($n = 6$) ($^*p < 0.05$; $^{**}p < 0.01$; $^{***}p < 0.001$).

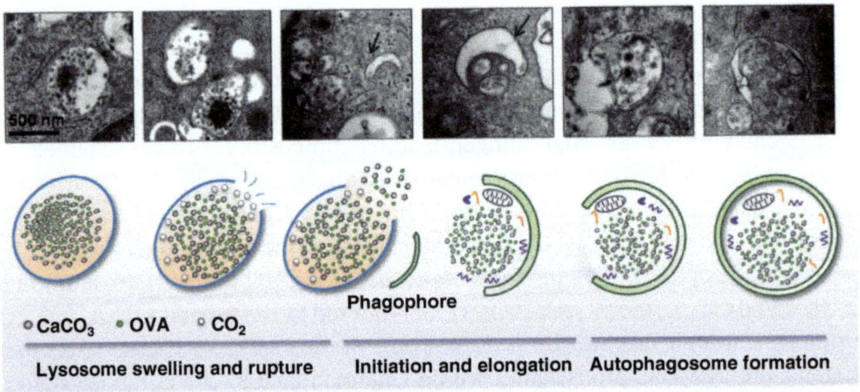

Figure 8.14 TEM images and the corresponding scheme illustrating lysosome escape and autophagy in DCs. The isolation/elongation of isolated membrane (bowl-like phagophore, indicated by black arrows) and the typical double-membrane autophagosomes.

8.5 Summary

With the aid of porous membrane emulsification technique, particles were prepared with uniform sizes. By adjusting the solidification, the particles can be further fabricated with almost homogeneous structures and well-defined properties, which set up the tailorable chassis for vaccine assembly and mechanism exploration. Furthermore, the uniformity also offered the potential for higher batch-to-batch stability of large-scale production and the bioeffects of the novel vaccine adjuvant and delivery systems. In this chapter, we discussed how rational design can exploit the intrinsic immunogenicity of particles through improved physiochemical properties and better processes for delivery. Particulate cancer vaccines still face challenges in translation and application, despite recent advances. As far as, we are concerned, there are a few challenges that need to be addressed: (i) comprehensive safety and efficacy assessments: synthetic particle-based cancer vaccines must be extensively researched and profiled for toxicity. The increased immune activation may also lead to an enhanced autoimmune response, but the safety profile of the raw materials is unclear. (ii) Clearer mechanisms: the biodistribution of the uniform particles, their interaction with the immune cells, and the inherent immunogenicity of these particles may have been unclear. The underlying mechanism should be revealed by new technologies, including spatiotemporal transcriptome, computer-aided simulations, and high-throughput data analysis. (iii) Reproducibility and scale-up production: multicomponent and multistructured synthetic particles usually result in complexity and increased cost of manufacture. We anticipate future efforts to develop new equipment, such as membrane-emulsification techniques, microfluidic devices, or methods for facile and effective control of the structures and properties of batches. Thus, an intensified multidisciplinary approach was necessary to integrate the high-end advances in immunology, chemistry, medicine, and engineering for clinical translation of particles in the vaccine industry.

References

1 Sharma, O., Sultan, A.A., Ding, H. et al. (2020). A review of the progress and challenges of developing a vaccine for Covid-19. *Frontiers in Immunology* 11: 585354.

2 Walker Bruce, D. and Burton Dennis, R. (2008). Toward an aids vaccine. *Science* 320 (5877): 760–764.

3 Zepp, F. (2010). Principles of vaccine design—lessons from nature. *Vaccine* 28: C14–C24.

4 Bowen, W.S., Svrivastava, A.K., Batra, L. et al. (2018). Current challenges for cancer vaccine adjuvant development. *Expert Review of Vaccines* 17 (3): 207–215.

5 Brito, L.A., Malyala, P., and O'hagan, D.T. (2013). Vaccine adjuvant formulations: a pharmaceutical perspective. *Seminars in Immunology* 25 (2): 130–145.

6 An, H.-W., Mamuti, M., Wang, X. et al. (2021). Rationally designed modular drug delivery platform based on intracellular peptide self-assembly. *Exploration* 1 (2): 20210153.

7 Wu, S., Xia, Y., Hu, Y. et al. (2021). Bio-mimic particles for the enhanced vaccinations: lessons learnt from the natural traits and pathogenic invasion. *Advanced Drug Delivery Reviews* 176: 113871.

8 Xia, Y., Song, T., Hu, Y. et al. (2020). Synthetic particles for cancer vaccines: connecting the inherent supply chain. *Accounts of Chemical Research* 53 (10): 2068–2080.

9 Wu, J. and Ma, G. (2020). Biomimic strategies for modulating the interaction between particle adjuvants and antigen-presenting cells. *Biomaterials Science* 8 (9): 2366–2375.

10 Guo, Y., Wang, Z., Shi, X. et al. (2022). Engineered cancer cell membranes: an emerging agent for efficient cancer theranostics. *Exploration* 2 (1): 20210171.

11 Bershteyn, A., Hanson, M.C., Crespo, M.P. et al. (2012). Robust igg responses to nanograms of antigen using a biomimetic lipid-coated particle vaccine. *Journal of Controlled Release* 157 (3): 354–365.

12 Wu, J., Fan, Q., Xia, Y. et al. (2015). Uniform-sized particles in biomedical field prepared by membrane emulsification technique. *Chemical Engineering Science* 125: 85–97.

13 Rosenthal, J.A., Chen, L., Baker, J.L. et al. (2014). Pathogen-like particles: biomimetic vaccine carriers engineered at the nanoscale. *Current Opinion in Biotechnology* 28: 51–58.

14 Ford Versypt, A.N., Pack, D.W., and Braatz, R.D. (2013). Mathematical modeling of drug delivery from autocatalytically degradable plga microspheres — a review. *Journal of Controlled Release* 165 (1): 29–37.

15 Yue, H., Wei, W., Fan, B. et al. (2012). The orchestration of cellular and humoral responses is facilitated by divergent intracellular antigen trafficking in nanoparticle-based therapeutic vaccine. *Pharmacological Research* 65 (2): 189–197.

16 Liu, Q., Jia, J., Yang, T. et al. (2016). Pathogen-mimicking polymeric nanoparticles based on dopamine polymerization as vaccines adjuvants induce robust humoral and cellular immune tesponses. *Small* 12 (13): 1744–1757.

17 Hamdy, S., Haddadi, A., Hung, R.W. et al. (2011). Targeting dendritic cells with nano-particulate plga cancer vaccine formulations. *Advanced Drug Delivery Reviews* 63 (10): 943–955.

18 Uyen, N.T.T., Hamid, Z.A.A., Tram, N.X.T. et al. (2020). Fabrication of alginate microspheres for drug delivery: a review. *International Journal of Biological Macromolecules* 153: 1035–1046.

19 Islam, M.A., Firdous, J., Choi, Y.-J. et al. (2012). Design and application of chitosan microspheres as oral and nasal vaccine carriers: an updated review. *International Journal of Nanomedicine* 7: 6077–6093.

20 Hu, L., Sun, Y., and Wu, Y. (2013). Advances in chitosan-based drug delivery vehicles. *Nanoscale* 5 (8): 3103–3111.

21 Liu, Y., Yin, Y., Wang, L. et al. (2013). Surface hydrophobicity of microparticles modulates adjuvanticity. *Journal of Materials Chemistry B* 1 (32): 3888–3896.
22 Wang, Y.-Q., Wu, J., Fan, Q.-Z. et al. (2014). Novel vaccine delivery system induces robust humoral and cellular immune responses based on multiple mechanisms. *Advanced Healthcare Materials* 3 (5): 670–681.
23 Holmgren, J. and Czerkinsky, C. (2005). Mucosal immunity and vaccines. *Nature Medicine* 11 (4): S45–S53.
24 El-Kamary, S.S., Pasetti, M.F., Mendelman, P.M. et al. (2010). Adjuvanted intranasal norwalk virus-like particle vaccine elicits antibodies and antibody-secreting cells that express homing receptors for mucosal and peripheral lymphsoid tissues. *The Journal of Infectious Diseases* 202 (11): 1649–1658.
25 Zhao, Y., Zhao, X., Cheng, Y. et al. (2018). Iron oxide nanoparticles-based vaccine delivery for cancer treatment. *Molecular Pharmaceutics* 15 (5): 1791–1799.
26 Dong, Z., Liu, W., Liu, K. et al. (2021). Effects on immunization of the physicochemical parameters of particles as vaccine carriers. *Drug Discovery Today* 26 (7): 1712–1720.
27 Hess, K.L., Medintz, I.L., and Jewell, C.M. (2019). Designing inorganic nanomaterials for vaccines and immunotherapies. *Nano Today* 27: 73–98.
28 Li, X., Wang, X., and Ito, A. (2018). Tailoring inorganic nanoadjuvants towards next-generation vaccines. *Chemical Society Reviews* 47 (13): 4954–4980.
29 Braydich-Stolle, L.K., Speshock, J.L., Castle, A. et al. (2010). Nanosized aluminum altered immune function. *ACS Nano* 4 (7): 3661–3670.
30 Jiang, H., Wang, Q., Li, L. et al. (2018). Turning the old adjuvant from gel to nanoparticles to amplify $CD8^+$ T cell responses. *Advanced Science* 5 (1): 1700426.
31 Cao, P., Han, F.Y., Grøndahl, L. et al. (2020). Enhanced oral vaccine efficacy of polysaccharide-coated calcium phosphate nanoparticles. *ACS Omega* 5 (29): 18185–18197.
32 Chiu, D., Zhou, W., Kitayaporn, S. et al. (2012). Biomineralization and size control of stable calcium phosphate core–protein shell nanoparticles: potential for vaccine applications. *Bioconjugate Chemistry* 23 (3): 610–617.
33 Pei, M., Liang, J., Zhang, C. et al. (2019). Chitosan/calcium phosphates nanosheet as a vaccine carrier for effective cross-presentation of exogenous antigens. *Carbohydrate Polymers* 224: 115172.
34 Henriksen-Lacey, M., Korsholm, K.S., Andersen, P. et al. (2011). Liposomal vaccine delivery systems. *Expert Opinion on Drug Delivery* 8 (4): 505–519.
35 De Serrano, L.O. and Burkhart, D.J. (2017). Liposomal vaccine formulations as prophylactic agents: design considerations for modern vaccines. *Journal of Nanobiotechnology* 15 (1): 83.
36 Zhou, S., Kawakami, S., Yamashita, F. et al. (2010). Intranasal administration of CpG DNA lipoplex prevents pulmonary metastasis in mice. *Cancer Letters* 287 (1): 75–81.
37 Hou, X., Zaks, T., Langer, R. et al. (2021). Lipid nanoparticles for mRNA delivery. *Nature Reviews Materials* 6 (12): 1078–1094.

38 Pardi, N., Tuyishime, S., Muramatsu, H. et al. (2015). Expression kinetics of nucleoside-modified mRNA delivered in lipid nanoparticles to mice by various routes. *Journal of Controlled Release* 217: 345–351.

39 Tenchov, R., Bird, R., Curtze, A.E. et al. (2021). Lipid nanoparticles—from liposomes to mRNA vaccine delivery, a landscape of research diversity and advancement. *ACS Nano* 15 (11): 16982–17015.

40 Zhang, N.-N., Li, X.-F., Deng, Y.-Q. et al. (2020). A thermostable mRNA vaccine against Covid-19. *Cell* 182 (5): 1271–1283. e1216.

41 Cornebise, M., Narayanan, E., Xia, Y. et al. (2022). Discovery of a novel amino lipid that improves lipid nanoparticle performance through specific interactions with mRNA. *Advanced Functional Materials* 32 (8): 2106727.

42 Miao, L., Li, L., Huang, Y. et al. (2019). Delivery of mRNA vaccines with heterocyclic lipids increases anti-tumor efficacy by sting-mediated immune cell activation. *Nature Biotechnology* 37 (10): 1174–1185.

43 Tian, M., Zhou, Z., Tan, S. et al. (2018). Formulation in DDA-MPLA-TDB liposome enhances the immunogenicity and protective efficacy of a DNA vaccine against mycobacterium tuberculosis infection. *Frontiers in Immunology* 9: 310.

44 Lau, K.K.S. and Gleason, K.K. (2006). Particle surface design using an all-dry encapsulation method. *Advanced Materials* 18 (15): 1972–1977.

45 Koppolu, B.P., Smith, S.G., Ravindranathan, S. et al. (2014). Controlling chitosan-based encapsulation for protein and vaccine delivery. *Biomaterials* 35 (14): 4382–4389.

46 Ilyinskii, P.O., Roy, C.J., O'neil, C.P. et al. (2014). Adjuvant-carrying synthetic vaccine particles augment the immune response to encapsulated antigen and exhibit strong local immune activation without inducing systemic cytokine release. *Vaccine* 32 (24): 2882–2895.

47 Yue, H. and Ma, G. (2015). Polymeric micro/nanoparticles: particle design and potential vaccine delivery applications. *Vaccine* 33 (44): 5927–5936.

48 Beiss, V., Mao, C., Fiering, S.N. et al. (2022). Cowpea mosaic virus outperforms other members of the secoviridae as in situ vaccine for cancer immunotherapy. *Molecular Pharmaceutics* 19 (5): 1573–1585.

49 Stranger-Jones Yukiko, K., Bae, T., and Schneewind, O. (2006). Vaccine assembly from surface proteins of staphylococcus aureus. *Proceedings of the National Academy of Sciences* 103 (45): 16942–16947.

50 Singh, M., Chakrapani, A., and O'hagan, D. (2007). Nanoparticles and microparticles as vaccine-delivery systems. *Expert Review of Vaccines* 6 (5): 797–808.

51 Atabekov, J.G., Nikitin, N.A., and Karpova, O.V. (2015). New type platforms for in vitro vaccine assembly. *Moscow University Biological Sciences Bulletin* 70 (4): 177–183.

52 Li, A.W., Sobral, M.C., Badrinath, S. et al. (2018). A facile approach to enhance antigen response for personalized cancer vaccination. *Nature Materials* 17 (6): 528–534.

53 Guan, X., Chen, J., Hu, Y. et al. (2018). Highly enhanced cancer immunotherapy by combining nanovaccine with hyaluronidase. *Biomaterials* 171: 198–206.

54 Xi, X., Zhang, L., Lu, G. et al. (2018). Lymph node-targeting nanovaccine through antigen-CpG self-assembly potentiates cytotoxic T cell activation. *Journal of Immunology Research* 2018: 3714960.

55 Zhou, X., Zhang, X., Yu, X. et al. (2008). The effect of conjugation to gold nanoparticles on the ability of low molecular weight chitosan to transfer DNA vaccine. *Biomaterials* 29 (1): 111–117.

56 Grayson, E.J., Bernardes, G.J.L., Chalker, J.M. et al. (2011). A coordinated synthesis and conjugation strategy for the preparation of homogeneous glycoconjugate vaccine candidates. *Angewandte Chemie, International Edition* 50 (18): 4127–4132.

57 Lu, L., Duong, V.T., Shalash, A.O. et al. (2021). Chemical conjugation strategies for the development of protein-based subunit nanovaccines. *Vaccine* 9 (6): 563.

58 Hoyle, C.E. and Bowman, C.N. (2010). Thiol–Ene click chemistry. *Angewandte Chemie, International Edition* 49 (9): 1540–1573.

59 Kaur, J., Saxena, M., and Rishi, N. (2021). An overview of recent advances in biomedical applications of click chemistry. *Bioconjugate Chemistry* 32 (8): 1455–1471.

60 Holay, M., Guo, Z., Pihl, J. et al. (2021). Bacteria-inspired nanomedicine. *ACS Applied Bio Materials* 4 (5): 3830–3848.

61 Zhou, J., Kroll, A.V., Holay, M. et al. (2020). Biomimetic nanotechnology toward personalized vaccines. *Advanced Materials* 32 (13): 1901255.

62 Somiya, M. and Kuroda, S. (2015). Development of a virus-mimicking nanocarrier for drug delivery systems: the bio-nanocapsule. *Advanced Drug Delivery Reviews* 95: 77–89.

63 Yan, H., Shao, D., Lao, Y.H. et al. (2019). Engineering cell membrane-based nanotherapeutics to target inflammation. *Advanced Science* 6 (15): 1900605.

64 Luo, G.F., Chen, W.H., Zeng, X. et al. (2021). Cell primitive-based biomimetic functional materials for enhanced cancer therapy. *Chemical Society Reviews* 50 (2): 945–985.

65 Jia, J.L., Zhang, W.F., Liu, Q. et al. (2017). Adjuvanticity regulation by biodegradable polymeric nano/microparticle size. *Molecular Pharmaceutics* 14 (1): 14–22.

66 Yue, H., Yuan, L., Zhang, W.W. et al. (2018). Macrophage responses to the physical burden of cell-sized particles. *Journal of Materials Chemistry B* 6 (3): 393–400.

67 Zhang, X., Song, C., Ma, G.H. et al. (2018). Mechanical determination of particle-cell interactions and the associated biomedical applications. *Journal of Materials Chemistry B* 6 (44): 7129–7143.

68 Mohsen, M.O., Heath, M.D., Cabral-Miranda, G. et al. (2019). Vaccination with nanoparticles combined with micro-adjuvants protects against cancer. *Journal for Immunotherapy of Cancer* 7: 114.

69 Yue, H., Wei, W., Yue, Z. et al. (2010). Particle size affects the cellular response in macrophages. *European Journal of Pharmaceutical Sciences* 41 (5): 650–657.

70 Hong, X.Y., Zhong, X.F., Du, G.S. et al. (2020). The pore size of mesoporous silica nanoparticles regulates their antigen delivery efficiency. *Science Advances* 6 (25): 4462.

71 Kang, S., Ahn, S., Lee, J. et al. (2017). Effects of gold nanoparticle-based vaccine size on lymph node delivery and cytotoxic T-lymphocyte responses. *Journal of Controlled Release* 256: 56–67.

72 Kim, S.Y., Noh, Y.W., Kang, T.H. et al. (2017). Synthetic vaccine nanoparticles target to lymph node triggering enhanced innate and adaptive antitumor immunity. *Biomaterials* 130: 56–66.

73 Zhang, Y.N., Lazarovits, J., Poon, W. et al. (2019). Nanoparticle size influences antigen retention and presentation in lymph node follicles for humoral immunity. *Nano Letters* 19 (10): 7226–7235.

74 Liu, H., Moynihan, K.D., Zheng, Y. et al. (2014). Structure-based programming of lymph-node targeting in molecular vaccines. *Nature* 507 (7493): 519–522.

75 Underhill, D.M. and Goodridge, H.S. (2012). Information processing during phagocytosis. *Nature Reviews Immunology* 12 (7): 492–502.

76 Yu, W., Liu, R., Zhou, Y. et al. (2020). Size-tunable strategies for a tumor targeted drug delivery system. *ACS Central Science* 6 (2): 100–116.

77 Hua, S., He, J., Zhang, F. et al. (2021). Multistage-responsive clustered nanosystem to improve tumor accumulation and penetration for photothermal/enhanced radiation synergistic therapy. *Biomaterials* 268: 120590.

78 Xu, F., Huang, X., Wang, Y. et al. (2020). A size-changeable collagenase-modified nanoscavenger for increasing penetration and retention of nanomedicine in deep tumor tissue. *Advanced Materials* 32 (16): e1906745.

79 Wang, Y.Q., Fan, Q.Z., Liu, Y. et al. (2016). Improving adjuvanticity of quaternized chitosan-based microgels for H_5N_1 split vaccine by tailoring the particle properties to achieve antigen dose sparing effect. *International Journal of Pharmaceutics* 515 (1–2): 84–93.

80 Lybaert, L., Ryu, K.A., Nuhn, L. et al. (2017). Cancer cell lysate entrapment in $CaCO_3$ engineered with polymeric Tlr-Agonists: immune-modulating microparticles in view of personalized antitumor vaccination. *Chemistry of Materials* 29 (10): 4209–4217.

81 Fromen, C.A., Rahhal, T.B., Robbins, G.R. et al. (2016). Nanoparticle surface charge impacts distribution, uptake and lymph node trafficking by pulmonary antigen-presenting cells. *Nanomedicine* 12 (3): 677–687.

82 Seydoux, E., Rodriguez-Lorenzo, L., Blom, R.A. et al. (2016). Pulmonary delivery of cationic gold nanoparticles boost antigen-specific CD4(+) *T* cell proliferation. *Nanomedicine* 12 (7): 1815–1826.

83 Han, R., Zhu, J., Yang, X. et al. (2011). Surface modification of poly(DL-lactic-*co*-glycolic acid) nanoparticles with protamine enhanced cross-presentation of encapsulated ovalbumin by bone marrow-derived dendritic cells. *Journal of Biomedical Materials Research. Part A* 96 (1): 142–149.

84 Xu, J., Lv, J., Zhuang, Q. et al. (2020). A general strategy towards personalized nanovaccines based on fluoropolymers for post-surgical cancer immunotherapy. *Nature Nanotechnology* 15 (12): 1043–1052.

85 Qiao, C., Liu, J., Yang, J. et al. (2016). Enhanced non-inflammasome mediated immune responses by mannosylated zwitterionic-based cationic liposomes for Hiv DNA vaccines. *Biomaterials* 85: 1–17.

86 Caster, J.M., Callaghan, C., Seyedin, S.N. et al. (2019). Optimizing advances in nanoparticle delivery for cancer immunotherapy. *Advanced Drug Delivery Reviews* 144: 3–15.

87 Jiang, Y., Huo, S.D., Mizuhara, T. et al. (2015). The interplay of size and surface functionality on the cellular uptake of Sub-10 Nm gold nanoparticles. *ACS Nano* 9 (10): 9986–9993.

88 Du, Y., Xia, Y., Zou, Y. et al. (2019). Exploiting the lymph-node-amplifying effect for potent systemic and gastrointestinal immune responses via polymer/lipid nanoparticles. *ACS Nano* 13 (12): 13809–13817. https://doi.org/10.1021/acsnano.9b04071.

89 Talamini, L., Violatto, M.B., Cai, Q. et al. (2017). Influence of size and shape on the anatomical distribution of endotoxin-free gold nanoparticles. *ACS Nano* 11 (6): 5519–5529.

90 Ni, D., Qing, S., Ding, H. et al. (2017). Biomimetically engineered demi-bacteria potentiate vaccination against cancer. *Advanced Science* 4 (10): 1700083.

91 Mammadov, R., Cinar, G., Gunduz, N. et al. (2015). Virus-like nanostructures for tuning immune response. *Scientific Reports* 5: 16728.

92 Kumar, S., Anselmo, A.C., Banerjee, A. et al. (2015). Shape and size-dependent immune response to antigen-carrying nanoparticles. *Journal of Controlled Release* 220 (Part A): 141–148.

93 Sharma, G., Valenta, D.T., Altman, Y. et al. (2010). Polymer particle shape independently influences binding and internalization by macrophages. *Journal of Controlled Release* 147 (3): 408–412.

94 Lin, X., Lin, X., and Gu, N. (2020). Optimization of hydrophobic nanoparticles to better target lipid rafts with molecular dynamics simulations. *Nanoscale* 12 (6): 4101–4109.

95 Niu, Y., Yu, M., Meka, A. et al. (2016). Understanding the contribution of surface roughness and hydrophobic modification of silica nanoparticles to enhanced therapeutic protein delivery. *Journal of Materials Chemistry B* 4 (2): 212–219.

96 Yang, Y., Lu, Y., Abbaraju, P.L. et al. (2017). Multi-shelled dendritic mesoporous organosilica hollow spheres: roles of composition and architecture in cancer immunotherapy. *Angewandte Chemie International edtion in English* 56 (29): 8446–8450.

97 Skwarczynski, M., Zhao, G., Boer, J.C. et al. (2020). Poly(amino acids) as a potent self-adjuvanting delivery system for peptide-based nanovaccines. *Science Advances* 6 (5): eaax2285.

98 Peng, S., Cao, F., Xia, Y. et al. (2020). Particulate alum via pickering emulsion for an enhanced Covid-19 vaccine adjuvant. *Advanced Materials* 32 (40): 2004210.

99 Min, Y.Z., Roche, K.C., Tian, S.M. et al. (2017). Antigen-capturing nanoparticles improve the abscopal effect and cancer immunotherapy. *Nature Nanotechnology* 12 (9): 877–882.

100 Yue, H., Wei, W., Yue, Z. et al. (2012). The role of the lateral dimension of graphene oxide in the regulation of cellular responses. *Biomaterials* 33 (16): 4013–4021.

101 Xia, Y., Wu, J., Wei, W. et al. (2018). Exploiting the pliability and lateral mobility of pickering emulsion for enhanced vaccination. *Nature Materials* 17 (2): 187–194.

102 Hui, Y., Yi, X., Wibowo, D. et al. (2020). Nanoparticle elasticity regulates phagocytosis and cancer cell uptake. *Science Advances* 6 (16): eaaz4316.

103 Palomba, R., Palange, A.L., Rizzuti, I.F. et al. (2018). Modulating phagocytic cell sequestration by tailoring nanoconstruct softness. *ACS Nano* 12 (2): 1433–1444.

104 Nguyen, T.L., Choi, Y., and Kim, J. (2019). Mesoporous silica as a versatile platform for cancer immunotherapy. *Advanced Materials* 31 (34): 1803953.

105 Sun, Q.X., Barz, M., De Geest, B.G. et al. (2019). Nanomedicine and macroscale materials in immuno-oncology. *Chemical Society Reviews* 48 (1): 351–381.

106 Fries, C.N., Curvino, E.J., Chen, J.L. et al. (2021). Advances in nanomaterial vaccine strategies to address infectious diseases impacting global health. *Nature Nanotechnology* 16 (4): 1–14.

107 Hong, S. and Zhang, Z. (2018). B cells are the dominant antigen-presenting cells that activate naive CD_4^+ T cells upon immunization with a virus-derived nanoparticle antigen. *Immunity* 49 (4): 695–708.

108 Irvine, D.J., Hanson, M.C., Rakhra, K. et al. (2015). Synthetic nanoparticles for vaccines and immunotherapy. *Chemical Reviews* 115 (19): 11109–11146.

109 Azzi, J., Yin, Q., Uehara, M. et al. (2016). Targeted delivery of immunomodulators to lymph nodes. *Cell Reports* 15 (6): 1202–1213.

110 Li, F., Nie, W.D., Zhang, F. et al. (2019). Engineering magnetosomes for high-performance cancer vaccination. *ACS Central Science* 5 (5): 796–807.

111 Song, T.T., Xia, Y.F., Du, Y.Q. et al. (2021). Engineering the deformability of albumin-stabilized emulsions for lymph-node vaccine delivery. *Advanced Materials* 33 (26): 2100106.

112 Wang, J., Wang, S., Ye, T. et al. (2020). Choice of nanovaccine delivery mode has profound impacts on the intralymph node spatiotemporal distribution and immunotherapy efficacy. *Advanced Materials* 7 (19): 2001108.

113 Liu, Q., Chen, X., Jia, J. et al. (2015). pH-Responsive poly(DL-lactic-*co*-glycolic acid) nanoparticles with rapid antigen release behavior promote immune response. *ACS Nano* 9 (5): 4925–4938.

114 Xi, X., Ye, T., Wang, S. et al. (2020). Self-healing microcapsules synergetically modulate immunization microenvironments for potent cancer vaccination. *Science Advances* 6 (21): eaay7735.

115 Xie, X., Hu, Y., Ye, T. et al. (2021). Therapeutic vaccination against leukaemia via the sustained release of co-encapsulated anti-Pd-1 and a leukaemia-associated antigen. *Nature Biomedical Engineering* 5 (5): 414–428.

116 Baranov, M.V., Kumar, M., Sacanna, S. et al. (2020). Modulation of immune responses by particle size and shape. *Frontiers in Immunology* 11: 607945.

117 Alexis, F., Pridgen, E., Molnar, L.K. et al. (2008). Factors affecting the clearance and biodistribution of polymeric nanoparticles. *Molecular Pharmaceutics* 5 (4): 505–515.

118 Bachmann, M.F. and Jennings, G.T. (2010). Vaccine delivery: a matter of size, geometry, kinetics and molecular patterns. *Nature Reviews Immunology* 10 (11): 787–796.

119 Jia, Y., Omri, A., Krishnan, L. et al. (2017). Potential applications of nanoparticles in cancer immunotherapy. *Human Vaccines & Immunotherapeutics* 13 (1): 63–74.

120 Xia, T., Rome, L., and Nel, A. (2008). Nanobiology: particles slip cell security. *Nature Materials* 7 (7): 519–520.

121 Ahmad, A., Khan, J.M., and Haque, S. (2019). Strategies in the design of endosomolytic agents for facilitating endosomal escape in nanoparticles. *Biochimie* 160: 61–75.

122 Zhou, S., Huang, Y., Chen, Y. et al. (2020). Engineering ApoE3-incorporated biomimetic nanoparticle for efficient vaccine delivery to dendritic cells via macropinocytosis to enhance cancer immunotherapy. *Biomaterials* 235: 119795.

123 Stead, S.O., Kireta, S., Mcinnes, S.J.P. et al. (2018). Murine and non-human primate dendritic cell targeting nanoparticles for in vivo generation of regulatory T-cells. *ACS Nano* 12 (7): 6637–6647.

124 Yang, R., Xu, J., Xu, L. et al. (2018). Cancer cell membrane-coated adjuvant nanoparticles with mannose modification for effective anticancer vaccination. *ACS Nano* 12 (6): 5121–5129.

125 Kim, D., Wu, Y., Shim, G. et al. (2021). Lipid nanoparticle-mediated lymphatic delivery of immunostimulatory nucleic acids. *Pharmaceutics* 13 (4): 490.

126 Gu, Z.L., Chen, S.H., Ding, Z.W. et al. (2019). The molecular mechanism of robust macrophage immune responses induced by pegylated molybdenum disulfide. *Nanoscale* 11 (46): 22293–22304.

127 Lee, K., Huang, Z.N., Mirkin, C.A. et al. (2020). Endosomal organization of CpG constructs correlates with enhanced immune activation. *Nano Letters* 20 (8): 6170–6175.

128 Wang, S., Ni, D., Yue, H. et al. (2018). Exploration of antigen induced $CaCO_3$ manoparticles for therapeutic vaccine. *Small* 14 (14): 1704272.

9

Applications of Uniform Microspheres and Super-porous Microspheres in Biochemical Engineering

9.1 Introductions

Biotechnology is in a rapid development stage and has formed an interdisciplinary and comprehensive system of research, development, and production, which is led by medical biotechnology, agricultural biotechnology, and industrial biotechnology. The rapid development of the upstream field of biotechnology has put forward higher requirements to the downstream field of biochemical engineering (Figure 9.1). Take bioseparation engineering as an example, as a key component of the downstream field of biotechnology, it plays a primary role in efficient purification of target proteins from complex systems. Among many bioseparation technologies, chromatography is one of the most effective and mildest separation methods and has been widely used in the purification of proteins, enzymes, vaccines, antibodies, polypeptides, nucleic acids, and other biological molecules. The development of chromatography depends on the chromatographic media. The key factor of chromatography is to select the chromatographic media with both good performance and reasonable price and to optimize the process conditions at the same time. For another example, cell culture engineering is the key stage of biochemical engineering and is the basis of large-scale production of vaccines, antibodies, therapeutic proteins, and other biological products. Cell culture carriers are the core materials of this process.

Microspheres are the most widely used biochemical engineering materials because of their rich structure and variety, easy functionalization, excellent pore structure, and good stability. Microspheres are spherical particles that have a good effect as packing material for columns, and the space between microspheres is relatively uniform when the column is packed, and the pressure resistance is far lower than that of other materials. Furthermore, microspheres usually have stable physical and chemical properties and have excellent surface properties and pore structures, which are suitable for purification according to various separation mechanisms. The well-designed microspheres also have good performance for maintaining bioactivity of biomolecules during the purification of biomolecules. In addition, microspheres can be produced on large scale with good stability from batch to batch.

Novel Membrane Emulsification: Principles, Preparation, Processes, and Bioapplications,
First Edition. Guanghui Ma.

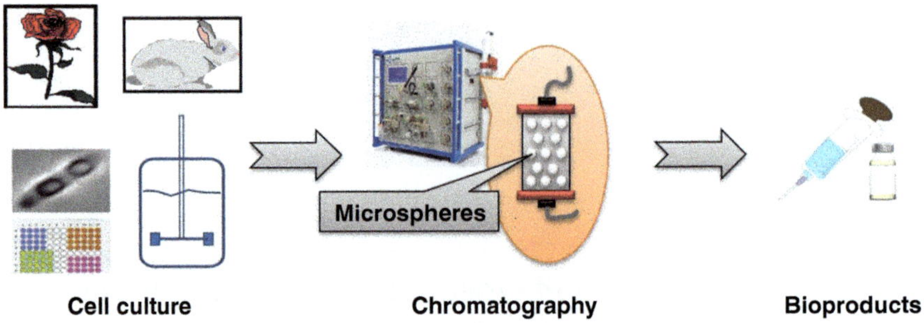

Figure 9.1 The role of uniform microspheres and super-porous microspheres in biochemical engineering.

Uniform microspheres have always been pursued in the field of biochemical engineering. In chromatography, the uniform-sized chromatographic media can improve the resolution, reduce the back pressure and, hence, greatly improve the separation efficiency. In cell culture, uniform microspheres are useful for cell proliferation in a homogeneous microenvironment. Our group has developed membrane emulsification technology for preparation of uniform microspheres used in biochemical engineering and now, this technology has received a great attention in various areas.

Super-porous microspheres have unique advantages in purification of superlarge biomolecules. The mass transfer of biomolecules in traditional chromatographic media is mainly by diffusion. For large biomolecules, some even reach tens of millions Da, and the size up to hundreds of nanometers or even microns, they can only bind to the surface of the traditional agarose-based media and cannot enter the inner pores, resulting in low mass transfer efficiency and unable to meet the purification requirements. Compared with traditional chromatographic media, super-porous microspheres have stronger permeability to biomolecules, which can flow in the inner pores of super-porous microspheres by both convection and diffusion, greatly increasing the separation speed and improving the mass transfer efficiency.

This chapter will introduce the applications of both uniform microspheres and super-porous microspheres in biochemical engineering. The preparation materials for microspheres will be discussed, and the microspheres prepared by various materials have their unique structures and functions, which play a special effect in applications.

9.2 Uniform Microspheres for Chromatographic Media

9.2.1 Significance of Particle Size Uniformity in Chromatography

Since the 1990s, the global market for biotechnology products has grown by more than 30% a year. Genetic engineering and cell engineering as representatives of life science and technology have maintained rapid growth. A variety of biological products, e.g. antibodies and vaccines have been available, which have made great contributions to enhance human health. The rapid growth in biological market and higher demands for the quality and safety of products present serious challenges to separation and purification in downstream fields. High effective separation and purification technology is the most reliable guarantee for large-scale production of biological products with high purity and activity. As one of the most important bioseparation technologies, chromatography has the advantages of high separation efficiency, good maintenance for bioactivity, easy scale-up and high degree of automation, and has been widely used in the downstream fields of biotechnology.

As the stationary phase of chromatography, chromatographic media is the core element of this technology. Microspheres are the most common appearance of chromatographic media. Particle size and distribution, pore size and distribution, ligand density, and other property parameters are closely related to the performance of chromatographic media, thus affecting the final separation effect. Among them, the size and uniformity of microspheres have a direct impact on chromatography process and being the most important factors influencing the final separation effect [1].

The effect of particle size should be considered comprehensively. On one hand, microspheres with small particle size have short paths of mass transfer, which is beneficial to biomolecules diffusing into the internal part of microspheres, thus both achieving a higher resolution and providing a larger surface area are helpful in increasing the loading capacity. However, the flow rate is slow and the back pressure increases if the particle size is too small. On the other hand, when microspheres with large particle size are used, the diffusion equilibrium is reached for a longer time, leading to a lower resolution, while the flow rate is fast with small resistance to the mobile phase, which is suitable for large-scale industrial separation and purification.

The uniformity of microspheres is closely related to the separation efficiency. Nonuniform microsphere brings a series of disadvantages, including the decrease of resolution, the decrease of void content, and the increase of back pressure. On the contrary, the more uniform the microspheres are, the higher resolution the chromatographic column has, and the better the separation effect of every component in the mixture is. Mainly, the effects of the uniformity of microspheres lie in the following two aspects as shown in Figure 9.2. (i) Reduction of the peak width. The multipath effect caused by nonuniformity can be overcome. The path lengths of solute molecules flowing through the column packed with nonuniform microspheres are different, resulting in the distribution of fluid velocity in the cross-section of the column. Therefore, solute molecules are unable to move rapidly in mobile phase to achieve the balance in the cross-section of the column. The direct consequences are the increase of the plate height, the broadening of the peak width, and the decrease of the resolution, all of which lead to a low column efficiency. However, uniform microspheres can significantly reduce the distribution of the fluid velocity of solute molecules in the cross-section of the column, thereby inhibiting the increase of the plate height and the expansion of the peak width. In addition, the longitudinal diffusion effect of solutes can also be weakened. Solutes diffuse in the moving direction of the mobile phase due to the concentration gradient in the column, forming a Gaussian distribution and resulting in the broadening

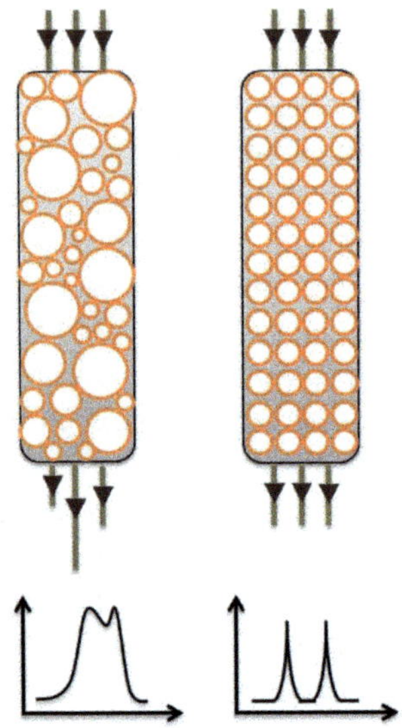

Figure 9.2 Scheme of the effects of particle size uniformity on separation efficiency. (a) Nonuninform and (b) uniform.

of the peak width. This condition is improved when uniform microspheres are used, as the plate height caused by longitudinal diffusion can be reduced effectively, and the peak width is reduced accordingly. (ii) Improvement of column permeability and reduction of column pressure. The column permeability is closely related to the column pressure. The worse the permeability is, the higher the pressure is. The permeability of the column packed with nonuniform microspheres is relatively worse than that packed with uniform microspheres because small-sized microspheres fill in the void of large-sized ones, and hence the column is easy to be blocked with a great increase of column pressure. However, the column packed with uniform microspheres has a relatively high void content, which is beneficial for its permeability. Under this condition, the mobile phase flows through the column more fluently and the column pressure thus decreases.

Therefore, microsphere with uniform and a controllable particle size is one of the basic requirements for highly efficient purification and is also a direction of effort for microsphere producers. The microspheres prepared using traditional suspension polymerization method are not uniform. In order to overcome this disadvantage, we used membrane emulsification technology to prepare uniform microspheres for chromatography, and good application results have been achieved.

9.2.2 Agarose Microspheres

Agarose is a kind of natural polysaccharide, which is refined from agar extracted from red seaweeds such as Gelidium or Gracilaria. It is a linear polysaccharide composed of 1,3-linked β-D-galactose and 1,4-linked 3,6-dehydrated-α-L-pyranogalactose residues linked alternately (Figure 9.3). Agarose dissolves in hot water to form agarose solution, and agarose gel is formed after the temperature of agarose solution is cooling down. The process of forming agarose solution into

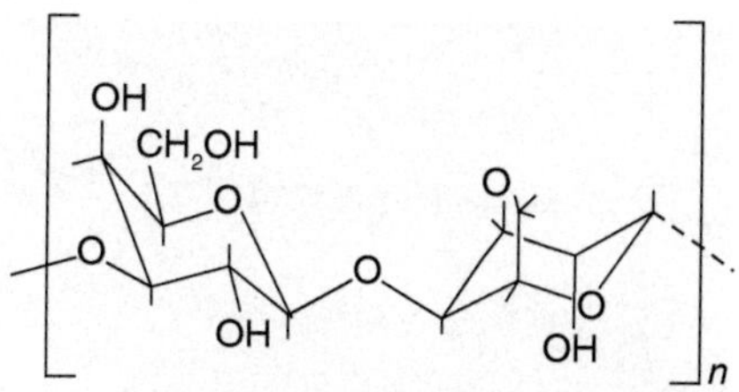

Figure 9.3 The structural formula of agarose.

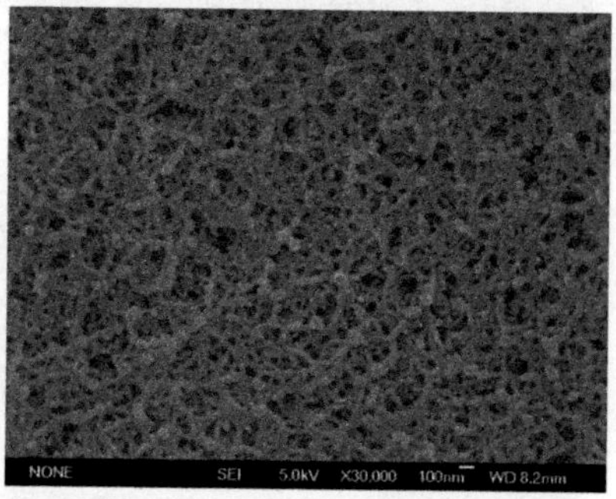

Figure 9.4 Scanning electron microscopy (SEM) photographs of agarose microspheres.

agarose gel is a reversible process in which hydrogen bond is the main driving force. Agarose hydrogels contain a large number of pores that allow the diffusion of macromolecules (Figure 9.4). In contrast to single-strand gel bundles formed by other natural polysaccharides, such as dextran, multistrand gel bundles are formed in agarose gel with larger pores. Agarose has a series of virtues, e.g. rich hydroxyl groups, good hydrophilicity, easy to be functionalized, high binding capacity, excellent pore structure, low nonspecific adsorption, good stability, etc. and is very suitable for preparation of chromatographic media. Agarose-based media is by far the most commonly used chromatographic media for ion-exchange chromatography, hydrophobic chromatography, and affinity chromatography. It has been used for purification of proteins, nucleic acids, peptides, polysaccharides and other substances in the fields of biopharmaceutical, food industry, and chemical industry [2, 3].

Traditionally, agarose microspheres are prepared using a mechanical stirring method and have a wide particle size distribution, which needs to be sieved subsequently. Therefore, it needs multisteps for preparation and wastes raw materials. Moreover, microspheres are still nonuniform even after sieving, leading to a low separation efficiency. Our group has successfully prepared uniform agarose microspheres using membrane emulsification method (Figure 9.5) [4–6]. Uniform agarose microspheres with a controllable particle size were prepared using both direct membrane emulsification method and rapid (premix) membrane emulsification method, respectively. In direct membrane emulsification method, agarose solution was forced to pass through a membrane with a certain pore size into the oil phase containing the emulsifier by applying trans-membrane pressure, to form uniform droplet. In rapid membrane emulsification method, a pre-emulsified water-in-oil suspension of agarose solution is forced to pass through the membrane

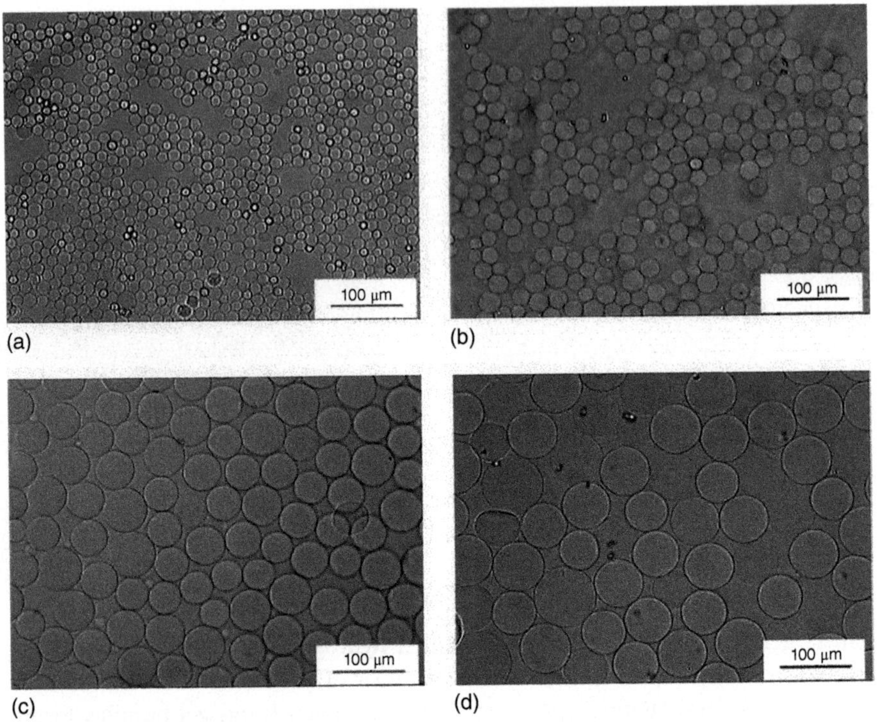

Figure 9.5 Optical microscopic photographs of agarose microspheres prepared by direct membrane emulsification technique with different pore sizes of membrane: (a) 4.7 μm; (b) 5.7 μm; (c) 10.2 μm; and (d) 19.6 μm. Source: Zhou et al. [5]/Elsevier.

to form small droplet. Then the W/O emulsion formed was cooled down and agarose microspheres were obtained. Microspheres produced using these methods have a much more uniform particle size distribution than those manufactured by traditional mechanical stirring methods. By optimizing the oil phase composition, types and amounts of emulsifiers, emulsification temperature, and other conditions, preparation of agarose microspheres with uniform and controllable particle size was achieved. By the rapid membrane emulsification process, uniform agarose microspheres with both small particle size and high agarose concentration were successfully prepared, for purpose of microspheres below 10 microns used in analytical chromatography of biomolecules. The effects of the trans-membrane pressure, trans-membrane times, properties of pre-emulsion, and composition of oil–water phase on both size and size distribution of microspheres were studied. Various rigid chromatographic media can be further prepared followed by cross-linking of agarose microsphere and modifying it further with various ligands.

Generally, the average particle size of agarose chromatographic media for industrial production is around 90 μm, because the agarose chromatographic media with a particle size of this range has better hydraulics properties and a lower operation back pressure, which is suitable for large-scale purification. Using the direct membrane emulsification technology, uniform agarose microspheres with a controllable particle size have been successfully prepared and used for bioseparation. However, the particle size of agarose microspheres prepared by this technology was generally between 15–60 μm, and it was difficult to prepare larger ones, because the trans-membrane pressure was not easy to be controlled when the membrane pore was large, and the large droplets were unstable and they easily coalesced on the bottom of the reactor, and then further were broken by agitation.

To prepare large-sized uniform agarose microspheres, we developed a series of strategies [7, 8]. For example, we used 6% agarose solution as water phase, a mixture of liquid paraffin and petroleum ether with different volume ratios as oil phase and PO-5S as emulsifier, uniform agarose microspheres with an average particle size of 90 μm were prepared. The effects of the membrane pore size, the oil phase composition, the emulsification temperature, and the trans-membrane pressure on the size and size distribution of agarose microspheres were investigated. The results showed that under the optimized conditions, i.e. the membrane pore size was 25.9 μm, the volume ratio of liquid paraffin to petroleum ether was 11 : 1, and the emulsification temperature was 65 °C, uniform large-sized agarose microspheres with a mean size of 93.3 μm and a Span value of 1.25 were prepared. Under this condition, good repeatability was achieved, and the relative standard deviation of each batch was only 1.34%. The microscopic photographs of agarose microspheres prepared by membrane emulsification and Sepharose 6FF were shown in Figure 9.6 [8], and it can be seen that agarose microspheres prepared by membrane emulsification method were more uniform than Sepharose 6FF, even though the latter has been sieved. The particle size distribution of agarose microspheres has a great effect on the pressure–velocity curve of the chromatographic media, and these agarose microspheres were cross-linked by an optimized two-step procedure. The pressure–velocity curves of cross-linked self-made chromatographic media, domestic media, and imported media were as shown in Figure 9.7 [8], and it can be seen that under the same pressure, the flow rates of three different chromatographic media were similar at the beginning, and after that the flow rate of self-made

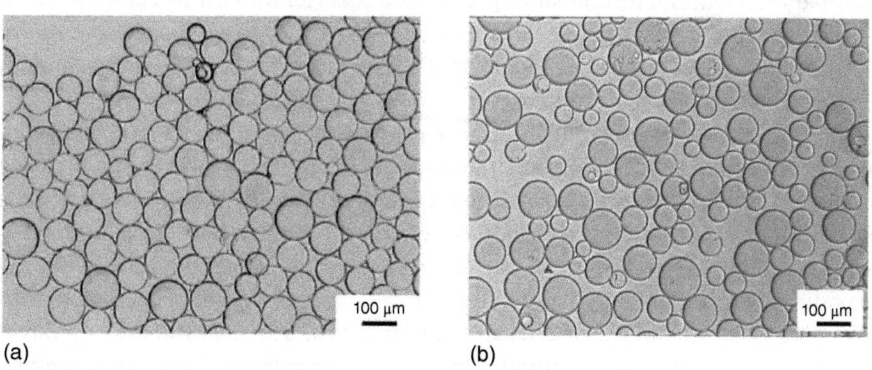

Figure 9.6 Optical microscopic photographs of agarose microspheres prepared by different methods. (a) Microspheres prepared by membrane emulsification; (b) Sepharose 6FF. Source: Lin et al. [8]/The Chinese Journal of Process Engineering.

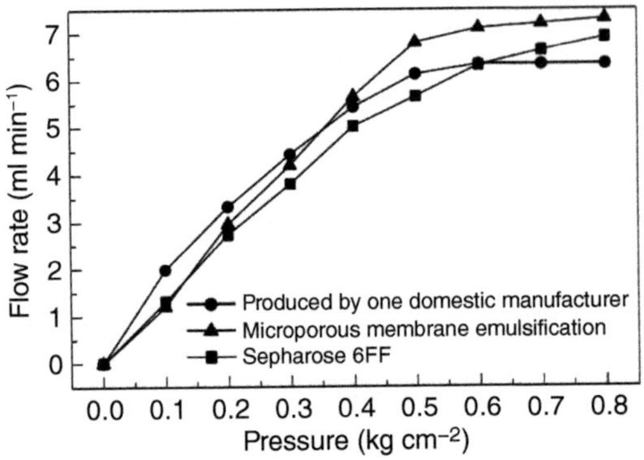

Figure 9.7 Comparison of pressure-flow rate curves of three kinds of agarose microspheres. Source: Lin et al. [8]/with permission from John Wiley & Sons, Inc.

uniform agarose microspheres was the highest, indicating that it had the best column permeability. The results showed that agarose microspheres prepared by membrane emulsification method had a better hydraulic flow performance without need of sieving process [8].

Furthermore, we have developed a direct membrane emulsification apparatus for scale-up of the preparation of large microspheres [9]. This apparatus has two improvements compared with the original one. One is that the arrangement mode of the membrane tube is changed from horizontal to vertical, and the emulsion droplets fall off from the surface of the membrane tube with the help of the gravity to avoid the damage of the shear force on the uniform droplets. The other is that the pump circulation is changed to mechanical stirring, and then the emulsion droplets are dispersed into the emulsion tank in a gentle way to avoid their secondary breaking caused by pump circulation. Also, the effect of membrane pore size, trans-membrane pressure, stirring speed during emulsification, oil phase composition, emulsifier, and emulsification temperature on the size and size distribution of agarose microsphere were studied, respectively [10]. Under optimized preparation conditions, uniform large-sized agarose microspheres were prepared with $d_{0.5}$ of 92.1 μm and Span value of 1.15, respectively. The obtained product can be used for bioseparation without further sieving and has a superior uniformity compared with commercial Sepharose 4FF. On this basis, the formation mechanism of uniform emulsion was deeply discussed, which laid a foundation for large-scale preparation of large-sized agarose microspheres.

This direct membrane emulsification method was also used for preparation of highly cross-linked uniform agarose microspheres. In order to solve the problems of low cross-linking efficiency and weak gel strength of agarose microspheres prepared by traditional cross-linking methods, a new method named precross-linking for preparing highly cross-linked agarose microspheres was proposed. By introducing cross-linking agent of allyl glycidyl ether into agarose solution, agarose with allyl groups was obtained and used to prepare uniform microspheres by direct membrane emulsification method. Then, after bromination and cross-linking reaction successively, highly cross-linked uniform agarose microspheres were prepared. This method can effectively improve the concentration of cross-linking agent in agarose microspheres, to 49.8 μmol g^{-1} gel, and the cross-linking agent was homogeneously distributed in microspheres. With combining direct membrane emulsification method, highly cross-linked uniform agarose microspheres were successfully prepared. The effects of the amount of cross-linking agent and the reaction conditions on the mechanical strength of agarose microspheres were systematically investigated, and the preparation process was optimized. Under optimized conditions, highly cross-linked uniform agarose microspheres were regularly spherical with a maximum flow rate of $1452\,\text{cm}\,\text{h}^{-1}$, which was 170% higher than that of commercial Sepharose 4FF.

In high-resolution chromatography, both strength and flow rate are also important factors for agarose microspheres besides uniformity due to their small particle size. In order to meet these requirements, our strategy is to prepare high-concentration agarose microspheres. Direct membrane emulsification method was used to prepare uniform agarose microspheres with the largest concentration of 6% agarose solution, and the uniformity of the microspheres decreased when the concentration of agarose solution continued to increase. We developed rapid membrane emulsification method to prepare high-concentration agarose microspheres. At the first step, larger agarose droplets were prepared by conventional mechanical stirring method, then they were pressed through the membrane rapidly, and broken into uniform smaller droplets under optimum conditions. We have successfully used this rapid membrane emulsification method to prepare high-concentration agarose microspheres with uniform small sizes [11, 12]. To improve the resolution further, dextran was introduced to the high-concentration agarose microsphere, which had both macroporous structures belonging to agarose and microporous structures belonging to dextran. Also, the effective cross-linking method was adopted to prepare high-resolution chromatographic media. Model protein mixtures were loaded onto this high-resolution column, and K_d-R_m curves were plotted, where K_d was the distribution coefficient and R_m was the viscosity radius of model protein, and the pore size distribution of chromatographic media was also estimated (Figures 9.8 and 9.9) [12]. By adjusting both the composition of two polysaccharides and the molecular weight of the dextran, the particle size and pore structure of the microspheres can be well controlled. With increasing the concentration of agarose solution or decreasing the concentration of dextran solution, the average pore size of the composite microspheres increased firstly and then decreased, and the pore size distribution became narrower. With the increase of the molecular weight of dextran, the microspheres had smaller pores and narrower pore size distribution. The composite microspheres have a higher resolution than pure agarose microspheres, and also the separation range is controllable. Compared with conventional analytical chromatographic media, the composite microspheres prepared by both 10% agarose /2% dextran T40 or 8% agarose /4% dextran T150 have a higher resolution, especially in the separation range of low molecular weight. At the same time, the mechanical strength of the composite polysaccharide microspheres can be improved by regulating the composition of microspheres. This composite polysaccharide microsphere with uniform small size has great potential for applications of high-resolution chromatography.

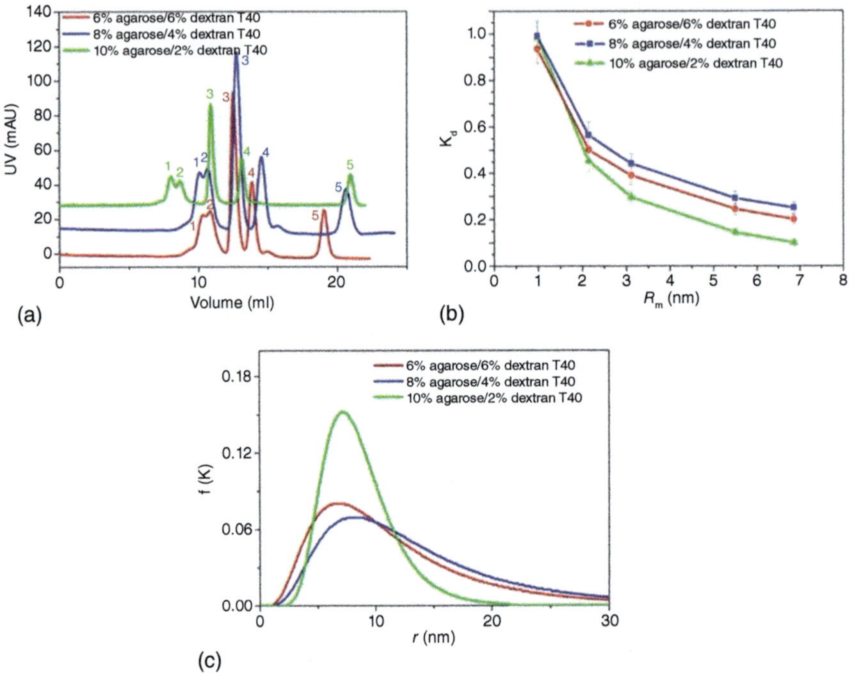

Figure 9.8 (a) Chromatogram of model proteins, (b) K_d-R_m curves, and (c) pore size distributions of polysaccharide composite microspheres with various compositions. Source: From Zhang et al. [12]/with permission from Springer Nature.

9.2.3 Konjac Glucomannan Microspheres

Konjac Glucomannan (KGM) is a neutral polysaccharide with the largest molecular weight formed by β-1,4 bond polymerization of mannan and glucose (Figure 9.10). It is generally derived from konjac, which is widely grown in tropical and subtropical regions of Asia. KGM has a unique network structure and physicochemical properties similar to that of dextran, for example, high hydrophilicity, biocompatibility, degradability, easy-to-absorb water, and gel formation. And the price of KGM is much lower than that of dextran; it has greater application potentials for bioseparation, enzyme immobilization, microcarriers for cell culture, and so forth.

Using KGM as raw material, we prepared uniform KGM microspheres by rotating membrane emulsification method combined with chemical cross-linking method. During the preparation, three kinds of 12% KGM solutions with different viscosities were used as dispersed phase (water phase), and a mixed oil phase composed of liquid paraffin and petroleum ether as continuous phase. The effects of emulsifier types on the stability of KGM emulsion, and the effects of the viscosity of water phase, the oil phase composition, and the rotation speed of membrane on preparation of KGM microspheres were investigated, respectively. The higher the viscosity of KGM water phase was, the lower the optimal viscosity of the oil phase was. The optimal preparations were obtained as the viscosity of KGM water phase was 1548 mPa·s, the volume ratio of liquid paraffin to petroleum ether was 5 : 1, the

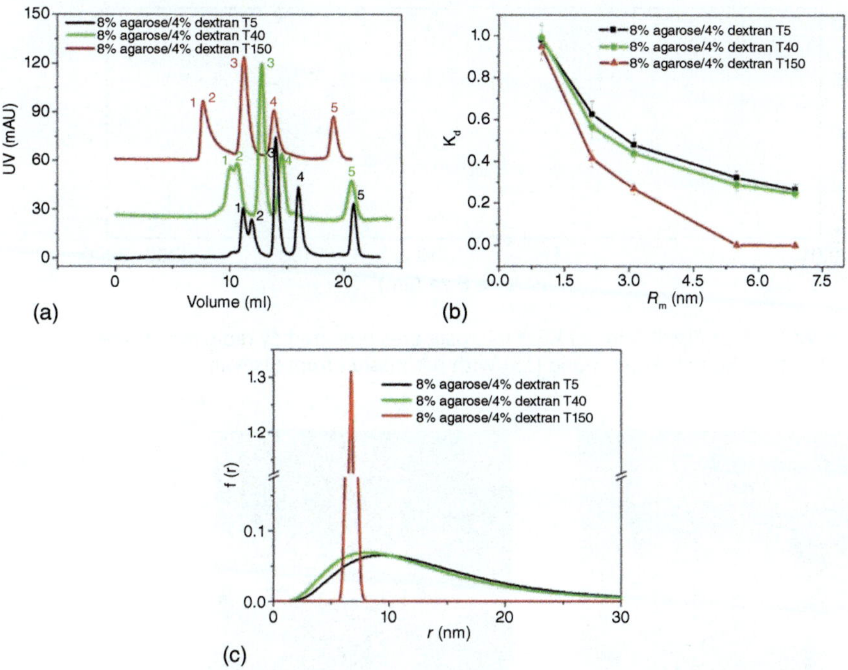

Figure 9.9 (a) Chromatogram of model proteins, (b) K_d-R_m curves, and (c) pore size distributions of polysaccharide composite microspheres using dextran with various molecular weights. Source: From Zhang et al. [12]/with permission from Springer Nature.

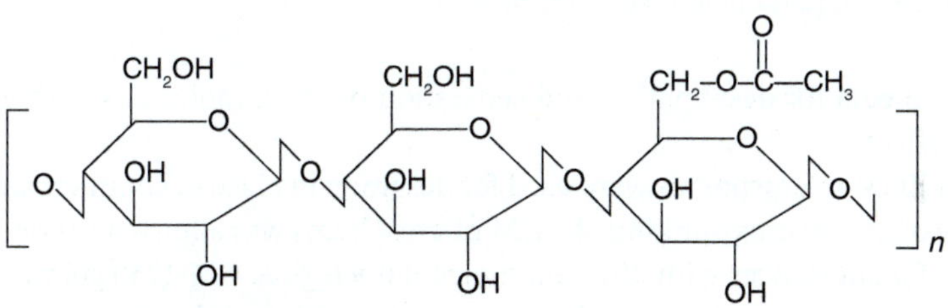

Figure 9.10 The structural formula of KGM.

rotational speed was 400 r·m^{-1}, and the emulsifier was 4% Span 80. Under these optimal conditions, uniform KGM microspheres were obtained with an average size of about 70 μm and a Span value <1.0 [13, 14].

We have also prepared uniform KGM microspheres with an average particle size of less than 10 μm by using rapid membrane emulsification method. The effects of the oil phase composition, the amount of emulsifier, oil–water ratio, and the stirring speed during the cross-linking procedure on the size and size distribution of microspheres were studied, respectively, and the optimum preparation conditions were determined by orthogonal experiments. Under the optimum condition, KGM microspheres with an average size of less than 10 μm and a Span value of 0.9 were obtained (Figures 9.11 and 9.12) [15]. The KGM microspheres can be used as

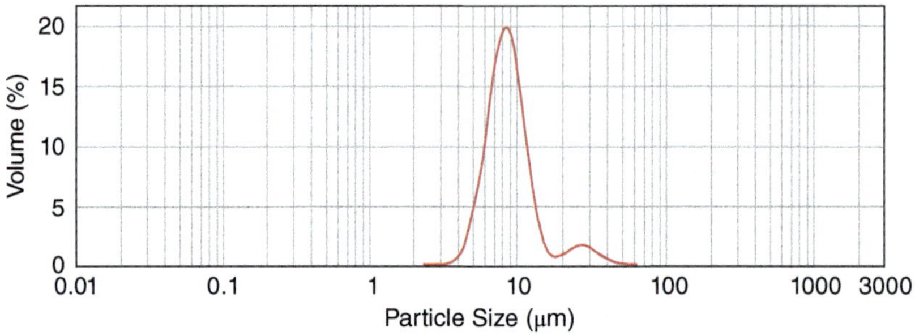

Figure 9.11 Size distribution of KGM microspheres prepared by rapid membrane emulsification. Source: From Xiong [15]/with permission from Elsevier.

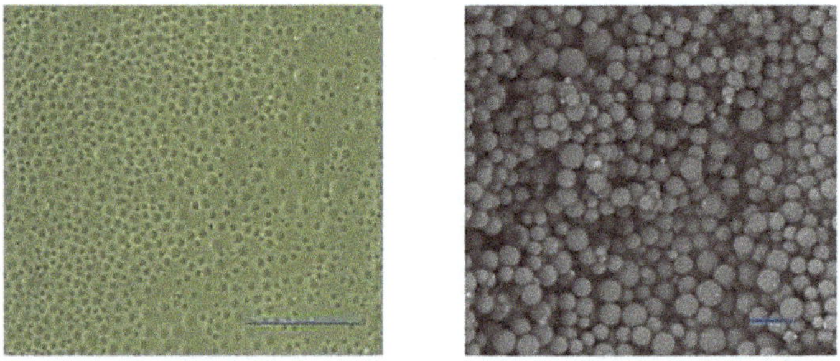

Figure 9.12 (a) Optical micrograph of KGM microspheres; (b) SEM of KGM microspheres (Bar: 10 μm). Source: Xiong [15]/Chinese Academy of Sciences.

chromatographic media for desalination and separation of small molecules with a good separation effect.

These low-cost KGM microspheres were used for desalting in place of commercial dextran gels. The degree of cross-linking of KGM microspheres was adjusted to control its selectivity for substances with different molecular weights. GM-1250 gel was found to be suitable for fractionation of globular proteins with a molecular weight range of 1–10 kDa, and GM-1225 gel with a fractionation range of 4–30 kDa. The desalting effect of filamentous hemagglutinin (FHA) on GM-1250 gel and Sephadex G25 was compared as shown in Figure 9.13 [16]. FHA is a large surface-associated adhesive protein of *Bordetella pertussis* with a molecular weight of 220 kDa, which is a component of some new acellular pertussis vaccines. The recoveries for FHA on both columns were over 95%, showing the GM-1250 has comparative desalting capacity with the widely used commercial product.

The stability of KGM microspheres stored in various aqueous solutions at 40 °C for 1 week was investigated by measuring the recovery of BSA before and after storage (Figure 9.14) [16]. KGM microspheres are excellent in chemical stability and can be used over the whole pH range (1–14). They can be stored in 0.1 M HCl or 1 M NaOH at 40°C for 1 week without any change in chromatographic properties. Moreover, the microspheres can be used in high concentrations of guanidine

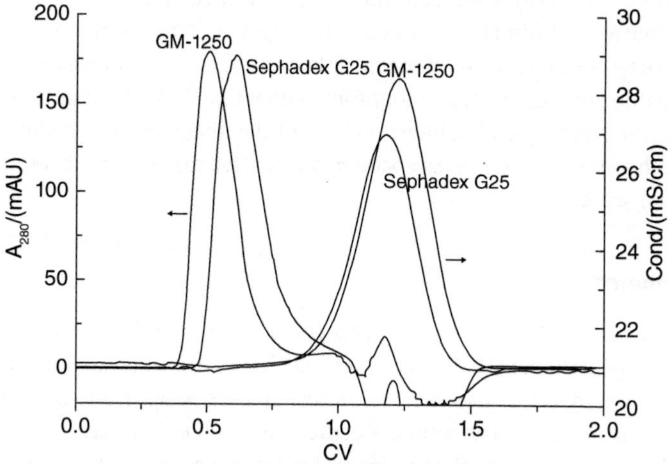

Figure 9.13 The desalting effect of filamentous hemagglutinin on Sephadex G25 and GM-1250 gel (CV represented the column volume). Source: From Xiong et al. [16]/with permission from Elsevier.

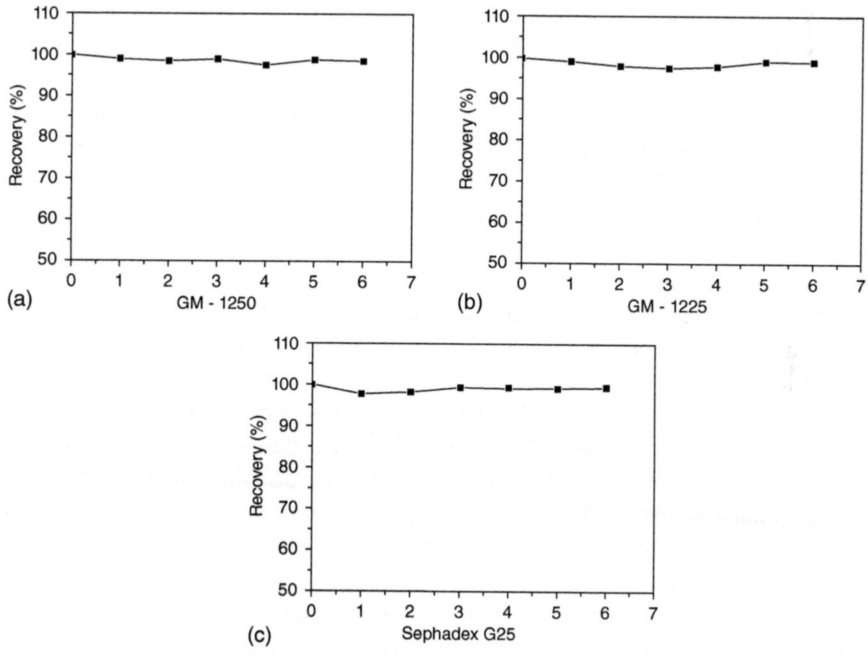

Figure 9.14 Stability of KGM gels measured as change in the recovery of proteins after storing in various solutions or after autoclaving. Storage solutions: 0, control; 1, 0.1 M HCl; 2, 1 M NaOH; 3, 6.0 M guanidine hydrochloride; 4, 8.0 M urea 5, 0.5 M HCl; 6, autoclaving. Storage test (1–5): the gel was kept in the solutions at 40 °C for one week (experiments 1–4) and one week at room temperature (experiment 5). Autoclaving: experiment 6, 121 °C for 20 minutes. Source: From Xiong et al. [16]/with permission from Elsevier.

hydrochloride and urea. The results showed that KGM microspheres are all stable in commonly used buffer solutions and denaturing agents, and insensitive to autoclaving as well. They could be reused for multiple cycles in the desalting experiment without noticeable change of chromatographic behavior. KGM microspheres showed almost the same stability as Sephadex G25, and the superior performance and significant cost advantage make it a prospective alternative material for protein desalting for industrial scale.

9.2.4 PST Microspheres

Micron-sized cross-linked copolymer microspheres (poly(styrene-divinylbenzene)) of styrene (ST) and divinylbenzene (DVB) have high mechanical strength, chemical stability, and hydrophobicity, and also offer a range of pore sizes from several nanometers to several hundred nanometers. Furthermore, their surfaces can be chemically modified for gel filtration chromatography, reversed phase chromatography, ion-exchange chromatography, hydrophobic chromatography, and affinity chromatography, so as to achieve efficient separation of biological components [17, 18]. Therefore, PST microspheres have been one of the most widely used polymeric supports for bioseparation. Among them, monodisperse PST microspheres have special properties, such as high resolution, high surface area, and high binding capacity, therefore, they are especially popular in high-performance liquid chromatography.

Compared with conventional seeded polymerization, by which uniform PST microspheres can be prepared, porous membrane emulsification technology has obvious advantages in large-scale preparation, such as reducing time, costs, and energy consumption. Our group [17–22] has used direct membrane emulsification method combined with suspension polymerization method to prepare uniform PST microspheres with an average particle size range of 20–40 μm and studied the structure control mechanism of microspheres as described in Chapter 2. The phase separation rule between polystyrene chains and pore-forming agent was obtained, and uniform PST microspheres with a controllable structure were successfully prepared, which has been successfully used as chromatographic media. To optimize scale-up manufacture process, the effects of the linear flow velocity in pipeline, the trans-membrane pressure on the particle size, and size distribution of microspheres were systematically studied, and the best preparation conditions were obtained. These micron-sized uniform PST porous microspheres were successfully obtained in scale-up production [23].

Recently, polypeptides have drawn increasing attention, and their applications are becoming more and more wide. The synthetic crude polypeptide samples contain various amino acid residues, small peptides, inorganic salts, and other impurities, which need to be removed by separation. Reversed-phase chromatography is one of the main methods for polypeptides separation. Because PST microspheres have many advantages, such as generally tolerating extreme pH (pH 1–14), good chemical stability, high mechanical strength, nonpolarity of the surface, and low production cost, it is an ideal chromatographic media for peptide separation.

Our group prepared highly cross-linked polystyrene (PST)-based anion-exchange chromatographic media with uniform size, high ion-exchange capacity, and high hydrophilicity after modification by a novel surface functionalization method [18]. The uniform PST microspheres prepared by direct membrane emulsion method were modified by two ways. One was conversing residual vinyl groups to epoxy groups followed by quaternization, and another was to functionalize aromatic ring matrix including nitration, reduction, and attachment of glycidyl trimethylammonium chloride. The morphology and porous structure of microspheres were observed by SEM and atomic force microscopy (AFM), respectively, and the results showed that the surface of modified PST microspheres became roughness but the particle size remained the same (Figures 9.15 and 9.16). Both hydrophilicity and biocompatibility of modified PST microspheres were greatly improved after being functionalized. The purity of exenatide was improved using modified PST column, and these modified PST microspheres had a great application potential for efficient separation of peptides (Figures 9.17 and 9.18).

Very fast separation of octreotide from synthetic peptide crude mixtures was obtained on a PST-based reversed-phase chromatography. Uniform-sized cross-linked PST microspheres with high specific surface areas were synthesized by rapid membrane emulsification method [24]. The average diameter of PST microspheres was 13.7 μm with a Span value of 0.74. The pore size and specific surface area were

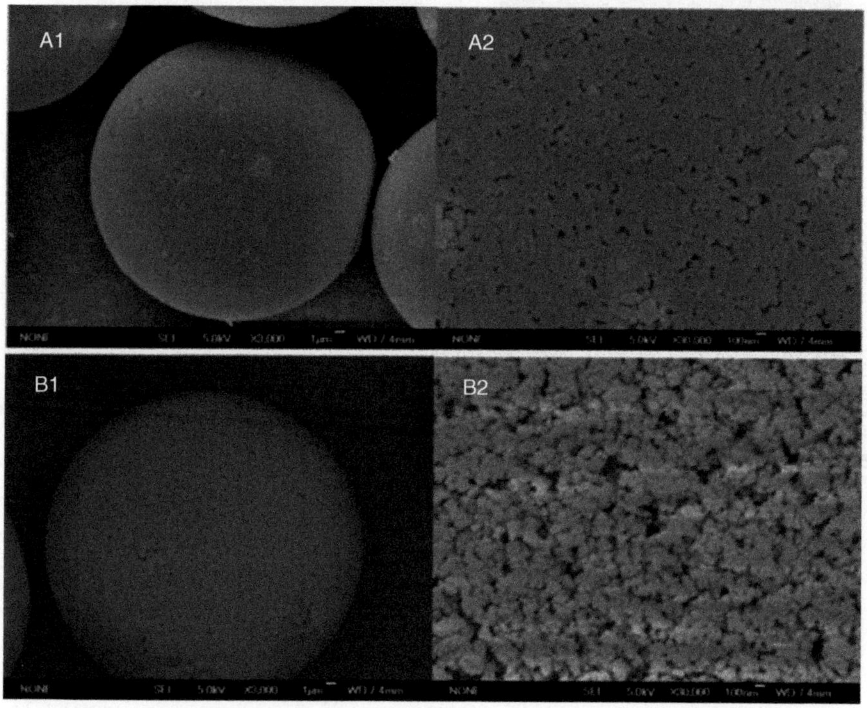

Figure 9.15 SEM images of PST before and after modification (A1, A2 before; B1, B2 after). Source: Li et al. [18]/Elsevier.

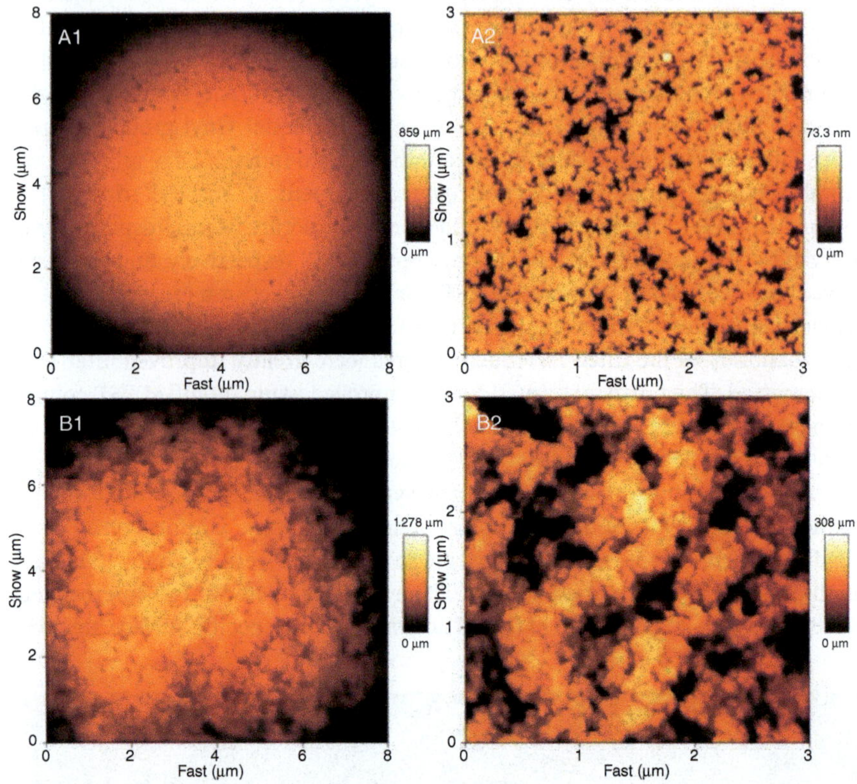

Figure 9.16 AFM images of PST before and after modification (A1, A2 before; B1, B2 after)Source: Li et al. [18]/Elsevier..

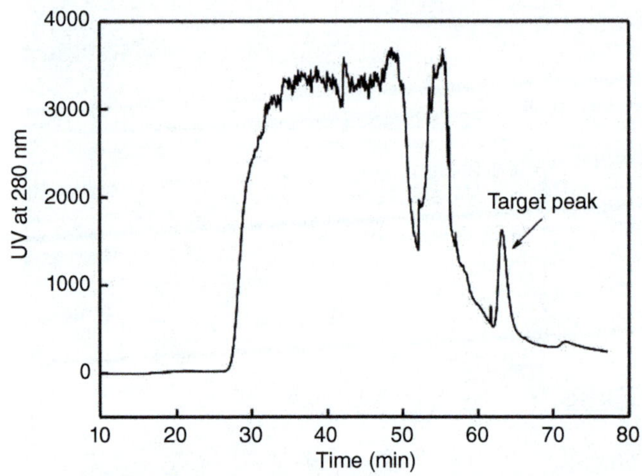

Figure 9.17 Separation of exenatide from peptide crudes using PST-based anion-exchange column. Source: From Li et al. [18]/with permission from Elsevier.

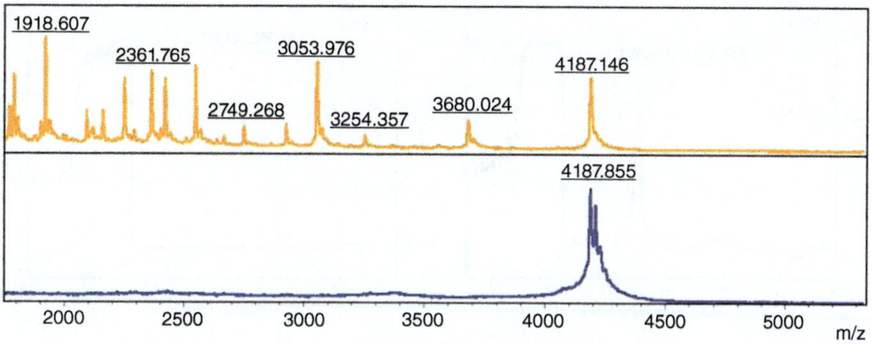

Figure 9.18 MS spectra of synthetic crude peptides (orange) and purified exenatide (blue). Source: From Li et al. [18]/with permission from Elsevier.

14 nm and 230 $m^2 \cdot g^{-1}$, respectively [25]. Compared with the traditional swelling method, the obtained microspheres prepared using one-step membrane emulsification method has both high specific surface area and uniform size beneficial for HPLC.

For the chromatography, the slurry of PST microsphere was transferred into a stainless steel reservoir and then packed into the column at a high pressure. The conditions of packing column were optimized and the theoretical plate number was up to 10970 m^{-1}. The linearity of pressure drop vs. flow rate confirmed a good throughput without deformation of PST microspheres even at a high pressure of 12 MPa, due to their good mechanical strength and good uniformity, exhibiting a low back pressure at high flow rate. The optimized PST media has a high loading capacity, high recovery yield, and purity. Under the optimized conditions, octreotide was purified from 42.89% to 99.99%, which was higher than that of C_{18} column with a purity of 96.19% (Figure 9.19) [25, 26]. Electrospray ionization mass spectrometry (ESI-MS) was used for analysis of purified peptides and the results showed that the purified octreotide had the same molecular mass as the proposed sequence (Figure 9.20) [26]. Great difference in impurity was determined by nuclear magnetic resonance (NMR) spectroscopy, and the data of purified octreotide by PST column was in keeping with that of octreotide standard (Figure 9.21) [26].

PST column can also be used to detect ligand leakage of nitrilotriacetic acid immobilized metal affinity chromatographic media during the separation process. In comparison, the detection limit of chemical derivatization method is 0.5 $mg \cdot l^{-1}$, and the detection limit of HPLC method is 1 $\mu g \cdot l^{-1}$, indicating that uniform PST microspheres have wide application in leakage detection of chromatography (Figure 9.22) [24].

PST microspheres have been used in preparative chromatography techniques, including both low-pressure liquid chromatography and high-performance liquid chromatography for separations of active components from natural products. Our group reported a successful purification of a natural product from crude extract using uniform PST microspheres [27]. An efficient low-pressure preparative chromatography for purification of icariin from crude extract was developed by

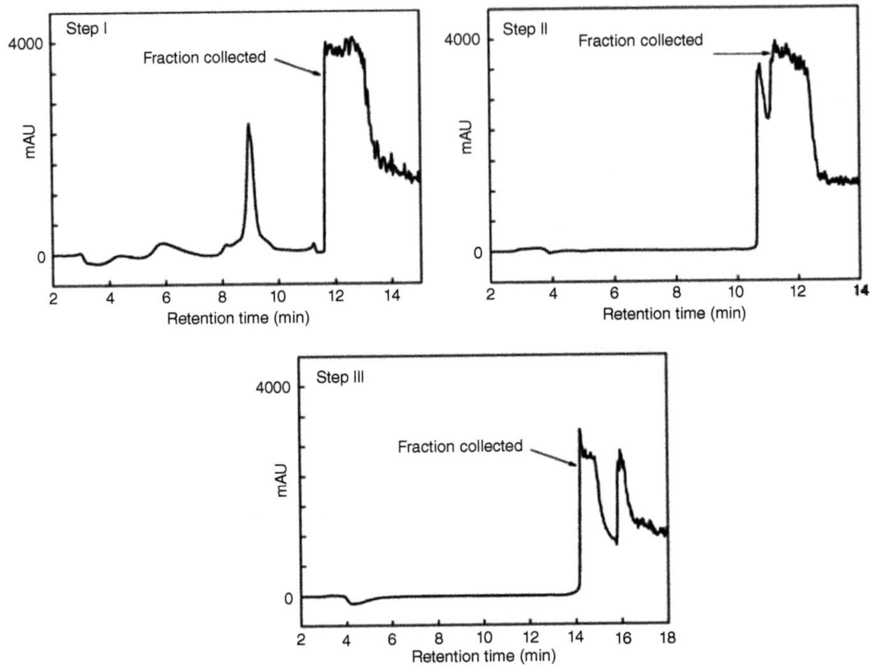

Figure 9.19 Three-step profiles of liquid chromatography purification of octreotide from synthesized crude peptides broth by PST column. Flow rate: 1 ml·min^{-1}. Detection UV: 220 nm. Sample loading amount: 50 mg. Mobile phase A: H_2O + 0.1% trifluoroacetic acid; B: acetonitrile + 0.1% trifluoroacetic acid. Step I: stepwise gradient 70% B in 15 minutes. Step II: linear gradient from 10% B to 100% B in 15 minutes. Step III: linear gradient from 40% B to 60% B in 18 minutes. Source: From Zhang et al. [26]/with permission from Elsevier.

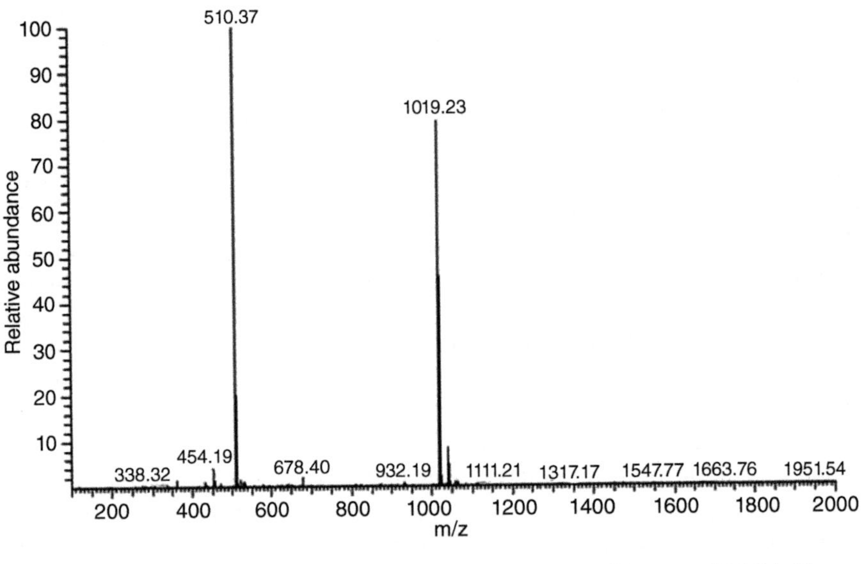

Figure 9.20 The ESI–MS of purified octreotide. Source: From Zhang et al. [26]/with permission from Elsevier.

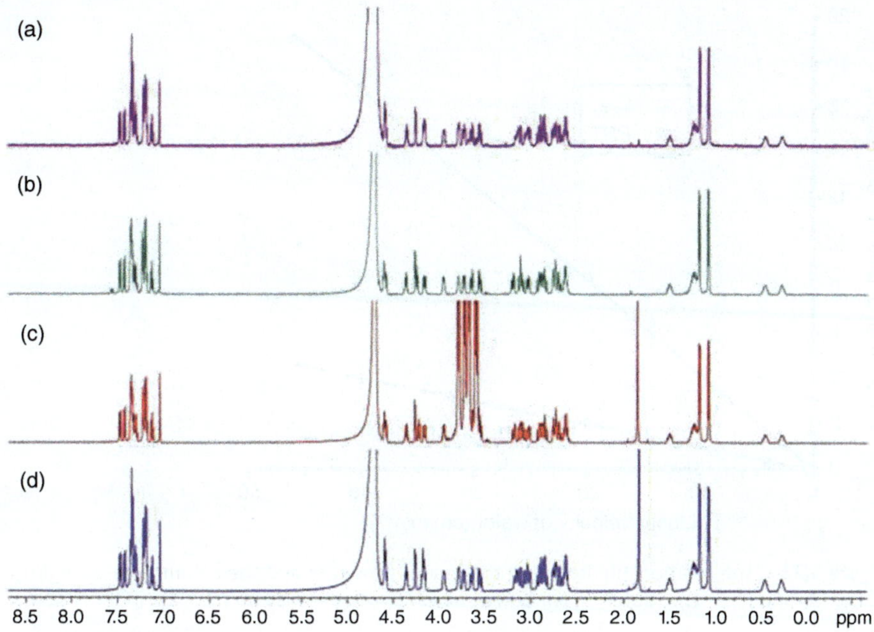

Figure 9.21 The spectra of ^1H NMR: (a) purified octreotide by PST column, (b) purified octreotide by C_{18} column, (c) synthetic peptide crude mixtures, and (d) octreotide standard. Source: From Zhang et al. [26]/with permission from Elsevier.

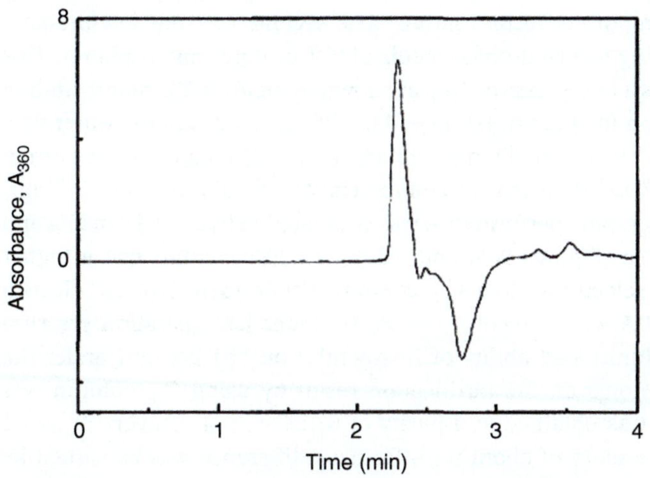

Figure 9.22 HPLC of nitrilotriacetic acid using the direct method on PST column. Source: From Zhang et al. [24]/DuraSpace/CC BY-NC 4.0.

using novel porous PST microspheres as packing material. 30 μm-average sized PST spherical microspheres with a narrow size distribution were prepared by combining microporous membrane emulsification technique and suspension polymerization process. The target product, namely icariin, was one of flavonoid components and considered as the principal compound responsible for the biological activities

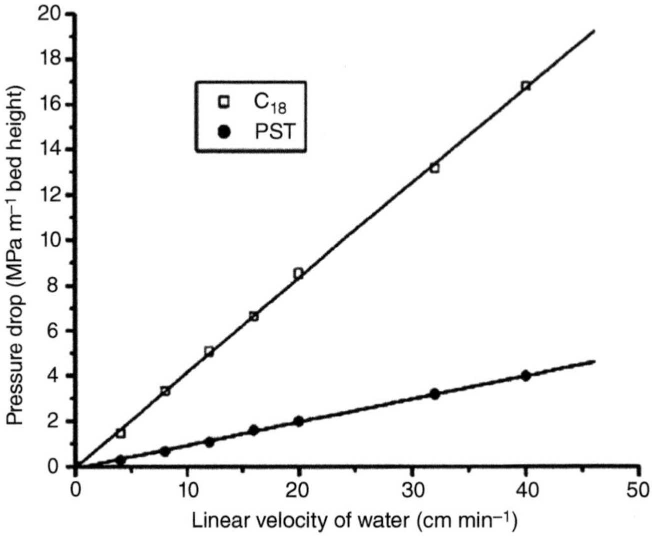

Figure 9.23 The relationship between the eluent flow-rate and the column pressure drop of C_{18} (□) and PST (●). Source: From Sun et al. [27]/with permission from Springer Nature.

of *Epimedium* in traditional Chinese medicines. A two-stage separation process was designed for icariin purification. The first was a preseparation step to remove a large amount of impurities from a crude extract of icariin with its content up to 20% by using microporous resin column. The second was the low-pressure chromatography by using reversed-phase media of PST uniform microspheres. This uniform PST media has a low-pressure drop as shown in Figure 9.23, demonstrating the relationship between the pressure drop and the linear velocity of the water flow through the column [27]. PST media has a much higher chromatographic permeability K of 1.68×10^{-9} m^2 than that of commercial C_{18} media of 4.0×10^{-10} m^2. The column chromatographic performance was compared between PST media and commercially C_{18} media (Figures 9.24 and 9.25) [27]. PST column had a higher resolution and better selectivity than C_{18} column. Crude extract of 20% icariin can be purified to 92.0% with a recovery of 99.9% under low operation pressure of 0.05 MPa with a column load ability of 2.5 mg·ml^{-1} on PST column under the optimized condition. However, the purification result by using C_{18} column was unsatisfactory, icariin was obtained at a purity of 63.1% with a recovery of 50.1% under the operation pressure of about 0.5 MPa. This difference was explained by the high resolution of uniform PST media and the interaction between the aromatic analytes and aromatic-rich PST surface, which was absent for C_{18} column. After the fraction purified by PST chromatography was crystallized and dried, the final purity of icariin after crystallization reached more than 98% with a total recovery of 93%. The purification process reported was batch-to-batch reproducible.

To further increase the specific surface area, we used divinylbenzene as the monomer instead of a mixture of styrene and divinylbenzene, and uniform polydivinylbenzene (PDVB) microspheres were prepared by direct membrane

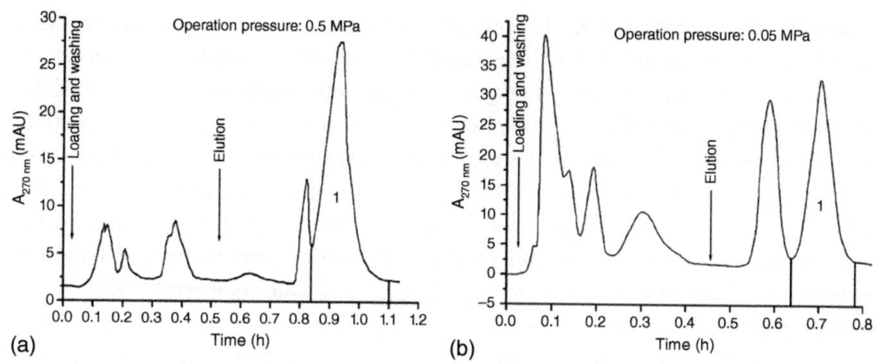

Figure 9.24 The profile of chromatographic purification of icariin from crude extract by C_{18} column (a) and PST column (b). (a) The wash/elution mobile phase was washing with 50% methanol and eluting with 70% methanol at 9 CVs·h^{-1}. (b) The wash/elution solvent was washing with 55% methanol and eluting with 60% methanol at 15 CVs·h^{-1}. The eluent was detected at 270 nm. Peak 1 represented icariin. Source: From Sun et al. [27]/with permission from Springer Nature.

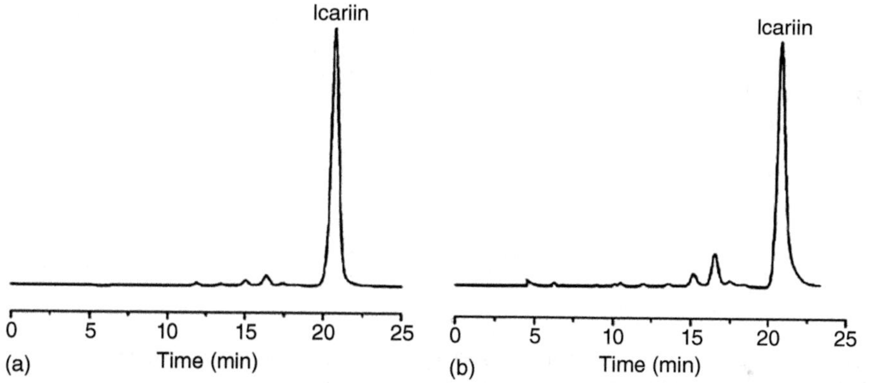

Figure 9.25 HPLC analysis of chromatographic fractions corresponding to peak 1 in Figure 9.24a,b. Chromatographic conditions: ODS-3 column (250 mm·4.6 mm I.D. 5 μm); isocratic elution: 30% acetonitrile in water at 0.6 ml·minutes^{-1}. (a) C_{18} column fractions; (b) PST column fractions. Source: From Sun et al. [27]/with permission from Springer Nature.

emulsification method [28, 29]. Uniform emulsion prepared by membrane emulsification method was synthesized using suspension polymerization method at 75 °C for 24 hours. The surface morphology, particle size, and size distribution of the microspheres were characterized, and the chromatographic performance was investigated. The results showed that the microspheres were porous with an average size of about 25 μm, an average pore diameter of 14 nm, a specific surface area of 520 m^2·g^{-1}, and good mechanical strength with a high resistance to a pressure of 40–50 MPa. A semipreparative HPLC column (300 mm × 10 mm I.D.) packed with PDVB microspheres was used to separate phenolic and polypeptide mixtures, respectively. The separation degrees were both greater than 1.5, and the maximum theoretical plate numbers were 13 000 and 21 000 m^{-1}, respectively. There was good reproducibility between batches. The detailed results are introduced as follows.

Figure 9.26 showed the chromatograms of phenolic mixtures by using columns packed with uniform PDVB microsphere and commercial C_{18} media, respectively [29]. It can be seen that the mixture of four kinds of phenols was well separated by both PDVB and C_{18} columns. The chromatographic parameters of both PDVB and C_{18} chromatographic media were listed in Table 9.1 [29]. It can be seen that PDVB semipreparative column had excellent separation performance for phenolic mixtures, especially for the isomer of hydroquinone and resorcinol. Additionally, the tailing factor (F) at the 10% height of each peak in the chromatogram of PDVB column ranged from 1.00 to 1.03, and all of peak shapes were symmetrical. The theoretical plate number (N) was up to 13000 m^{-1}. The results showed that PDVB media had a very promising application prospect in reversed-phase chromatography for purification of small molecules with a good separation effect.

Secondly, uniform PDVB microspheres had better separation results for polypeptide mixtures than commercial C_{18} media. The chromatograms of four polypeptides on PDVB column and C_{18} column were shown in Figure 9.27 [29]. The separation

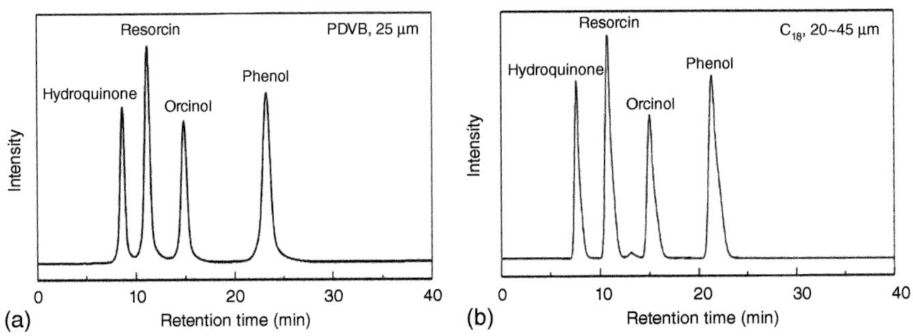

Figure 9.26 Chromatograms of phenol and its derivatives obtained by different packings. Source: From Wang et al. [29]/with permission from Elsevier.

Table 9.1 Theoretical plate number (N), tailing factor (F (10%)), and resolution (R_s) of phenol and its derivatives.

	PDVB			C_{18}		
Peak	N **(m^{-1})**	**Tailing factor, F (10%)**	R_s	N **(m^{-1})**	**Tailing factor, F (10%)**	R_s
1	5 100	1.027	—	4 300	1.677	—
2	6 400	1.025	2.61	5 500	1.707	3.31
3	8 800	1.014	3.44	8 100	1.708	3.75
4	13 000	1.001	6.29	11 000	1.770	4.66

Note: Peaks 1, 2, 3, and 4 represented hydroquinone, resorcin, orcinol, and phenol, respectively.
Source: From Wang et al. [29]/with permission from Elsevier.

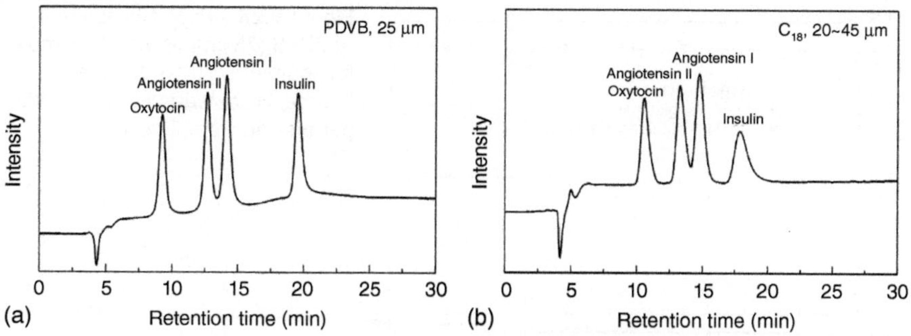

Figure 9.27 Chromatograms of four standard peptides obtained by different packings. Source: From Wang et al. [29]/with permission from Elsevier.

Table 9.2 Theoretical plate number, tailing factor (10%), and resolution of peptides.

	PDVB			**C_{18}**		
Peak	**N (m^{-1})**	**Tailing factor, F (10%)**	**R_s**	**N (m^{-1})**	**Tailing factor, F (10%)**	**R_s**
1	5 400	1.069	—	4 700	1.374	—
2	9 100	1.181	3.60	6 100	—	2.28
3	11 000	1.092	1.51	6940	—	1.16
4	21 000	1.347	5.44	4 700	1.501	1.96

Note: Peaks 1, 2, 3, and 4 represented oxytocin, angiotensin II, angiotensin I, and insulin, respectively.
Source: From Wang et al. [29]/with permission from Elsevier.

degrees of four polypeptides, the theoretical plate numbers, and tailing factor of each peak from two columns are listed in Table 9.2 [29]. The results showed that PDVB column could completely separate four polypeptides. The separation degree was greater than 1.5, and the tailing factors of each peak ranged from 1.00 to 1.35. The peak shape was symmetrical, and the theoretical plate number was up to 21000 m^{-1}. However, the chromatograms of C_{18} column had a wider peak shape, and both the separation degree and the theoretical plate number were much lower than those of PDVB column. Especially for angiotensin I and angiotensin II, there was only a slight difference of two amino acid residues between them, and it was difficult to separate them with C_{18} column, while a complete separation was achieved using PDVB column, indicating that PDVB microspheres have a good application prospect in separation of angiotensin and other polypeptide drugs.

Batch repeatability is one of important parameters to evaluate the performance of chromatographic media and column in separation. If the batch repeatability is not satisfied, no matter how good the separation results of this media are, it is not suitable for large-scale production. Here, PDVB column was used to separate the polypeptide mixture for evaluation of batch repeatability. Samples were injected for four consecutive times under the same conditions, and the batch repeatability

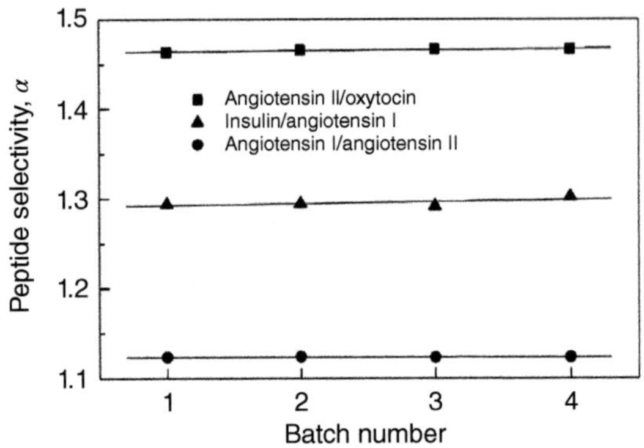

Figure 9.28 Peptide selectivity of PDVB (25 μm) vs. batch number for four consecutive batches. Source: From Wang et al. [29]/with permission from Elsevier.

of PDVB chromatographic column was investigated with the results shown in Figure 9.28 [29]. It can be seen that the separation effect of the polypeptide mixtures was almost identical after four successive injections, indicating that PDVB column has very stable performance and good batch repeatability, which is suitable for large-scale separation.

9.2.5 PGMA Microspheres

PGMA microspheres have excellent mechanical strength and chemical stability, and are easy to be functionalized due to their rich epoxy groups. Therefore, it is very promising for chromatographic media and it can be coupled with various ligands, such as amino group, phenyl group, and affinity group, used for ion exchange, hydrophobic, and affinity chromatographic media. We prepared uniform porous PGMA microspheres by direct membrane emulsification technology (see Figure 2.4, Chapter 2) [30]. That is, direct membrane emulsification – suspension polymerization was employed by using glycidyl methacrylate (GMA) as the main monomer. Then, polyethylene glycol (PEG) with a series of molecular weights was used to modify PGMA microspheres for eliminating irreversible protein adsorption, and a hydrophobic interaction chromatographic media, namely PGMA–PEG4000 was obtained. This media has been used in hydrophobic interaction chromatography for both purifications of kallikrein and lysozyme refolding by chromatography, and good results have been obtained [30–32].

The separation results of kallikrein by hydrophobic interaction chromatography of PGMA–PEG4000 microspheres were shown in Figure 9.29, which were compared with that of Octyl Sepharose FF in Figure 9.30 [31, 32]. The separation process of kallikrein using PGMA–PEG4000 was different from that of Octyl Sepharose FF. Kallikrein was adsorbed on Octyl Sepharose FF at a loading condition of 15% ammonium sulfate saturation and purified by gradient elution. However, at higher concentrations of ammonium sulfate (e.g. 30%), kallikrein was not adsorbed to PGMA–PEG4000 but flew directly through the column. Therefore, PGMA–PEG4000 had a weaker hydrophobicity than that of Octyl Sepharose FF. As shown in Figure 9.29, P_1 was the flow-through part in PGMA–PEG4000 column, containing kallikrein with a specific activity of 119.1 U·mg^{-1} and activity recovery

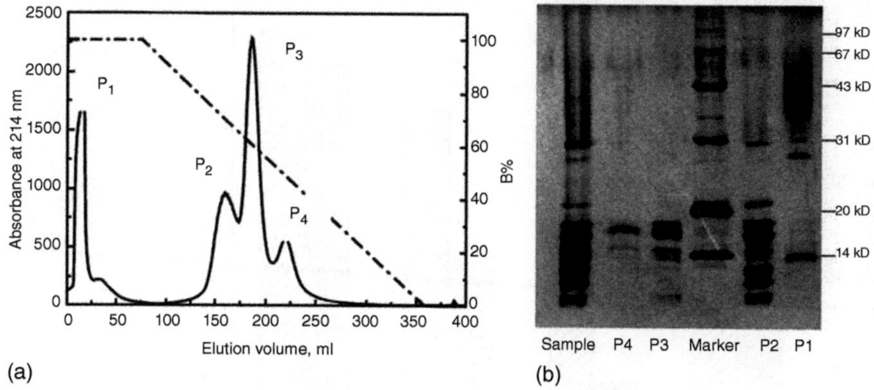

Figure 9.29 Chromatogram and electrophoresis of kallikrein purified on PGMA–PEG4000 (30% ammonium sulfate saturation). Source: Wang [31]/Europe PMC.

Figure 9.30 Chromatogram of kallikrein purified on Octyl Sepharose FF (15% ammonium sulfate saturation). Source: From Li et al. [32]/The Chinese Journal of Process Engineering.

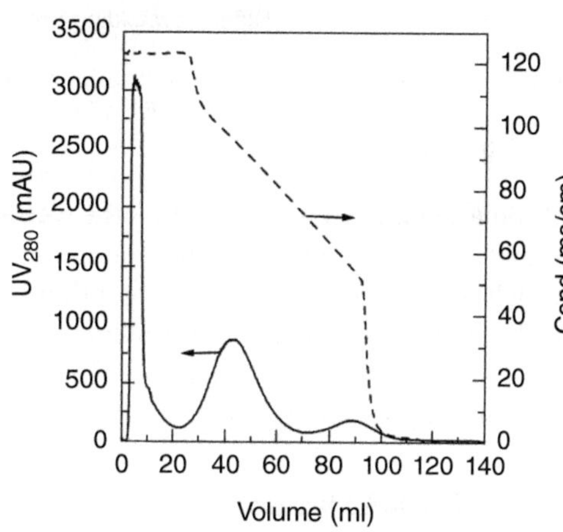

of 95%, while the elution of P_2, $P_{3,}$ and P_4 did not contain kallikrein. Also, these three high-resolution peaks in elution of PGMA–PEG4000, probably belonged to other enzymes, e.g. elastase and trypsin, which provided a perspective for copurification of pig pancreatic enzymes in one-step hydrophobic interaction chromatography by further optimization of chromatographic conditions. As shown in Figure 9.30, for Octyl Sepharose FF, P_1 was the flow-through part, while P_2 and P_3 were the elution part, respectively, and only P_2 contained kallikrein with a specific activity of 220 U·mg^{-1} and activity recovery of 91%, and the other two peaks did not contain kallikrein. Comparatively, PGMA–PEG4000 showed outstanding advantages of simple process, high flow rate, high resolution, and large capacity. Its flow rate was four times higher than that of Octyl Sepharose FF, showing a remarkable processing ability.

Lysozyme refolding was performed using PGMA–PEG4000 chromatography on a denaturant-glycerol double gradient elution model and compared with refolding by dilution method, Phenyl Sepharose HP chromatography and Butyl Sepharose 4FF

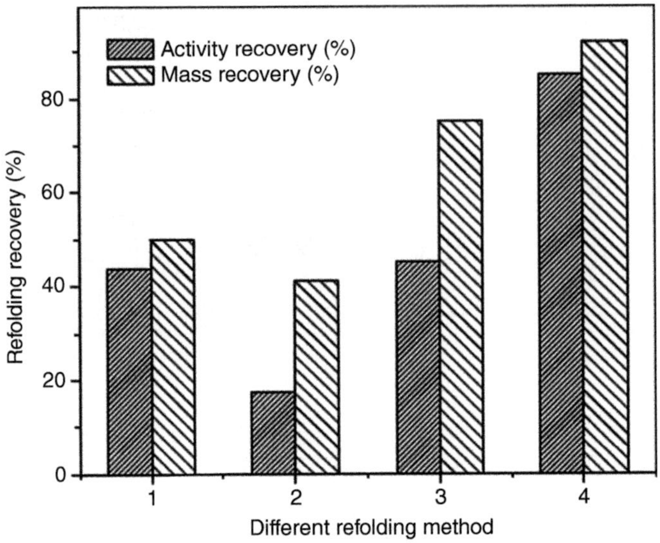

Figure 9.31 Recovery of refolding lysozyme with dilution or HIC. 1. dilution, 2. Phenyl Sepharose HP, 3. Butyl Sepharose 4FF, and 4. PGMA-PEG4000. Source: From Wang [31]/Chinese Academy of Sciences.

chromatography. It was found that the recovery yield of protein reached 92% and the activity yield reached 85% when refolding with PGMA-PEG4000 chromatography, which was obviously superior to other methods (Figure 9.31) [31]. This mild hydrophobic interaction chromatographic media has long, flexible PEG chains, which can provide a more useful environment for protein folding, therefore, it has outstanding advantages in protein folding.

9.2.6 PHEMA Microspheres

PHEMA microspheres have hydroxyl groups that can be functionalized further easily, so can be used in gel filtration chromatography, hydrophobic interaction chromatography, ion-exchange chromatography, affinity chromatography, and so forth. In addition to hydroxyl group, PHEMA molecules also contain C–C skeletons, so they have the potential to be used as reversed-phase chromatographic media. At the same time, the hydroxyl group in PHEMA can form hydrogen bond with the polar group in the solute molecule, thus improving the separation selectivity. We used membrane emulsification method to prepare uniform PHEMA microspheres with both controllable particle size and pore size, the broad size distribution and low separation efficiency brought by traditional suspension polymerization method are avoided. In addition, the polarity of PHEMA microspheres was controlled by changing the composition of oil phase, and PHEMA microspheres with suitable hydrophobicity were prepared and used as reversed-phase chromatographic media with polar groups (Figure 9.32) [33–35].

Uniform PHEMA microspheres were used as chromatographic media to separate icariin from the extract of *Epimedium brevicornum* Maxim, and the results were

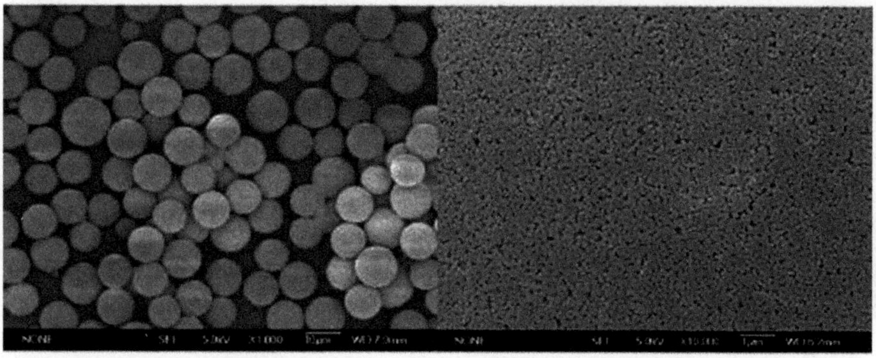

Figure 9.32 SEM image of PHEMA microspheres used as chromatographic packing materials (right is their surface). Source: Qu [33]/American Chemical Society.

compared with those of uniform PST microspheres (Figure 9.33 and Figure 9.34) [33]. It showed that under the same conditions, PHEMA media had higher adsorption selectivity for polar molecules, and lower concentration of organic solvent in the mobile phase was needed. The purity of icariin was 91% and the recovery yield was 98% after one-step elution with 30% ethanol solution on a PHEMA30 (particle size of 30 μm) column. For PST30 (particle size of 30 μm) column, when 35% ethanol solution was used as the mobile phase, icariin was strongly retained with an obvious tailing phenomenon. When 40% ethanol solution was used as the mobile phase, the retention time of icariin was greatly reduced and the purification was difficult with a low purity of the final product. Only gradient elution can obtain the best purification result. Icariin mainly existed in peak 2 after being washed with 35% ethanol solution followed by being eluted with 40% ethanol solution. Using PST30 column, the purity of icariin was 90% and the recovery was 86%, less than that of PHEMA30. The reason for the difference of separation efficiency between these two columns was due to the structure of microspheres and their interaction with icariin during chromatography. As a kind of chromatographic media with polarity, PHEMA has proper hydrophilicity and there was a hydrogen bond between PHEMA and icariin in ethanol–water system. Icariin was separated through moderate hydrophobic interaction with PHEMA and strong retention was avoided, obtaining a higher recovery yield. More importantly, the concentration of ethanol required for PHEMA column was lower, reducing the consumption of organic solvent and separation cost, and more friendly for the environment.

9.2.7 Silica Microspheres

Silica microspheres have been widely used for chromatographic media. We developed a novel emulsion strategy by combining membrane emulsification and sol–gel conversion to prepare uniform silica microspheres with controllable particle size [36]. Carboxyl modification was performed on both smooth and golf ball-like silica microspheres to evaluate the effect of surface structure on protein separation.

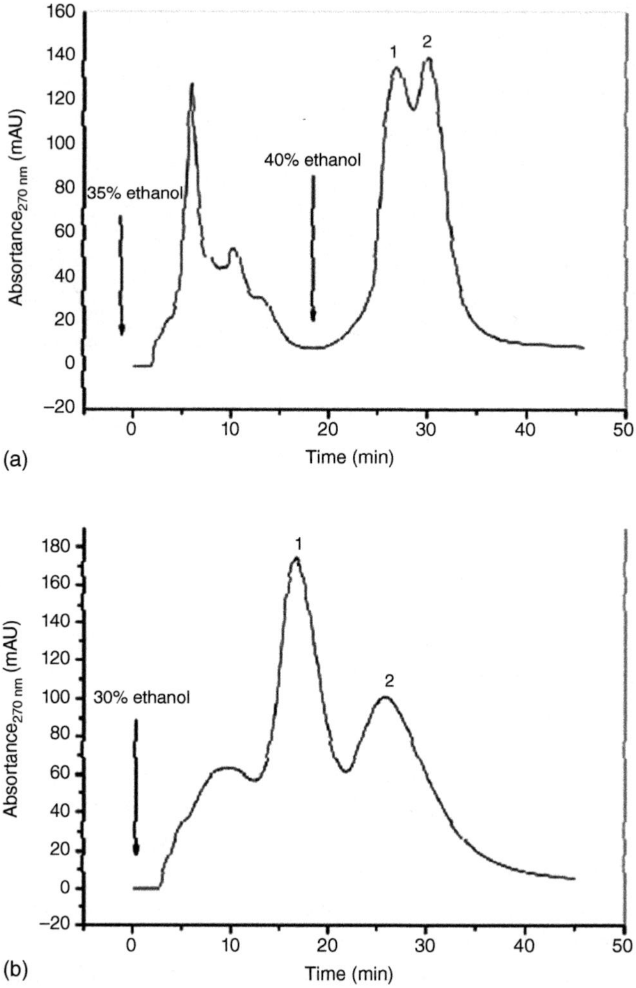

Figure 9.33 Chromatograms of icariin isolated from the extract solution of *Epimedium brevicornum* Maxim on PST30 (a) and PHEMA30 (b) column. Load: 5 mg sample in 5 ml 30% ethanol solution. Flow rate: 2 ml·min^{-1}. Source: From Qu [33]/American Chemical Society.

Dicarboxylic groups were coupled to silica microspheres via Si–OH groups followed by chelating with metal ions, i.e. Ni^{2+}, and Ni-chelate-IDA-SiO_2 was prepared. Complex structures greatly increased the amount of carboxyl groups and protein adsorption by improving specific surface area of microspheres. Figure 9.35 shows that the golf ball-like silica microsphere had a higher amount of adsorbed BHb than smooth one [36]. The reason was probably that the former had a higher surface area due to a large number of pits on the surface, and a higher modification degree was obtained. Blocking hydroxyl groups with 1% BSA solution can effectively reduce the nonspecific adsorption of silica microspheres and improve the desorption efficiency. The amount of adsorbed BHb was 299 mg·g^{-1} and 454 mg·g^{-1}, and the desorption efficiency was 70.4% and 65.8% for smooth and golf-like silica

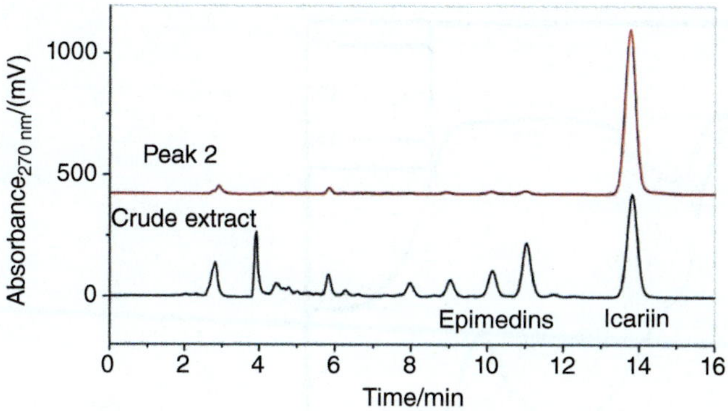

Figure 9.34 HPLC analysis of the crude extract solution and the target peak 2 eluted from the PHEMA30 column. Conditions: column, Inertsil ODS-3, 250 × 4.5 mm I.D.; mobile phase, ACN/water = 30/70, (V/V); flow rate, 1 ml·min^{-1}; and sample volume: 50 μl; detection at 270 nm. Source: From Qu [33]/American Chemical Society.

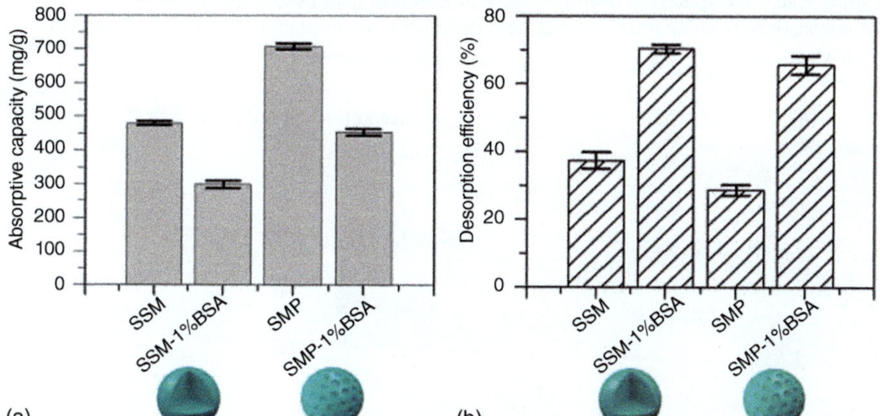

Figure 9.35 Protein separation performance of smooth and golf ball-like silica microspheres. (a) Adsorption capacity of BHb on Ni-chelate-IDA-SiO_2; (b) Desorption efficiency of BHb on Ni-chelate-IDA-SiO_2 (SSM: Ni-chelate smooth silica microspheres; SSM-1%BSA: SSM with 1% BSA blocked silanol; SMP: Ni-chelate golf ball-like silica microspheres; and SMP-1%BSA: SMP with 1%BSA blocked silanol). Source: From Hu [36]/American Chemical Society.

microspheres, respectively, showing their good protein adsorption and desorption performance.

Selective adsorption properties of modified silica microspheres were studied using a mixture of BSA and bovine hemoglobin (BHb) as the model proteins. In Figure 9.36, the UV spectrum of the protein mixture, the supernatant after adsorption and desorption showed that the protein mixture had a high UV adsorption value at 406 nm, and it decreased for the supernatant after adsorption by silica microspheres, indicating that proteins were adsorbed on the silica microsphere [36]. There was an obvious UV adsorption peak at 406 nm in the

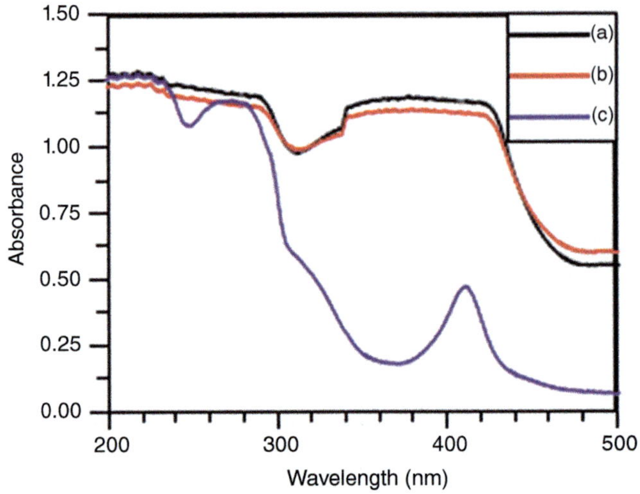

Figure 9.36 UV spectrum of the mixture of BSA and BHb before (a) and after (b) adsorption by silica microspheres and desorption by imidazole (c). Source: From Hu [36]/American Chemical Society.

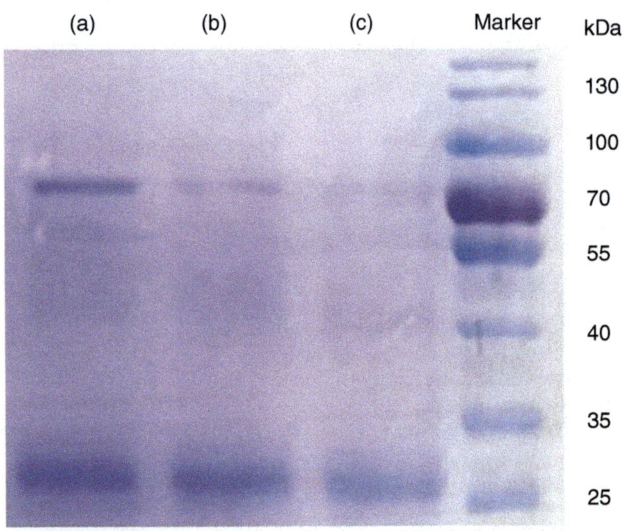

Figure 9.37 Electrophoresis of the mixture of BSA and BHb before (a) and after (b) adsorption by silica microspheres and desorption by imidazole (c). Source: Hu [36]/Chinese Academy of Sciences.

desorbed solution, indicating that the modified silica microsphere could selectively adsorb BHb from the protein solution. Electrophoresis results showed that both the protein mixture and the supernatant after adsorption contained both BSA and BHb, while the desorbed solution contained only BHb, further confirming that modified silica microspheres can selectively adsorb BHb from the protein mixture (Figure 9.37) [36].

9.3 Super-Porous Microspheres for Vaccine Separation

9.3.1 Significance of Super-Porous Microspheres for Vaccine Separation

With the development of biotechnology, more and more large biomolecules, such as virus-like particle (VLP) vaccine, adeno-associated virus (AAV), and mRNA benefit mankind in many ways, which also bring new challenges to separation media [37]. Due to their large molecular size, the mass transfer in separation media is limited because the pore size of traditional media is relatively small (10–30 nm). Due to the lack of large interconnecting channels, a considerable portion of pores inside particles is not easily accessible [38]. This results in lower loading capacity. Also, a lot of diffusive pores requires very long diffusional path. This causes band spreading that is detrimental to the resolution of a chromatography column, and low activity yield as well because the biomolecules are fragile [39, 40].

Super-porous media for liquid chromatography provides answers to the drawbacks of traditional media. They have large pores in the order of hundreds of nanometers [41–43]. POROS® perfusion media were the first super-porous commercial products introduced in the late 1980s. This first-generation super-porous medium was synthesized in two steps. The small nanosized subparticles were prepared first and then assembled into large particles with transecting super-pores ranging from 500 to 800 nm in size. With a particle size between 10 and 50 microns, these super-pores permit convective flow inside the particles in additional to flow in the interstitial spaces between the particles (Figures 9.38 and 9.39). This reduces column pressure drop considerably and thus allows much faster flow. These transecting super-pores divide the particle into many small regions, so that the diffusion path in each region is much shorter, which greatly reduces the mass transfer resistance. The existence of diffusive pores ensures a sufficient load capacity for biomolecules. This kind of chromatography using perfusion media is known as perfusion chromatography [44–47].

9.3.2 Preparation Methods for Super-Porous Microspheres

To make the preparation of super-porous microspheres more convenient and controllable, different methods were developed, including high internal phase emulsion

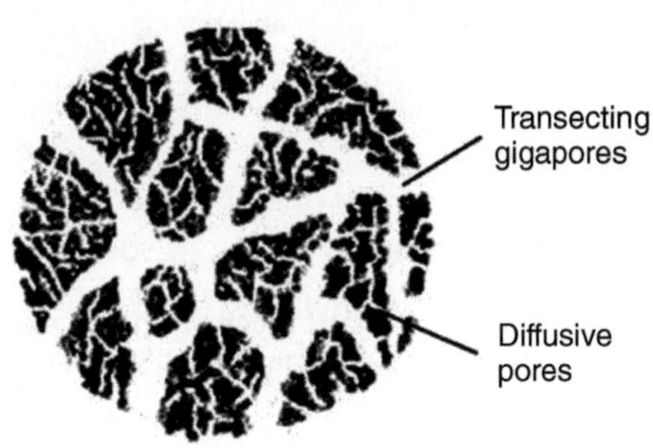

Figure 9.38 Schematic image of super-porous microspheres. Source: Gu et al. [38]/Elsevier.

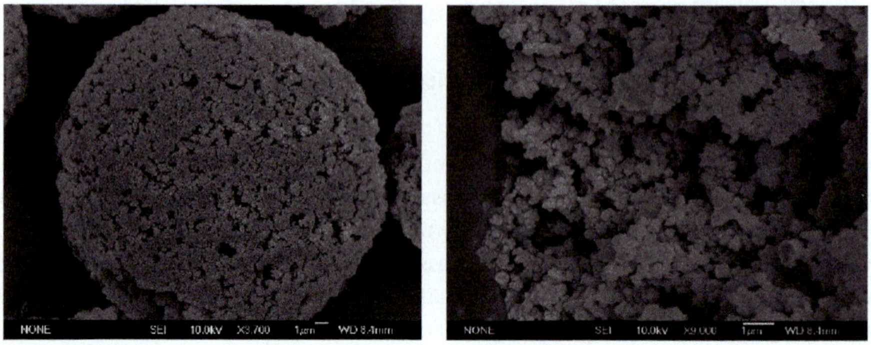

Figure 9.39 SEM photographs of POROS R1 particles. Source: Gu et al. [38]/Elsevier.

(HIPE) and liquid–solid mixed porogen. However, due to the high viscosity of HIPE, it is difficult to be dispersed and to form uniformed particles. As for the method using liquid–solid mixed porogen, it is difficult to incorporate and clean the solid porogen.

In order to overcome the problems of these cumbersome steps, our group developed a convenient one-step method to prepare super-porous polymeric microspheres, which were called gigaporous (or super-porous) microspheres [48]. We ingeniously utilized the swelling property of the reverse micelles to form a bi-continuous phase emulsion as shown in Figure 9.40. Enough amount of surfactant was added in the monomer phase to form reverse micelles. When the monomer phase was dispersed in the aqueous phase, these reverse micelles absorbed water into the polar cores spontaneously, and then the continuous aqueous channels were formed. The aqueous channels became large pores after polymerization. The nanopores in the polymer skeleton can be formed by porogen added in the monomer phase. Super-porous P(ST–DVB) microspheres, P(GMA–DVB) microspheres, and agarose microspheres were prepared by using this preparation strategy.

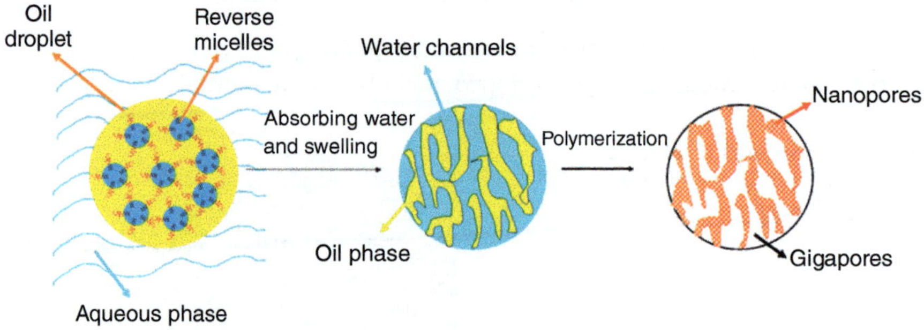

Figure 9.40 Schematic mechanism of the formation of super-pores in polymeric particles. Source: [48]/with permission from Elsevier.

9.3.2.1 Super-Porous P(ST–DVB) Microspheres

In preparation of super-porous P(ST–DVB) microspheres, it was found that the main factors, which affected the pore structure were the type and content of the surfactant in the monomer phase [49]. As shown in Figure 9.41, when the concentration of the optimized surfactant (Span 80) increased from 30% to 40% (based on total amount of monomer styrene and cross-linker divinylbenzene), the pore size of the microspheres became larger, around 500 nm (Figure 9.42). However, if the concentration of Span 80 exceeded 50%, the polymeric beads were broken. Besides the surfactant, the diluent also affected the pore size. If good solvent for monomer was added, the pore size of the polymeric particles became smaller because it restrained the water-adsorbing amount of the oil phase [48].

9.3.2.2 Super-Porous P(GMA–DVB) Microspheres

P(ST–DVB) microspheres are highly hydrophobic and have no functional group, which limits their application. Therefore, our group further applied the reverse micelles swelling method to prepare the poly(glycidyl methacrylate) [P(GMA)]

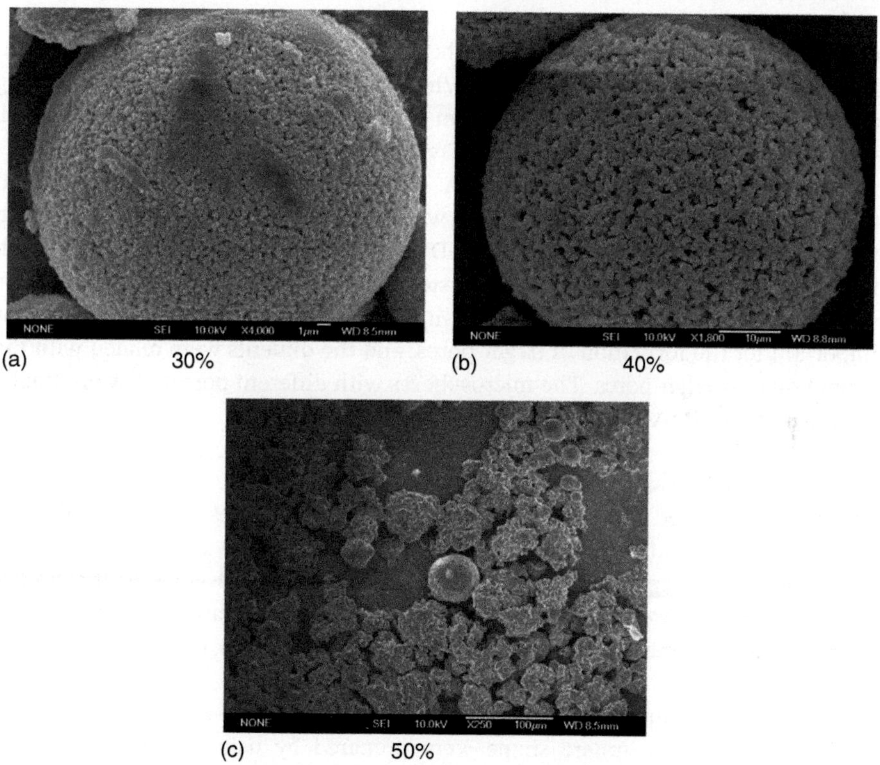

Figure 9.41 Effect of amount of Span 80 on the morphology of microspheres (The cross-linking degree was 13.8%). Source: Zhou et al. [49]/Elsevier. (a) 30%, (b) 40%, and (c) 50%.

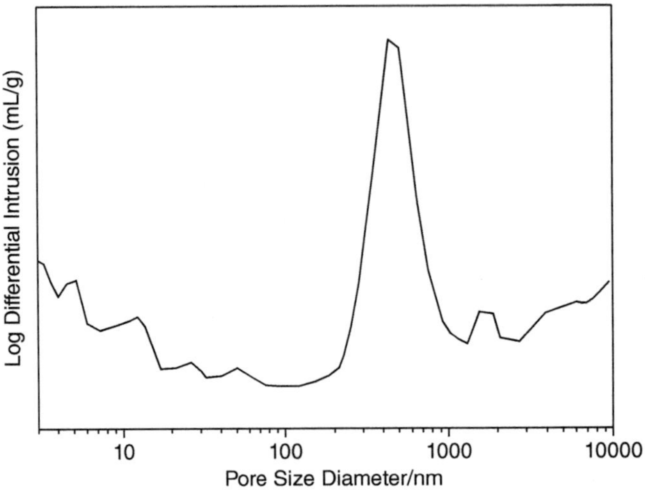

Figure 9.42 Pore size distribution curve of the microspheres prepared by 40% Span 80. Source: From Zhou et al. [49]/with permission from Informa UK Limited.

microspheres with epoxy groups [50]. The epoxy group can be modified by several ligands under moderate conditions. When a slightly hydrophilic cross-linking agent EGDMA was used, the maximum pore size of 140–200 nm was obtained, which was much smaller than P(ST–DVB) microspheres. When using DVB as a cross-linking agent instead of EDMA, higher concentration of surfactant and diluents could be added in the oil phase without break of the particles, resulting in the increase of the pore size of P(GMA–DVB) microspheres. This was related with higher hydrophobicity of GMA–DVB system. The necessity of co-operation of the surfactant and diluents was confirmed in GMA–DVB system. The surfactant was important for the formation of larger pores, and the diluents were related with the formation of smaller pores. The microspheres with different pore sizes were finally obtained in the GMA–DVB system as shown in Figure 9.43.

9.3.2.3 Super-Porous Agarose Microspheres

The above prepared super-porous P(ST–DVB) microspheres and P(GMA–DVB) microspheres are both hydrophobic particles that need to be hydrophilized for application in protein separation. Therefore, our group proposed whether the similar method (micelles swelling method) could be used in preparation of super-porous agarose microspheres in order to form large pores besides the gel structure of agarose [51].

As shown in Figure 9.44, highly cross-linked and super-porous agarose microspheres with regular sphere shape were prepared by using mixed surfactants. And the morphology of the microspheres varied when mixed surfactant with different HLB was employed. When a mixture of Triton X-100 and SDS (volume ratio 9 : 1, HLB 17.0) was used, the microspheres with large pores (about 100 nm) were obtained (Figure 9.44b). However, the surface of the microspheres became smoother when more SDS was added (Figure 9.44c).

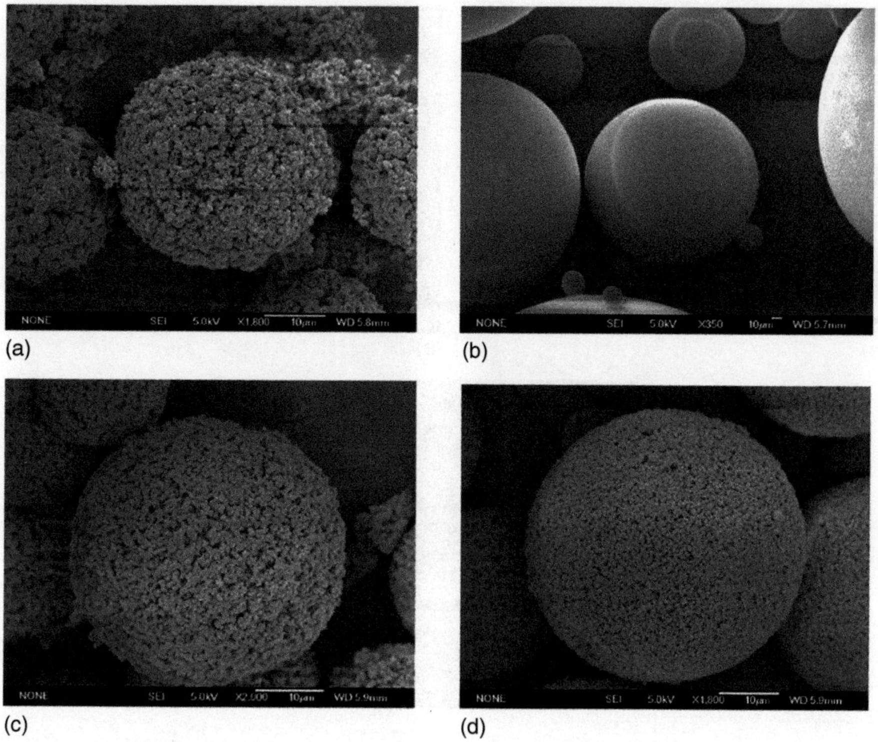

Figure 9.43 Effect of the co-operation of surfactant and diluents on the morphology of particles (Cross-linking agent DVB, concentration of cross-linking agent 27.5%. IO: isooctane; MP: 4-methyl-2-pentanol). Source: Zhou et al. [50]/Elsevier. (a) Span 80, 1.6 g; (b) IO/MP = 3:2, 1.0 g; (c) Span 80, 1.6 g, IO/MP = 3:2, 1.0 g; and (d) Span 80, 1.6 g, IO/MP = 1 : 1, 1.0 g.

9.3.2.4 Application of Membrane Emulsification Technology in the Preparation of Super-Porous Polymeric Microspheres

Based on the reverse micelles swelling method, super-porous polymeric microspheres with controllable large pores were successfully prepared. This method is convenient and has good reproducibility. However, a new problem appeared in the preparation process. Some particles aggregated weakly. Though the aggregation could be easily dispersed at about 0.01 MPa pressure, great effort, such as ultrasonic dispersion or grinding, had to be made to redisperse the aggregate in scaling up. The weak aggregation of microspheres is mainly due to the disappearance of the interface (or the boundary) between the oil phase and the water phase, which is because the water channels in oil droplet connect with the outside aqueous phase. Another reason is due to Ostwald ripening. According to the mechanism of Ostwald ripening, the more uniform is the droplet size, the weaker is the final aggregation degree. Therefore, we used premixed membrane emulsion method to solve this problem. With the improvement of the uniformity of particle size, the aggregation degree of microspheres decreased, but it was not completely avoided. This was because the connection between the internal and external water

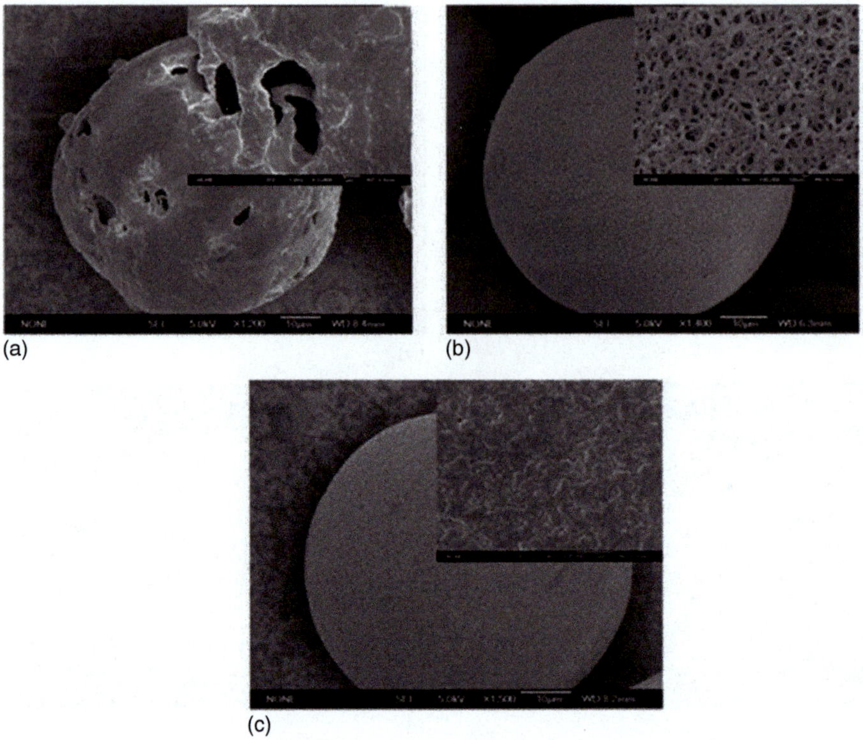

Figure 9.44 Effect of mixed surfactants on the morphology of the highly cross-linked and super-porous agarose microspheres. (a) PO-500 mixed with SDS (volume ratio 9 : 1, HLB8.4) (×1200); (b) Triton X-100 mixed with SDS (volume ratio 9 : 1, HLB 17.0) (×1400); and (c) Triton X-100 mixed with SDS (volume ratio 4 : 1, HLB 19.6) (×1500). Source: Zhao et al. [51]/Elsevier.

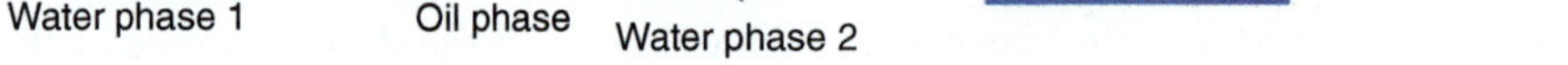

Figure 9.45 Two-step method to prepare the aqueous two-phase system (ATPS). Source: Zhang et al. [52]/MDPI/CC BY 4.0.

phases of oil droplets has not been solved. If a new interface could be rebuilt, the aggregation might be prevented. Considering the pore performing and aggregation mechanism, an aqueous two-phase system (ATPS) was chosen. The process was shown schematically in Figures 9.45 and 9.46 [52].

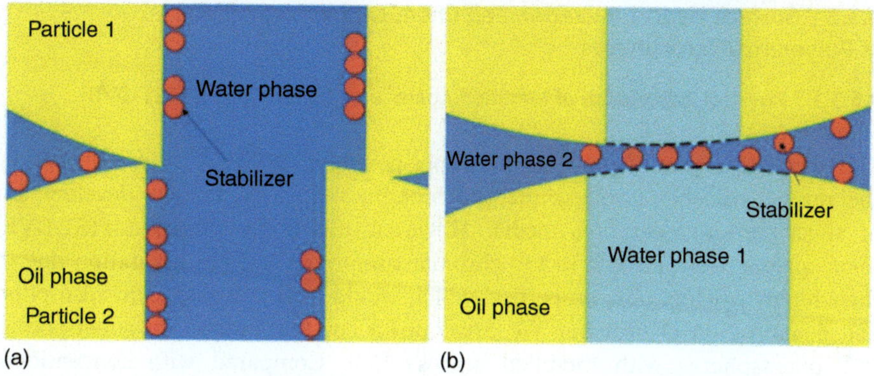

Figure 9.46 The schematic photographs of boundary loss (a) and boundary rebuilding (b). Source: Zhang et al. [52]/MDPI/CC BY 4.0.

The composition of oil phase remained unchanged. Water dissolved with the stabilizer poly(vinyl alcohol) (PVA) was used as the continuous phase. Another component, such as PEG or dextran, was added into the water phase in order to establish an ATPS. A phase diagram was drawn to guide the construct the ATPS between the inner pores and outside, and the concentration of each phase could be calculated. After preparing each aqueous phase, the oil phase was dispersed into one of the aqueous phases to form a first emulsion. A certain amount of time was needed to make the micelle swell. Then, the first emulsion was added into another aqueous phase to obtain the ATPS. As shown in Figure 9.46, an interface was rebuilt, which facilitated the stabilizer absorbed on it. The resulted emulsion was treated by the premixed membrane emulsion method for 3–5 times, and then started the process of polymerization. Combining the ATPS and the premixed membrane emulsion method, the weak aggregation of microsphere was well solved and uniform-sized super-porous microspheres were successfully prepared (Figure 9.47).

Figure 9.47 Well-dispersed super-porous microspheres prepared by combining the ATPS and the premixed membrane emulsion method.

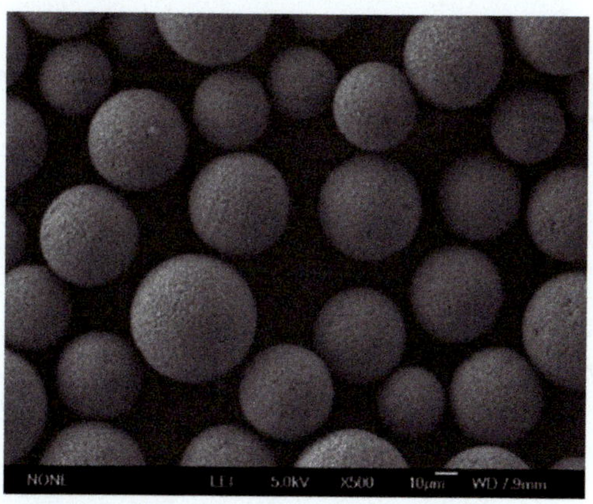

9.3.3 Surface Hydrophilization and Chemical Derivatization of Polymeric Microspheres

9.3.3.1 Physical Adsorption of Modified Agarose on Super-Porous P(ST-DVB) Microspheres

The successful preparation of the super-porous P(ST-DVB) microspheres by the reverse micelles swelling method brought new wish for the development of super-porous separation media. However, the firstly prepared P(ST-DVB) microspheres were limited in the chromatography for protein separation due to their highly hydrophobic properties, which would lead to nonspecific adsorption and denaturation of proteins. We developed a coating method for super-porous PST microspheres with modified agarose [53]. Compared with conventional macroporous polymeric microspheres, the large pores of the super-porous PST microspheres facilitated the coating with hydrophilic macromolecules such as agarose. Aromatic groups were introduced on agarose chains, which made it possible to adsorb agarose on polymeric microspheres. Then, the cross-linking of agarose network was carried out. After coating, the hydrophilicity and biocompatibility of super-porous PST microspheres were greatly enhanced (Figure 9.48). The nonspecific adsorption of proteins was nearly reduced to zero through the optimization of both the phenoxy content and the cross-linking conditions of adsorbed layer. After hydrophilization, the super-porous structure could be well maintained and convective flow of mobile phase through the super-porous structure in the medium was confirmed by its low backpressure and the Blake–Kozeny equation [54].

9.3.3.2 Chemical Modification with Poly(Vinyl Alcohol) of Super-Porous P(ST-DVB) Microspheres

For the above physical adsorption of modified agarose, there is the problem of coating desorption over time. In order to obtain a more stable hydrophilic coating with efficient masking and long lifetime, we coated the super-porous PST microspheres with cross-linked PVA hydrogel through *in situ* radical polymerization of vinyl acetate (VAc) and *N,N′*-methylene bis(acrylamide) (MBA) [55]. Then

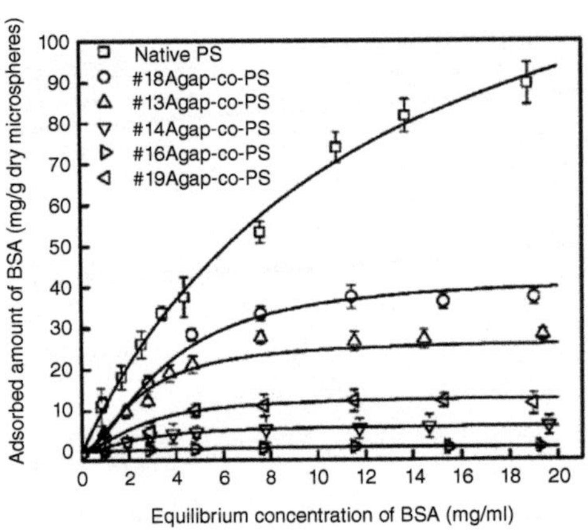

Figure 9.48 Adsorption isotherms of BSA on super-porous PST microspheres before and after coating. Source: Qu et al. [53]/with permission from American Chemical Society.

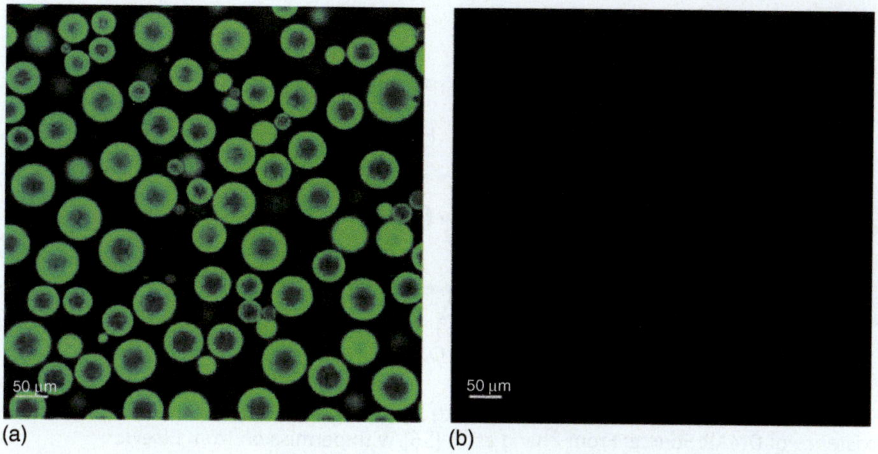

Figure 9.49 Preparation of cross-linked PVA hydrogel-coated super-porous PST Microspheres. LSCM images of (a) PST microspheres and (b) PST-PVA after being incubated with FITC-BSA in 0.05 M phosphate buffer (pH 7.4, 25 °C). Source: Qu et al. [55]/American Chemical Society.

the cross-linked PVA hydrogel-coated PST microspheres were obtained through alcoholysis of the PVAc on the PST microspheres.

The cross-linked PVA gave a dense covalent coating on PST microspheres, where the super-porous structure of microspheres still remained well. Both C1s XPS spectra and FT-IR spectra results confirmed that the PST microspheres were effectively coated with cross-linked PVA hydrogel. The adsorption isotherms of BSA on PS-PVA microspheres showed that the nonspecific adsorption amount of BSA on PST particles decreased significantly, which was attributed to the good shielding of hydrophobic PST microspheres surfaces by cross-linked PVA hydrogel. LSCM images gave a straight insight into the incremental hydrophilicity of PST microspheres after coating. Figure 9.49 illustrates that strong fluorescence from FITC-BSA was visualized on PST microspheres, whereas weak fluorescence was observed in PST-PVA microspheres which indicates the hydrophilicity and biocompatibility of PS-PVA were improved significantly. While the nonspecific adsorption of protein was still more than 5.0 $mg\,g^{-1}$ dry microspheres due to the hydrogen bond formed by the hydroxyl groups of PVA and carboxyl group of proteins.

9.3.3.3 Surface Hydrophilization of Super-Porous PGMA Microspheres

Surface hydrophilization of the super-porous P(GMA-DVB) microspheres is more convenient due to its high content of epoxy groups [56]. The epoxy groups can easily react with hydroxyl groups of the hydrophilic molecules through ester bond, as shown in Figure 9.50 [57, 58]. We coated dextran by covalent cross-linking on P(GMA-DVB) microspheres. We found that the main factors affected coating, included physical adsorption of dextran, 4-dimethylaminopyridine (DMAP) amounts, reaction time, and temperature [56]. DMAP was a basic catalyst for opening the epoxy groups to react with the adsorbed dextran. In the present of

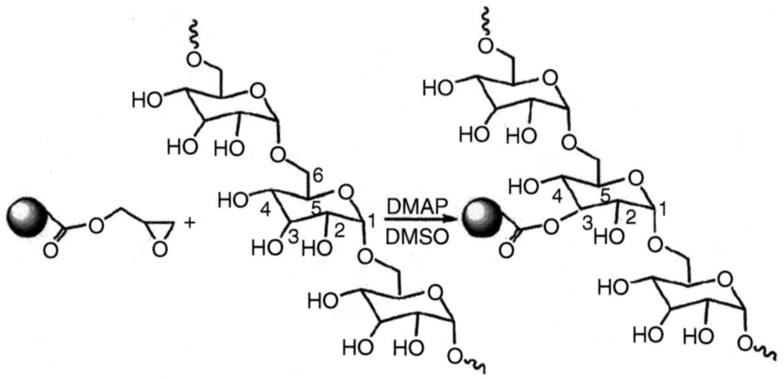

Figure 9.50 Chemical reaction between dextran and P(GMA-DVB) microspheres at the existence of DMAP. Source: From Zhang et al. [56]/with permission from Elsevier.

DMAP, the epoxy groups in the pore surface reacted with the hydroxyl groups of dextran via transesterification (Figure 9.50).

9.3.4 Applications in Biomolecules Separation

Based on the successful hydrophilization of super-porous polymeric microspheres, they become good chromatographic "base supports" for different types of chromatography, such as ion exchange and affinity chromatography, since the coatings can be easily derivatized by classical methods. Ion-exchange chromatography (IEC) medium has been widely used in protein separation, and nearly 75% protein purification includes the step of IEC. Therefore, the ion-exchange medium (DEAE-AP) was used as an example to study the basic properties and applications of super-porous microspheres as chromatographic separation medium.

9.3.4.1 Excellent Flow Hydrodynamics

We compared the relationship between flow velocity and the backpressures of DEAE-FF, DEAE-AP (DEAE derivatized hydrophilic super-porous PST), and super-porous PST columns (Figure 9.51). The linear relationship was obtained on DEAE-AP column for flow velocity up to 3612 $cm\,h^{-1}$. However, for DEAE-FF column, the linear relationship was only obtained for flow velocity up to 1668 $cm\,h^{-1}$ due to DEAE-FF resins' low mechanical stability. Moreover, the backpressure of DEAE-AP column was much lower than that of DEAE-FF column under the same flow velocity, which is the remarkable feature of perfusion chromatography.

Height equivalent of theoretical plate (HETP) is a typical criterion to describe the overall column efficiency. The experiments were conducted at flow velocities ranging from 180 to 2528 $cm\,h^{-1}$ under a nonretained condition. With increasing the flow velocity, the difference of HETP values between the two columns (DEAE-FF and DEAE-AP) was further augmented (Figure 9.52). For DEAE-FF columns, HETP rapidly increased with the increase of flow velocity, which is in accordance with the Van Deemter equation. It indicated that the mass transfer resistance in

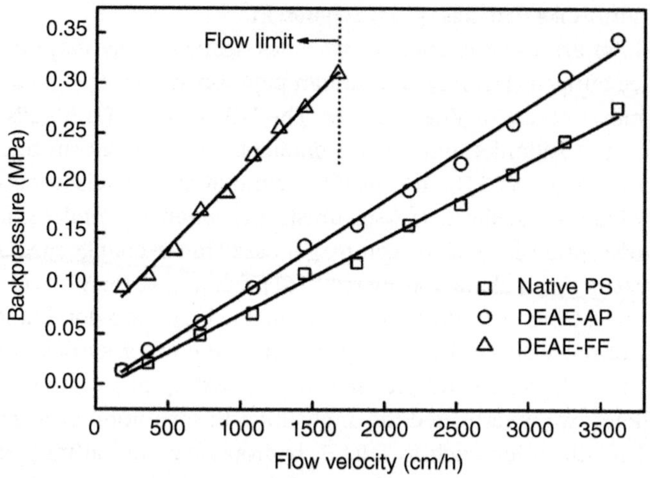

Figure 9.51 Relationship between the column backpressure and flow velocity. Column, 100 mm × 4.6 mm I.D.; mobile phase, high-purity water. Source: From Qu et al. [59]/with permission from John Wiley & Sons, Inc.

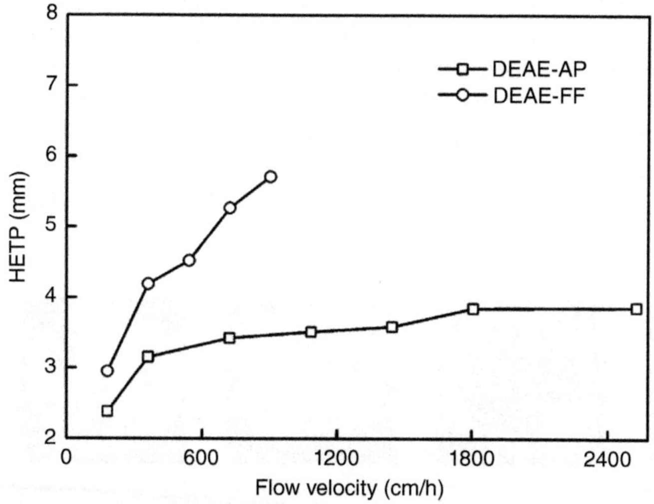

Figure 9.52 Column efficiency versus flow velocity. Column, 100 × 4.6 mm I.D.; sample, 2 mg ml^{-1} BSA; and mobile phase, 1 mol l^{-1} NaCl in 20 mM Tris-HCl buffer, pH 8.0. Source: From Qu et al. [59]/with permission from John Wiley & Sons, Inc.

the DEAE-FF beads was dominated by the intraparticle term. However, the HETP of DEAE-AP column only increased slightly with flow velocity up to 2528 cm h^{-1}, that is, HETP was nearly independent of flow velocity in a wide range of flow rate. The presence of flow-through large pores in DEAE-AP microsphere could induce intraparticle convective mass transfer and significantly lessened the stagnant mobile phase mass transfer of particles by reducing the diffusive distance. These results were also confirmed by our group's another work [60].

9.3.4.2 Application in Virus-Like Particles (VLPs) Separation

Virus-like particles (VLPs) are used as effective vaccines against many diseases. Hepatitis B virus surface antigen (HBsAg) and human papillomavirus (HPV) vaccine are currently the most successful VLP vaccines. The HBsAg VLPs (HB-VLPs) are composed of about 100 subunits and have a diameter of about 22 nm and molecular weight of more than 3000 kDa. It is well known that protein adsorption capacity on most ion-exchange media decreases drastically when the molecular mass is over 100 kDa. Adsorption capacity of agarose gel-based ion-exchange media for most proteins can easily reach about 100 mg ml^{-1} [61], while that for VLPs are usually less than 1 mg ml^{-1}. Obviously, the pore size of the media plays a dominant role in the accessibility of the VLPs to the binding sites on the interior surfaces of the microspheres. Based on the successful preparation and hydrophilic coating of super-porous microspheres, we investigated the application in separation of VLPs after suitable chemical modification such as DEAE, hydrophobic, and affinity of the super-porous microspheres.

Our group [62] examined the pore size effect of chromatographic media on the chromatographic behavior of HB-VLPs, in terms of binding capacity, uptake kinetics, and structural stability. Two commercial agarose media (DEAE-FF and DEAE-Capto) and two super-porous particles (DEAE-AP-120 nm and DEAE-AP-280 nm) possess the same DEAE ligands density but different pore size.

According to the confocal laser scanning microscopy observation (Figure 9.53), adsorption of the VLPs in DEAE-FF and DEAE-Capto was mostly confined to a thin shell on the surface of the beads, leaving the underlying pore space and the binding sites inaccessibly, while the large pores in super-porous media enabled the VLPs to access to the interior pore spaces by diffusion transport efficiently. The former led to

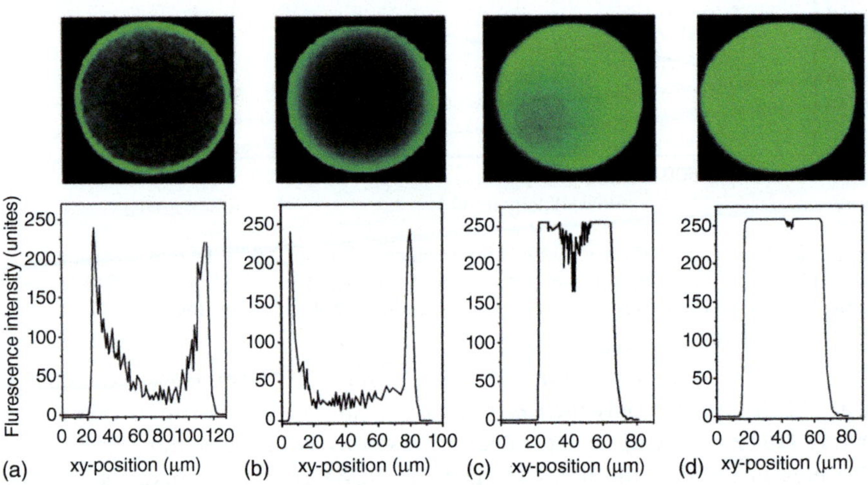

Figure 9.53 Experimental and simulated uptake curves of HB-VLPs (c0 = 0.12 mg ml^{-1}) to DEAE-FF, DEAE-Capto, DEAE-AP-120 nm, and DEAE-AP-280 nm. Source: Yu et al. [62]/Elsevier.

the low binding capacity and slow mass transfer rate of VLPs in the pore of agarose bead, which would also limit the chromatographic purification rate.

For practical chromatographic application, a high dynamic binding capacity (DBC) is more significant for media than high static adsorption capacity. During chromatographic process, both the properties of chromatographic media, such as pore size and ligand density, and the chromatographic conditions such as protein loading quantities, residence time, and solution environment, have profound effects on the interactions between protein stationary phase, as well as protein–protein interactions. What changes would be brought by super-porous media? As shown in Figure 9.54, the four anion-exchange media was individually fed with different volume of 0.08 mg ml^{-1} HB-VLPs solution to get four different HB-VLPs loading quantities of 0.5, 1.0, 1.5, and 2.0 mg protein ml^{-1}-media, respectively. With an increase in the loading quantities, changes in recovery of HB-VLPs from DEAE-FF and DEAE-Capto column generally showed a decreasing trend. However, the phenomena observed with DEAE-AP-120 nm and DEAE-AP-280 nm were distinctly different. The total recovery of HB-VLPs eluted from DEAE-FF column decreased from 36.8 to 18.3%, while that from DEAE-AP-280 nm column increased from 44.04 to 69.21% and that from DEAE-AP-120 nm column increased from 28.45 to 55.52%. The purification behavior of the large biomolecules in super-porous media is completely different from that in conventional agarose media.

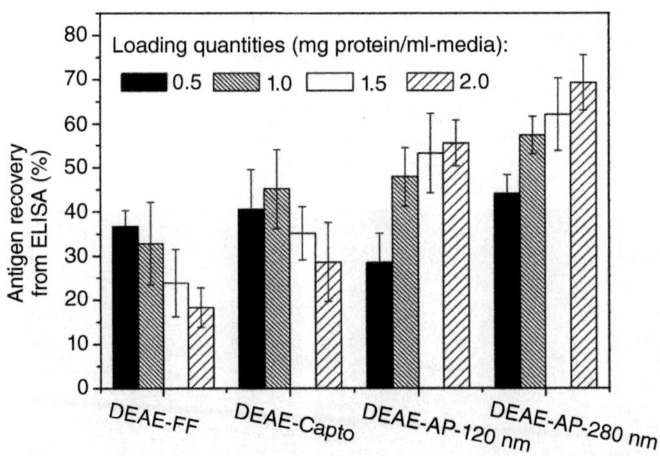

Figure 9.54 Effect of loading quantities (0.5, 1.0, 1.5, and 2.0 mg protein ml^{-1}-media) on ELISA-recovery of HB-VLPs eluted from four anion-exchange media. Conditions: a chromatographic column (50 mm × 10 mm I.D.) packed with each of the anion-exchange media, including DEAE-FF, DEAE-Capto, DEAE-AP-120 nm, and DEAE-AP-280 nm was loaded with 25, 50, 75, and 100 ml HB-VLPs (0.08 mg ml^{-1}) solutions at a flow rate of 80 cm hours^{-1} to yield protein loading quantities of 0.5, 1.0, 1.5, and 2.0 mg protein ml^{-1}-media, respectively. Then the adsorbed HB-VLPs were eluted with 1.0 M NaCl in pH 7.0, 20 mM sodium phosphate buffer, and the elution peak were polled and the antigen was detected by ELISA. The recovery of HB-VLPs was calculated by comparing the total antigen amount in the eluted fraction determined by ELISA with the initial total antigen in the loading samples assayed by ELISA. Source: From Yu et al. [62]/with permission from Elsevier.

To elucidate the relationship between HB-VLPs stability and pore size of media at varying loading quantities more clearly, HPSEC (high-performance size-exclusion chromatograph) analyses on the collected elution peak from the four IEC media at each loading quantity were carried out to investigate the aggregation and/or disassembly of HB-VLPs during the chromatographic process (Figure 9.55). Peak A corresponds to the whole HB-VLPs (Figure 9.55c) and Peak B represents the HB-VLPs disassembly (Figure 9.55d). These results gave direct proof that super-porous media could effectively reduce the disassembly of HB-VLPs.

When HB-VLPs were adsorbed in DEAE FF and DEAE-Capto, which have pore size close to the size of VLPs, multipoint adsorption was promoted on the surface of the pores. In case of adsorption and desorption of HB-VLPs in anion-exchange media, however, this multipoint adsorption between ion-exchange ligand and HB-VLPs might induce undesired irreversible disassembly of VLPs during adsorption–desorption process. The disassembly would be more significant

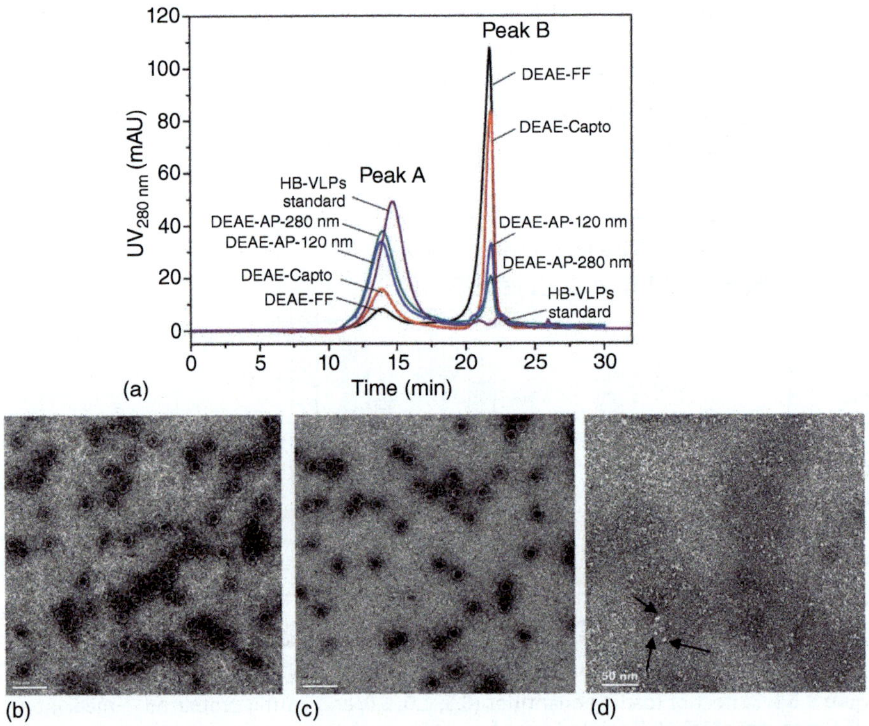

Figure 9.55 (a) HPSEC assay of HB-VLPs eluted by 1.0 M NaCl in 20 mM sodium phosphate buffer (pH 7.0) from DEAE-FF, DEAE-Capto, DEAE-AP-120 nm, and DEAE-AP-280 nm at the same loading quantity (2.0 mg ml^{-1}). The HPSEC profile of HB-VLPs standard was also compared. Conditions: TSK G5000 PW (300 mm × 7.5 mm); eluted by 50 mM PBS (pH 7.0) at room temperature (20 °C) at a flow rate of 0.6 ml minutes^{-1}; 100 μl of injection volume with a protein concentration in each sample adjusted to ca. 0.11 mg ml^{-1} prior to HPSEC analyses. (b) TEM image of purified HB-VLPs. Standard Bar, 100 nm; (c) TME image of HP-VLPs in Peak A after HPSEC assay. Bar, 100 nm; (d) TME image of HP-VLPs in Peak B after HPSEC assay. Bar, 50 nm. Source: Yu et al. [62]/Elsevier.

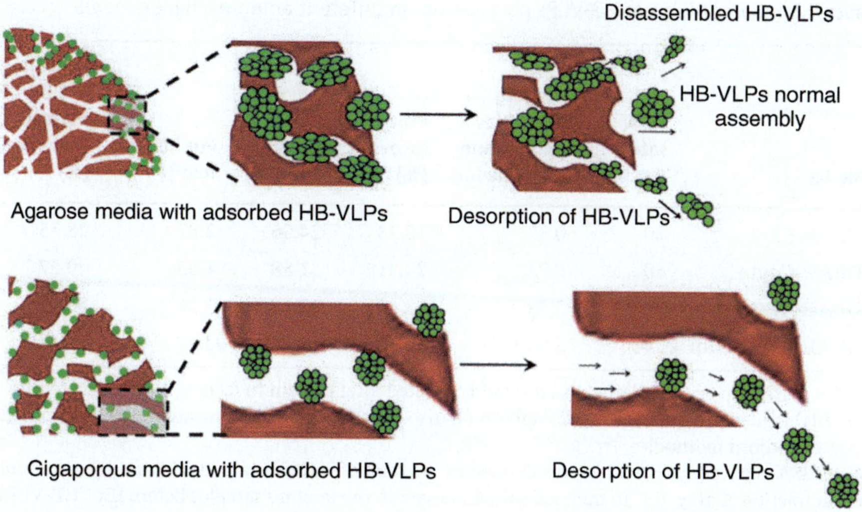

Figure 9.56 Schematic demonstration of HB-VLPs adsorption and disassembly during desorption process in agarose gel-based media and in super-porous media. Source: From Yu et al. [62]/with permission from Elsevier.

when the HB-VLPs were more crowdedly accumulated in the narrow pores due to the intense force of electrostatic interactions from all directions. The adsorption–desorption and disassembly of HB-VLPs in agarose media was schematically demonstrated in Figure 9.56. On the contrary, when HB VLPs were adsorbed on super-porous media, which pore size is about ten times larger than the size of VLPs, much less contact points were formed between protein and inner surface of the pores, so the disassembly of HB-VLPs was mostly avoided and much higher recovery was obtained.

To take full advantages of super-porous media in high dynamic bind capacity and high HB-VLPs recovery, higher loading quantity and flow rate were applied to DEAE-AP-120 nm and DEAE-AP-280 nm. Table 9.3 listed the detailed conditions for chromatographic purification of HB-VLPs on DEAE FF, DEAE-Capto, DEAE-AP-120 nm, and DEAE-AP-280 nm media. As expected, the IEC process on super-porous media exhibited significantly better purifying performance for HB-VLPs than on agarose media. Large pores in DEAE-AP-280 nm promoted the highest HB-VLPs recovery of 68.33% and the highest purification fold of 3.47 due to the lowered interactions between the HB-VLPs and the binding sites of the media. In addition, super-porous media DEAE-AP-280 nm gained about 12.9 times increase in static adsorption capacity, 8.0 times increase in DBC, and 11.4 times increase in effective pore diffusivity. Beyond increasing the binding capacity and enhancing the mass transfer, the super-porous structure also significantly improved the stability of the VLPs during intensive adsorption-desorption process by lowing the multipoint interaction between the VLPs and binding sites in the pores. At 2.0 mg ml^{-1}-media loading quantity, about 85.5% VLPs were correctly self-assembled after the chromatography with DEAE-AP-280 nm media; oppositely

Table 9.3 Comparison of HB-VLPs purification on different anion-exchange media.

Media	Flow rate (cm h^{-1})	Loading quantities (mg protein ml^{-1}-media)	Protein recovery[a] (%)	ELISA-recovery[b] (%)	Purification fold[c] (-)	Percentage of HB-VLPs normal assembly[d] (%)
DEAE-FF	30	0.37	10.73	24.06	2.87	28.35
DEAE-Capto	60	0.98	20.91	31.88	1.53	40.57
DEAE-AP-120 nm	240	2.98	13.46	43.18	3.26	56.17
DEAE-AP-280 nm	240	2.98	20.40	68.33	3.47	71.42

a) Protein recovery was defined as the ratio of eluted total protein in faction A after IEC to the total loaded protein before IEC (refer to Figure 9.55a). Protein concentration was determined by Bradford method.
b) ELISA-recovery of HB-VLPs was defined as the ratio of eluted total HB-VLPs antigen amount in fraction A after IEC to the total initial antigen in the loading samples before IEC. HB-VLPs antigen concentration was determined by ELISA.
c) Purification factor was defined as the ratio of the antigen concentration in fraction A after IEC to the initial antigen concentration before IEC. HB-VLPs antigen concentration was determined by ELISA.
d) Percentage of HB-VLPs normal assembly was resolved from HPSEC assay of eluted HB-VLPs in fraction A after IEC.

Source: From Yu et al. [62]/with permission from Elsevier.

about 85.2% VLPs lost their normal assembly with DEAE-FF due to irreversible disassembly. These results demonstrate the remarkable advantages of super-porous media in the purification of large biomolecules.

9.4 Uniform Microspheres for Cell Culture

Cell culture on microcarrier is one of cell production techniques, which integrates adherent culture and suspension culture, and is suitable for large-scale culture of various cells. Microcarriers, as one of the main components of this technology, are small suspended particles with the diameter of 60–300 μm, on which surface the cell can be adhered to and cultured. A good microcarrier should have the following characteristics: no cytotoxicity, good adhesion ability, a density slightly higher than that of the culture medium, uniform particle size, transparency for easy observation of cell proliferation, good batch stability, and mechanical stability. Uniform microspheres are helpful for their homogeneous mixing in the reactor and provide a similar microenvironment for cell growth.

We developed a positively charged microcarrier based on KGM microspheres and studied its effect on cell proliferation of Vero cell line [63]. KGM microspheres were named A-KGM-1225 according to the KGM concentration and the amount of the cross-linker. Vero cells were incubated in the microcarrier suspension culture system for several days using A-KGM-1225 and Cytodex-1, respectively (Figure 9.57) [63]. The number of Vero cells proliferated in the A-KGM-1225

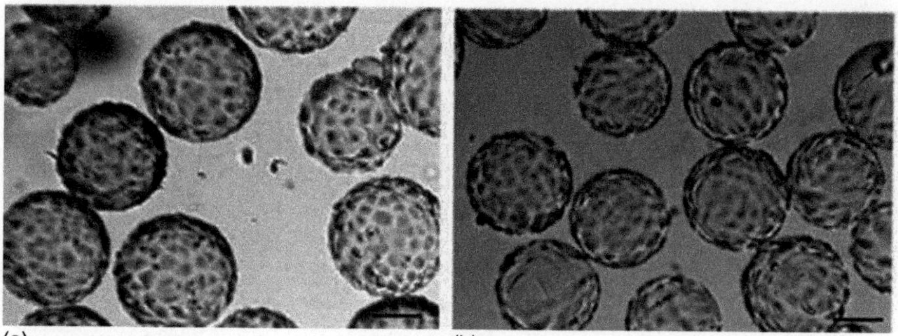

Figure 9.57 Morphology of Vero cell line grown on (a) A-KGM-1225 (b) Cytodex-1 microcarriers. Vero cells were inoculated at 1.0×10^5 cells/mL onto 2.0 mg/mL A-KGM-1225 and Cytodex-1 microcarriers, respectively. Photos were taken under inverted microscope at 72 hours after inoculation, bar: 100 μm. Source: Sun et al. [63]/with permission from Elsevier.

suspension culture system reached maximums after 8 days, and then declined slightly, which indicated that all possible areas of the microcarrier had been occupied by the cells. While in the Cytodex-1 suspension culture system, the cells reached maximums on 6 days. Also, both the size and its uniformity were important to the cell growth. With the same gel structure and ligand density, the cell density proliferated on large-sized KGM microcarrier was higher than that on small-sized KGM microcarrier. For KGM microcarrier with a particle size of 60–100 μm, the cell proliferation was poor, but for large-sized KGM microcarrier, the cell proliferation became better, which was similar to that of commercial microcarrier.

Microcarrier culture technology is also a promising technique for stem cell proliferation. Stem cells are highly susceptible to the biochemical and biophysical properties of the environment. Therefore, the properties of microcarriers are important for rapid proliferation and stemness maintenance of stem cells. For example, we prepared KGM microcarriers with good rigidity controllability in order to investigate the effect of microcarrier's rigidity on mesenchymal stem cells (MSCs) proliferation. Controlling the KGM concentration in the aqueous phase can modulate the skeleton density of microcarriers, and thus the rigidity. We prepared the microcarriers with KGM concentrations of 8, 10, 12, and 16% with the rigidity ranged from 0.18 to 9.73 MPa. We found that the proliferation and stemness of MSCs were well maintained when the rigidity was in the range of 2–4 MPa (12%-KGM microcarrier). With the optimal KGM microcarriers rigidity, MSCs proliferated about 27 times after 11 days (Figure 9.58), which was 1.7 times as much as commercial product Cytodex-1 (Figure 9.59), and the differentiation potential was 3.2 times as much as Cytodex-1. The KGM microcarriers with a controllable rigidity provide a promising select for MSCs large-scale culture and better application.

Agarose microspheres prepared by direct membrane emulsification method have been used as bacteria culture. During the preparation, agarose solution mixed with

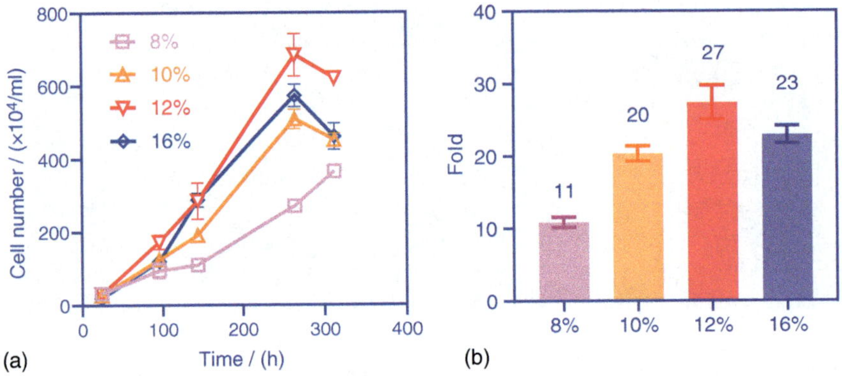

Figure 9.58 Effect of KGM microcarriers rigidity on MSCs proliferation. (a) Cell growth curve of MSCs on KGM microcarriers with different rigidity. (b) Proliferation fold of MSCs on different KGM microcarriers. Source: From Yan et al. [64]/with permission from Springer Nature.

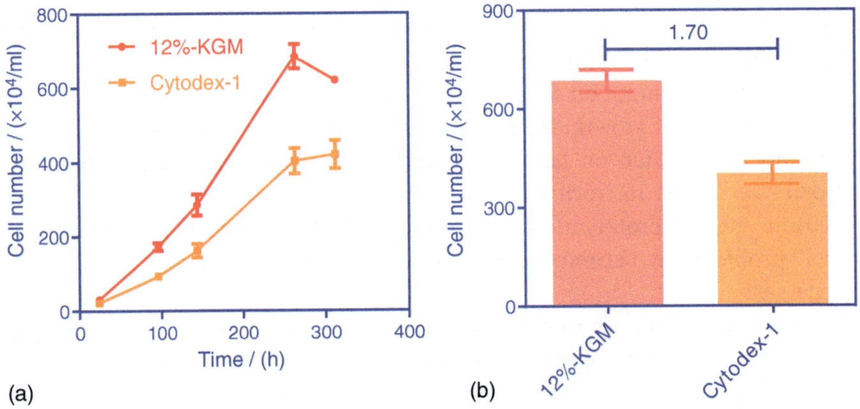

Figure 9.59 Comparation of cell proliferation between 12%-KGM microcarriers and commercial microcarriers Cytodex-1. (a) Cell growth curve of MSCs on 12%-KGM microcarriers and Cytodex-1. (b) Maximum cell numbers of MSCs on 12%-KGM microcarriers and Cytodex-1. Source: From Yan et al. [64]/with permission from Springer Nature.

a proper dilution of bacteria was used as the water phase and dispersed into the oil phase through the microporous membrane to form uniform-sized emulsion. After cooling, uniform bacteria-embedded agarose microspheres were obtained with a CV value of 12%. The particle size of agarose microspheres can be well controlled within a range of 20–60 μm using membranes with different pore sizes. There were several advantages to these uniform agarose microspheres used for bacterial culture. Firstly, it ensured that the same amount of bacteria was embedded in each microsphere, resulting in the same rate of proliferation during the cell culture process, which was beneficial for bacteria culture. Secondly, the bacteria-embedded process that was mixed with agarose solution and passed through the membrane into the oil phase was milder than that of traditional mechanical stirring method, in which the coated bacteria can well maintain both their biological activity and stability in agarose microspheres. Microspheres prepared were packed into a cylindrical container for continuous culture with a constant speed, and the amount of embedded bacteria increased after 14 days of culture (Figure 9.60) [6]. It showed that there were a lot of dark spots in microspheres and some of the microspheres became non-transparent, indicating the bacterial cells grew successfully. Using flow cytometry analysis, distinctive light-scattering signatures of dot plot data were obtained, and screening of agarose microspheres was performed depending on the optical density to separate microspheres embedded with different amounts of bacteria successfully (Figure 9.61) [6]. Compared with dot plot data of agarose microspheres without incubation (Figure 9.61a), there were four signal areas named P1, P2, P3, and P4 in dot plot data of agarose microspheres incubated after 14 days, representing the bacteria growth (Figure 9.61b). Furthermore, it showed that agarose microspheres collected from P3 area had a higher growth percent of bacteria compared with P1, P2, and P4 areas, which was promising for further culture and identification.

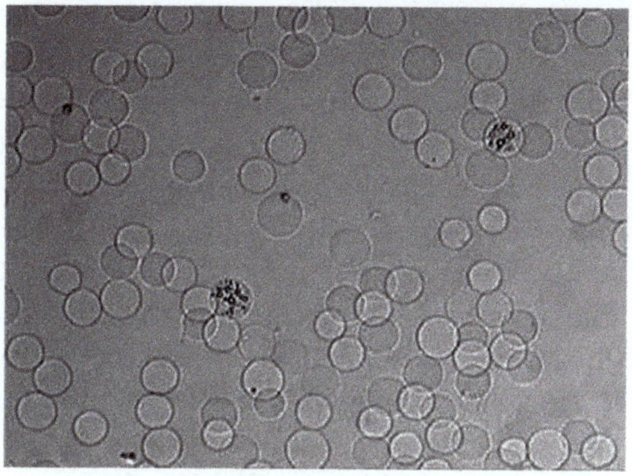

Figure 9.60 Optical micrographs of agarose microspheres encapsulating cells after incubation. Source: Zhou et al. [6]/with permission from American Chemical Society.

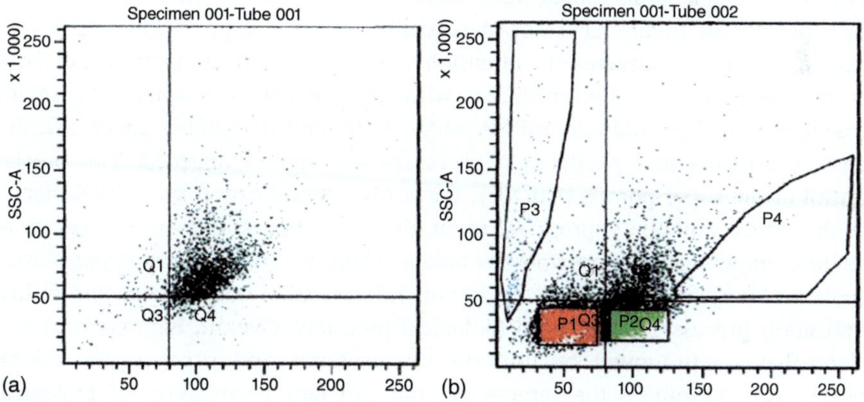

Figure 9.61 Distinctive light-scattering signatures of dot plot data of agarose microspheres (a) Dot plot data of agarose microspheres without incubation; (b) Dot plot data of agarose microspheres incubated after 14 days. Source: From Zhou et al. [6]/American Chemical Society.

9.5 Summary

This chapter mainly focuses on the application of both uniform microspheres and super-porous microspheres in the fields of biochemical engineering. For uniform microspheres, e.g. agarose microsphere, KGM microsphere, PST microsphere, PGMA microsphere, PHEMA microsphere, and silica microsphere, they have unique advantages in improving both the resolution and the column permeability in chromatography, and also for reliable cell culture. With the rapid development of biotechnology, both the performance and preparation technology of these microspheres are facing more demands and greater challenges. First of all, it is necessary to establish a suitable membrane emulsification process for special raw materials. For example, high-performance chromatographic media with characteristics of both high resolution and high flow rate is required for efficient and rapid purification of target protein. Uniform small-sized high-concentration agarose microspheres are able to meet these requirements, and therefore, the difficulties of high trans-membrane for high viscosity system and high constant temperature maintaining are vital for both the membrane emulsification technology and its equipment design. Secondly, attention should be paid to the scale-up of both the membrane emulsification process and the device. Many relevant factors such as process stability and balance between the scale-up size of the equipment and the cost are need to be considered. Thirdly, it is necessary to develop a green and energy-saving membrane emulsification technology to prepare uniform microspheres. For example, further simplifying the preparation process and reducing the amount of solvent. Our goal is to establish a high-level membrane emulsification technology and corresponding equipment for the purpose of the high-end manufacture in high-performance microspheres for biochemical engineering. For super-porous microspheres, they have the characteristics of fast flow rate, high yield, and high resolution, which can effectively maintain the structure and activity of biological macromolecules. Especially since the outbreak of COVID-19, the super-porous media has been successfully used in the purification of mRNA vaccine, which has attracted the attention of researchers all over the world. With the rapid development of biomedicines with large molecular weights and complex structures (such as antibody, mRNA, AAV, HPV, and exosomes), more efficient separation media and purification processes are urgently needed. The precise control of the super-porous structure, the surface modification specially designed for the target, and the appropriate separation process are still the key points of the development of the super-porous media. There is no doubt that super-porous media will play an increasingly important role for laboratory studies and industrialization process of large-sized biological products. Overall, high-performance microspheres represented by uniform microspheres and super-porous microspheres can guarantee the large-scale and efficient preparation of biological products.

References

1 Ma, G.H. and Su, Z.G. (2014). Process engineering for size and structure control of microspheres. *CIESC Journal* 65 (7): 2574–2587.

2 Huang, Y.D., Bi, J.X., Zhao, L. et al. (2010). Regulation of protein multipoint adsorption on ion-exchange adsorbent and its application to the purification of macromolecules. *Protein Expression and Purification* 74 (2): 257–263.

3 Zhou, W.B., Bi, J.X., Zhao, L. et al. (2007). A highly efficient hydrophobic interaction chromatographic absorbent improved the purification of hepatitis B surface antigen (HBsAg) derived from Hansenula polymorpha cell. *Process Biochemistry* 42 (5): 751–756.

4 Zhou, Q.Z. (2008). *Preparation and Application of Uniform-Sized Controllable Agarose Gel Beads*. Beijing: Institute of Process Engineering, Chinese Academy of Sciences.

5 Zhou, Q.Z., Wang, L.Y., Ma, G.H., and Su, Z.G. (2007). Preparation of uniform-sized agarose beads by microporous membrane emulsification technique. *Journal of Colloid and Interface Science* 311 (1): 118–127.

6 Zhou, Q.Z., Liu, X.Y., Liu, S.J. et al. (2008). Preparation of uniformly sized agarose microcapsules by membrane emulsification for application in sorting bacteria. *Industrial & Engineering Chemistry Research* 47 (17): 6386–6390.

7 Lin, N. (2009). *Preparation of Large Size Agarose Gel Beads by Microporous Membrane Emulsification Technique*. Beijing: Beijing University of Chemical Technology.

8 Lin, N., Wu, J., Zheng, G.J. et al. (2009). Preparation of large size agarose beads by microporous membrane emulsification technique. *The Chinese Journal of Process Engineering* 9 (5): 953–961.

9 Zhao, X. (2014). *Preparation of High-Performance Agarose Microsphere for Potential Bioseparation Media*. Beijing: Institute of Process Engineering, University of Chinese Academy of Sciences.

10 Zhao, X., Wu, J., Gong, F.L. et al. (2014). Preparation of uniform and large sized agarose microspheres by an improved membrane emulsification technique. *Powder Technology* 253: 444–452.

11 Zhang, H.W. (2018). *Preparation of Uniform-Sized Polysaccahride Complex Microspheres by Premix Membrane Emulsification Technique and Its use in Chromatography*. Beijing: China University of Mining & Technology.

12 Zhang, H.W., Zhao, L., Huang, Y.D. et al. (2018). Uniform polysaccharide composite microspheres with controllable network by microporous membrane emulsification technique. *Analytical and Bioanalytical Chemistry* 410 (18): 4331–4341.

13 Li, X.Q. (2014). *Preparation of Uniform Large-Sized Polysaccaride Beads by Rotating Membrane Emulsification*. Beijing: Institute of Process Engineering, University of Chinese Academy of Sciences.

14 Li, X.Q., Gong, F.L., and Ma, G.H. (2014). Preparation of uniform konjac glucomannan gel microspheres by rotating membrane emulsification. *The Chinese Journal of Process Engineering* 14 (2): 296–303.

15 Xiong, Z.D. (2015). *The Structure Design and Control for Preparation of Konjac Glucomannan Microspheres.* Beijing: Institute of Process Engineering, University of Chinese Academy of Sciences.

16 Xiong, Z.D., Zhou, W.Q., Sun, L.J. et al. (2014). Konjac glucomannan microspheres for low-cost desalting of protein solution. *Carbohydrate Polymers* 111: 56–62.

17 Huang, Y.D., Zhang, R.Y., Li, J. et al. (2014). A novel gigaporous GSH affinity medium for high-speed affinity chromatography of GST-tagged proteins. *Protein Expression and Purification* 95: 84–91.

18 Li, Q., Zhao, L., Zhang, R.Y. et al. (2016). Functional hydrophilic polystyrene beads with uniformly size and high cross-linking degree facilitated rapid separation of exenatide. *Journal of Chromatography B* 1017–1018: 129–135.

19 Hao, D.X., Gong, F.L., Hu, G.H. et al. (2009). The relationship between heterogeneous structures and phase separation in synthesis of uniform PolyDVB microspheres. *Polymer* 50 (14): 3188–3195.

20 Hao, D.X., Gong, F.L., Wei, W. et al. (2008). Porogen effects in synthesis of uniform micrometer-sized poly(divinylbenzene) microspheres with high surface areas. *Journal of Colloid and Interface Science* 323 (1): 52–59.

21 Zhu, L. (2013). *Preparation of Polystyrene Particles with Different Physical Properties and Their Effect on the Biological Responses.* Beijing: Beijing University of Chemical Technology.

22 Zhu, L., Gong, F.L., Yuan, Q.P., and Ma, G.H. (2013). Preparation of uniform and heterogeneous structure polystyrene particles by premix membrane emulsification. *The Chinese Journal of Process Engineering* 13 (2): 275–280.

23 Qin, L., Lin, H., Gong, F.L. et al. (2007). Preparation of uniform-sized porous PST-DVB microspheres by combining membrane emulsification with suspension polymerization. *The Chinese Journal of Process Engineering* 7 (3): 526–530.

24 Zhang, C.X. (2015). *Preparation of the Polystyrene-Divinyl Benzene Microsphere and the Application for Chromatographic Separation of Peptide.* Beijing: Beijing University of Chemical Technology.

25 Zhang, C.X., Li, Q., Huang, Y.D. et al. (2015). Chromatographic separation of octreotide with prepared polydivinylbenzene microspheres. *The Chinese Journal of Process Engineering* 15 (3): 482–488.

26 Zhang, C.X., Zhang, R.Y., Li, Q. et al. (2015). Rapid octreotide separation from synthetic peptide crude mixtures by chromatography on poly(styrene-*co*-divibylbenzene)-based reversed phases. *Separation and Purification Technology* 154: 351–358.

27 Sun, H.H., Li, X.N., Ma, G.H., and Su, Z.G. (2005). Polystyrene-type uniform porous microsphere enables high resolution and low-pressure chromatography of natural products – a case study with icariin purification. *Chromatographia* 61: 9–15.

28 Wang, J. (2009). *Preparation of Uniform Porous Polydivinylbenzene Microspheres and Their Applications in Separation and Purification of Peptides*. Beijing: Institute of Process Engineering, Chinese Academy of Sciences.

29 Wang, J., Qu, H.H., Hao, D.X. et al. (2009). Preparation of uniform polydivinylbenzene microspheres and their application in reversed-phase chromatographic separation of small molecular compounds and peptides. *The Chinese Journal of Process Engineering* 9 (4): 763–769.

30 Wang, R.W., Zhang, Y., Ma, G.H., and Su, Z.G. (2006). Preparation of uniform poly(glycidyl methacrylate) porous microspheres by membrane emulsification-polymerization technology. *Journal of Applied Polymer Science* 102 (5): 5018–5027.

31 Wang, R.W. (2006). *Preparation and Application of A New Hydrophobic Interaction Chromatography Media: Preparation, Modification and Application of Poly(glycidyl mathacrylate) Microsphere*. Beijing: Institute of Process Engineering, Chinese Academy of Sciences.

32 Li, C.F., Wang, R.W., Liu, S.Z. et al. (2005). Rapid purification of kallikrein by the combination of hydrophobic interaction chromatography and salt precipitation. *The Chinese Journal of Process Engineering* 5 (5): 550–553.

33 Qu, H.H. (2008). *Strategy for the Design of Uniform PHEMA Microspheres and Their Chromatographic Application*. Beijing: Institute of Process Engineering, Chinese Academy of Sciences.

34 Qu, H.H., Gong, F.L., Ma, G.H., and Su, Z.G. (2007). Preparation and characterization of large porous poly(HEMA-*co*-EDMA) microspheres with narrow size distribution by modified membrane emulsification method. *Journal of Applied Polymer Science* 105: 1632–1641.

35 Qu, H.H., Gong, F.L., Wang, J. et al. (2008). Strategies for the design of hydrophilic microspheres with controlled size and microstructure. *Industrial & Engineering Chemistry Research* 47 (17): 6399–6404.

36 Hu, Y.N. (2022). *Synthesis and Structure Regulation of Silica Microspheres by Membrane Emulsification*. Beijing: Institute of Process Engineering, Chinese Academy of Sciences.

37 Ladisch, M.R. (2001). *Bioseparations Engineering: Principles, Practice, and Economics*. New York: Wiley.

38 Gu, T.Y., Zhou, W.Q., Ma, G.H., and Su, Z.G. (2005). Rigid gigaporous chromatographic media and their potential impact on downstream processing. *China Particuology* 3 (6): 349–353.

39 Afeyan, N.B., Gordon, N.F., Mazsaroff, I. et al. (1990). Flow-through particles for the high-performance liquid chromatographic separation of biomolecules: perfusion chromatography. *Journal of Chromatography A* 519: 1–29.

40 Fulton, S.P., Afeyan, N.B., Gordon, N.F., and Regnier, F.E. (1991). Very high-speed separation of proteins with a 20 μm reversed-phase sorbent. *Journal of Chromatography A* 547: 452–456.

41 Frey, D.D., Schweinheim, E., and Horvath, C. (1993). Effect of intraparticle convection on the chromatography of biomacromolecules. *Biotechnology Progress* 9: 273–284.

42 Tallarek, U., Vergeldt, F.J., and Van As, H. (1999). Stagnant mobile phase mass transfer in chromatographic media: intraparticle diffusion and exchange kinetics. *Journal of Physical Chemistry B* 103: 7654–7664.

43 Gustavsson, P.E. and Larsson, P.O. (1996). Superporous agarose, a new material for chromatography. *Journal of Chromatography A* 734: 231–240.

44 Afeyan, N.B., Fulton, S.P., and Regnier, F.E. (1991). Perfusion chromatography packing materials for proteins and peptides. *Journal of Chromatography A* 544: 267–279.

45 McCoy, M., Kalghatgi, K., Regnier, F.E., and Afeyan, N. (1996). Perfusion chromatography-characterization of column packings for chromatography of proteins. *Journal of Chromatography A* 743: 221–229.

46 Whitney, D., McCoy, M., Gordon, N., and Afeyan, N. (1998). Characterization of large-pore polymeric supports for use in perfusion biochromatography. *Journal of Chromatography A* 807: 165–184.

47 Garcia, M.C., Marina, M.L., and Torre, M. (2000). Perfusion chromatography: an emergent technique for the analysis for food proteins. *Journal of Chromatography A* 880: 169–187.

48 Zhou, W.Q., Li, J., Wei, W. et al. (2011). Effect of solubilization of surfactant aggregates on pore structure in gigaporous polymeric particles. *Colloids and Surfaces A: Physicochemical and Engineering Aspects* 384: 549–554.

49 Zhou, W.Q., Gu, T.Y., Su, Z.G., and Ma, G.H. (2007). Synthesis of microporous poly(styrene-divinyl benzene) microspheres by surfactant reverse micelles swelling method. *Polymer* 48: 1981–1988.

50 Zhou, W.Q., Gu, T.Y., Su, Z.G., and Ma, G.H. (2007). Synthesis of microporous poly(glycidyl methacrylate) microspheres by surfactant reverse micelles swelling method. *European Polymer Journal* 43: 4493–4502.

51 Zhao, X., Huang, L., Wu, J. et al. (2019). Fabrication of rigid and microporous agarose microspheres by pre-crosslinking and surfactant micelles swelling method. *Colloids and Surfaces B: Biointerfaces* 182: 110377.

52 Zhang, D.L., Zhou, W.Q., Li, J. et al. (2016). The construction of an aqueous two-phase system to solve weak-aggregation of gigaporous poly(styrene-divinyl benzene) microspheres. *Polymers* 8 (5): 142.

53 Qu, J.B., Zhou, W.Q., Wei, W. et al. (2008). An effective way to hydrophilize gigaporous polystyrene microspheres as rapid chromatographic separation media for proteins. *Langmuir* 24: 13646–13652.

54 Sun, G.Y., Shi, Q.H., and Sun, Y. (2004). Novel biporous polymeric stationary phase for high-speed protein chromatography. *Journal of Chromatography A* 1061 (2): 159–165.

55 Qu, J.B., Huan, G.S., Chen, Y.L. et al. (2014). Coating gigaporous polystyrene microspheres with cross-linked poly(vinyl alcohol) hydrogel as a rapid protein chromatography matrix. *ACS Applied Materials and Interfaces* 6: 12752–12760.

56 Zhang, R.Y., Li, Q., Li, J. et al. (2012). Covalently coating dextran on microporous polyglycidyl methacrylate microsphere enabled rapid protein chromatographic separation. *Materials Science and Engineering C* 32: 2628–2633.

57 Zhang, X., Yan, W.D., Yang, H.F. et al. (2009). Graft polymerization from pore wall of three-dimensionally ordered microporous cross-linked polystyrene via atom transfer radical polymerization. *European Polymer Journal* 45: 1359–1366.

58 Yin, R.X., Tong, Z., Yang, D.Z., and Nie, J. (2012). Glucose-responsive insulin delivery microhydrogels from methacrylated dextran/concanavalin a: preparation and in vitro release study. *Carbohydrate Polymers* 89: 117–123.

59 Qu, J.B., Wan, X.Z., Zhai, Y.Q. et al. (2009). A novel stationary phase derivatized from hydrophilic gigaporous polystyrene-based microspheres for high-speed protein chromatography. *Journal of Chromatography A* 1216: 6511–6516.

60 Zhai, Y.Q., Zhou, W.Q., Wei, W. et al. (2012). Functional gigaporous polystyrene microspheres facilitating separation of poly(ethylene glycol)-protein conjugate. *Analytica Chimica Acta* 712: 152–161.

61 Tao, Y.Y., Almodovar, E.X.P., Carta, G. et al. (2011). Adsorption kinetics of deamidated antibody variants on microporous and dextran-grafted cation exchangers. III. microscopic studies. *Journal of Chromatography A* 1218: 8027–8035.

62 Yu, M.R., Li, Y., Zhang, S.P. et al. (2014). Improving stability of virus-like particles by ion-exchange chromatographic supports with large pore size: advantages of gigaporous media beyond enhanced binding capacity. *Journal of Chromatography A* 1331: 69–79.

63 Sun, L.J., Xiong, Z.D., Zhou, W.Q. et al. (2015). Novel konjac glucomannan microcarriers for anchorage-dependent animal cell culture. *Biochemical Engineering Journal* 96: 46–54.

64 Yan, X.R., Li, J., Na, X.M. et al. (2022). Mesenchymal stem cells proliferation on konjac glucomannan microcarriers: effect of rigidity. *Chinese Journal of Polymer Science* 40 (9): 1080–1089.

10

Membrane Emulsification Equipment and Industrialization

10.1 Introduction

The principle of innovative membrane emulsification technology is based on use of pressure to force the dispersed phase to permeate through a microporous membrane into the continuous phase, forming uniform emulsion [1–3]. After years of in-depth research, our team has obtained a series of innovative achievements on preparation techniques and applications, involving biomedicine, biopharmaceutical, biochemical engineering, biomaterial, fine chemistry, and personal care products [4–11]. Responding to the specific requests from the emulsion manufacturing industry, three subdivided membrane emulsification technologies including cross-flow membrane emulsification(CME) technology, also called direct membrane emulsification technique [4–7], premix membrane emulsification (PME) technology [8–10], also called rapid membrane emulsification technique, and rotary membrane emulsification (RME) technology [12, 13] have been developed successfully, which have been introduced in the previous nine chapters.

With development of membrane emulsification technology in preparation and applications, how to help the scientific achievements from laboratory to industry is a big challenge. Membrane emulsification is a new emerging technology for preparing uniform emulsions, microspheres, and microcapsules. It has been considered that industrialization is difficult because it is a very sophisticated technique that needs very carefully controlled pressure and seal between membrane and the module, a slight pressure vibration, membrane crack, etc., will result in failure of production. So, it is urgent to design an efficient scale-up membrane emulsification equipment to have this membrane emulsification become a universal emulsion manufacturing technology. Then, it becomes possible to industrialize and commercialize the novel uniform microsphere products by membrane emulsification technology.

Critical factors affecting the droplet size uniformity of emulsion have been clarified based on long-term professional experience of membrane emulsification and uniform droplet formation model (Chapter 1) by our team, as shown in Table 10.1. According to critical factors listed in this Table, we have developed membrane emulsification equipment for three subdivided membrane emulsification technologies, and successfully achieved a series of industrial applications.

Novel Membrane Emulsification: Principles, Preparation, Processes, and Bioapplications,
First Edition. Guanghui Ma.

Table 10.1 Critical factors affecting the size uniformity of emulsion.

Operation parameter	Transmembrane pressure
	Flow rate of dispersed phase
	Flow rate of continuous phase
Phase parameter (dispersed phase and continuous phase)	Interfacial tension, contact angle
	Density
	Viscosity
	Type and concentration of emulsifier

10.2 Cross-flow Membrane Emulsification Equipment

At the beginning of CME technology development, the hydrophilic glass porous membrane was used. Generally, CME was used to prepare oil-in-water (O/W) emulsions and hydrophobic particles. It involves using pressure to force an oil phase to permeate through a hydrophilic membrane into a water phase containing stabilizer and/or emulsifier to form uniform O/W emulsion. After in-depth study by our team, CME has been developed for preparation of uniform water-in-oil (W/O) emulsion by hydrophobically modified glass membrane or developed hydrophobic polymer membrane. Now, O/W, W/O, and double emulsions and particles solidified from them with uniform and controllable size of 5–100 μm are able to be prepared. The formation process of droplets by CME is shown in Figure 10.1.

According to Torque balance model and force analysis on a droplet at outlet of membrane pore in Chapter 1, it was found the transmembrane pressure worked on a droplet and flow shear force caused by circulation of continuous phase are critical affecting factors for the formation of uniform droplet. Based on these critical factors and operation parameters listed in Table 10.1, we have developed an automatic CME equipment of large scale. Automatic equipment can eliminate destabilization of transmembrane pressure caused by manual operation, and maintain flow of continuous phase accurately, for improving the success percentage of manufacture and batch-to-batch stability.

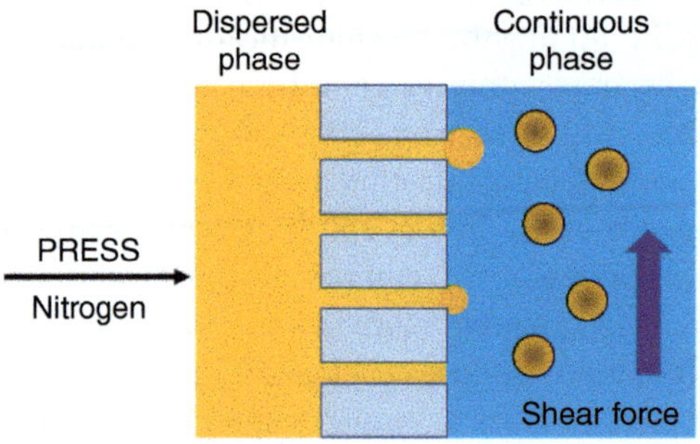

Figure 10.1 Cross-flow membrane emulsification method. Source: From Hu et al. [14]/with permission from Springer Nature.

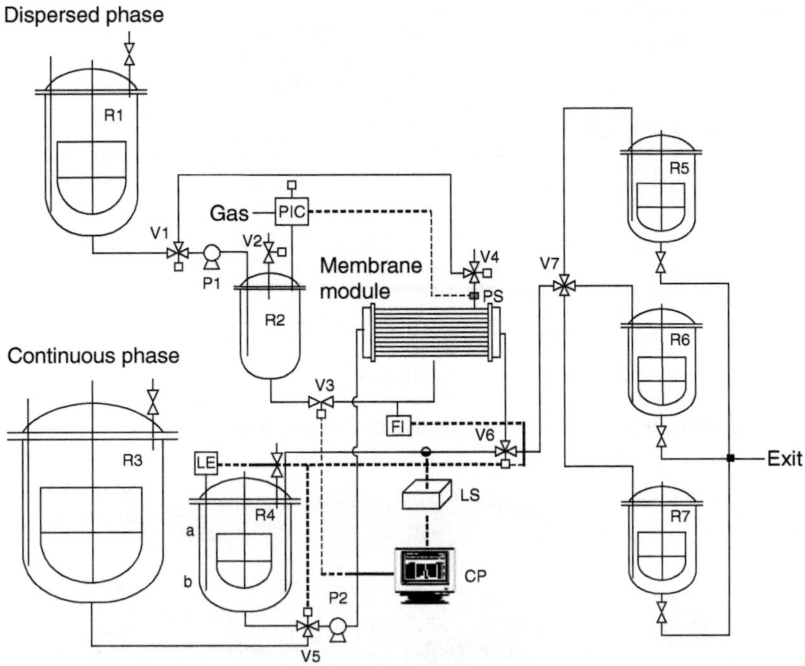

Figure 10.2 Process flow diagram of high throughput membrane emulsification equipment. R: tank; PIC: pressure indicator controller; V: valve; P: pump; FI: flow instrument; LE: liquidometer; LS: particle size detection; CP: control panel.

As shown in Figure 10.2, the large-scale high throughput equipment consists of pressure supply system, automatic control emulsification system, and data integration system.

As shown in Figure 10.2, high throughput membrane emulsification equipment has the following characteristics: (i) to maintain the stability of transmembrane pressure and the size uniformity of emulsion, CME equipment should balance the loss and supply of pressure constantly during the emulsification process based on pressure monitoring data; (ii) for maintaining the growth and torque balance of uniform droplets, the circulating pump can adjust the flow rate of continuous phase based on real-time data of flow instrument; and (iii) for improving the stability of emulsification process, the real-time data of liquidometer, particle size detection, and control system should be monitored and integrated.

To further increase the capacity of CME, magnification effect of equipment was explored as shown in Figure 10.3. By increasing effective membrane area and membrane linear velocity proportionally, the capacity of CME equipment increased substantially without influence on droplet size of emulsion. It provided a solution to enlarge the production scale of CME equipment by increasing the effective membrane area. To meet the demand for different products and purposes, our team

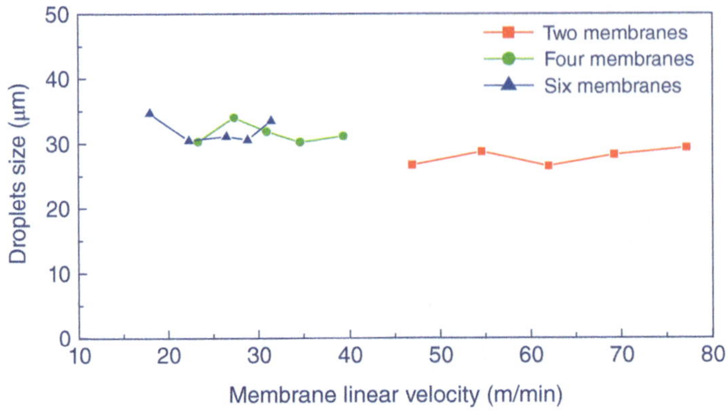

Figure 10.3 Magnification verification of effective membrane area.

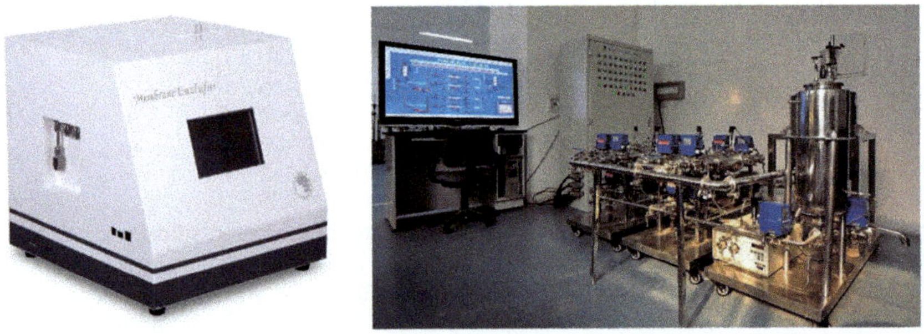

Figure 10.4 A series of novel cross-flow membrane emulsification equipment [1–3]. (a) Experimental equipment. (b) Industrial equipment (capacity over 100 l).

designed a series of novel CME equipment involving experimental equipment and industrial equipment (capacity over 100 l) as shown in Figure 10.4.

10.3 Premix Membrane Emulsification Equipment

PME technology involves using high pressure to force a coarse pre-emulsion to pass through microporous membrane and yielding smaller uniform emulsion droplets. PME with its superiorities of fast manufacturing speed and low energy consumption is not only appropriate for preparing O/W and W/O emulsions but also more suitable for manufacturing oil-in-water-in-oil (O/W/O), water-in-oil-in-water (W/O/W), and solid-in-oil-in-water (S/O/W) multiple emulsions. At present, both emulsions and particles solidified from them with uniform and controllable size of 0.1–50 μm, are able to be prepared by PME. The formation process of droplets by PME is shown in Figure 10.5.

Figure 10.5 Premix membrane emulsification method.

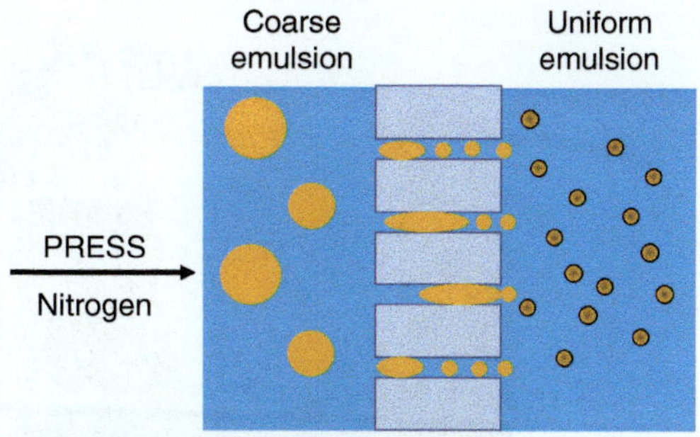

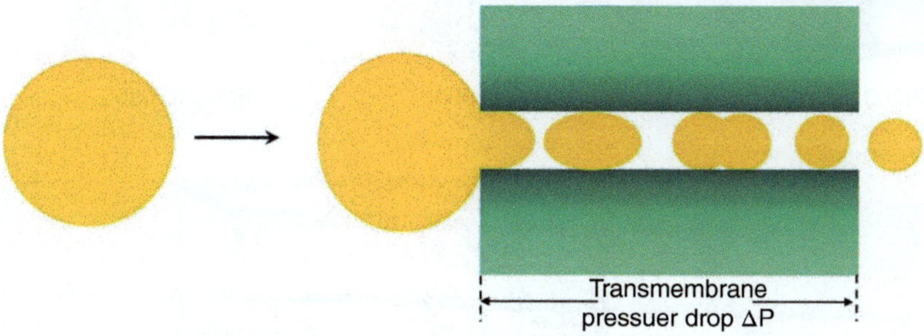

Figure 10.6 Disruption of coarse emulsion by microporous membrane.

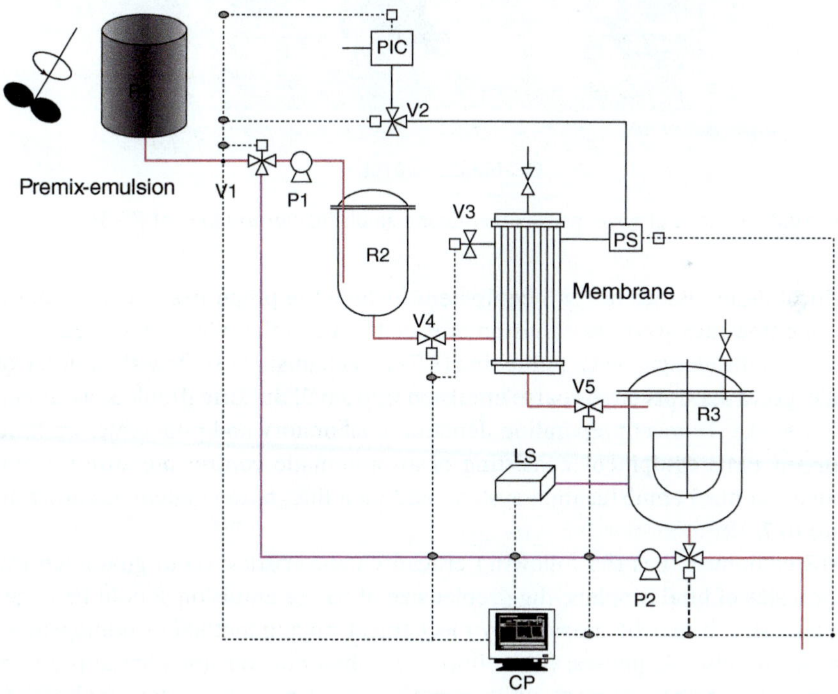

Figure 10.7 Process flow diagram of premix membrane emulsification equipment. R: tank; PIC: pressure indicator controller; V: valve; P: pump; LS: particle size detection; CP: control panel.

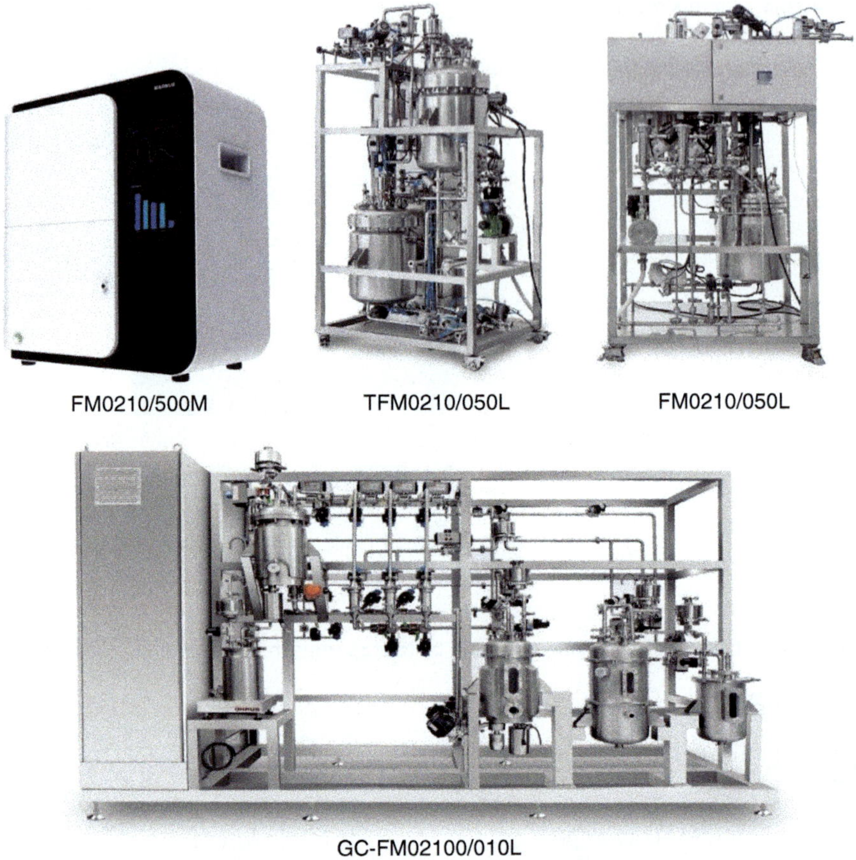

Figure 10.8 A series of novel premix membrane emulsification equipment [1–3].

Critical design issues of PME equipment include the properties of coarse emulsion, and flow rate of coarse emulsion pressed through microporous membrane.

Based on the elongative end pinching (EEP) mechanism [15], flow shear force of parallel pores disrupts large coarse emulsion into small uniform droplets, as shown in Figure 10.6. To meet preparation demand of laboratory and pilot scale, we have developed PME equipment consisting of an automatic control pressure system, automatic control emulsification system, and data integration system as shown in Figure 10.7.

PME equipment has the following crucial characteristics: (i) to guarantee the uniform size of final droplets, the droplet size of coarse emulsion should be larger than pore size, it may be prepared by mechanical stirring method or homogenizer from two immiscible phases; (ii) to improve stability of transmembrane pressure during manufacturing process, an automatic control pressure system replenishes the ullage of pressure in real-time based on pressure monitoring data; (iii) for controlling droplet size and uniformity of emulsion, transmembrane pressure should be adjusted based on pore size of microporous membrane and targeted particle size of product; and (iv) for repeatability and controllability of products, automatic control system is equipped with a data analysis module, which can be used for data storing and for reference of different manufacturing processes.

To meet the demand for different products, our team designed and produced a series of novel PME equipment including lab-scale equipment (FM0210/500M), pilot and industrial types of equipment (TFM0210/050L, FM0210/050L, and GC-FM02100/010L) as shown in Figure 10.8.

10.4 Rotary Membrane Emulsification Equipment

Compared with CME and PME, RME technology is developed for preparing microspheres with large particle size of 100 μm and above. As shown in Figure 10.9, dispersed phase permeates through the pores of a rotary cylindrical porous membrane into aqueous phase to form droplets under centrifugal force. Emulsion prepared by rotation is solidified by physical or chemical methods to obtain microspheres and microcapsules with large particle size. Both emulsions and particles solidified from them with uniform and large sizes of 0.1–2 mm are able to be prepared by RME.

In process of RME, the pore size uniformity of membrane and the rotation speed of membrane module have great influence on droplet size. RME equipment has the following crucial characteristics: (i) for great uniformity of prepared emulsion, micropores with a narrow size distribution should distribute uniformly on outer cylindrical membrane; (ii) to ensure every droplet acted by parallel centrifugal

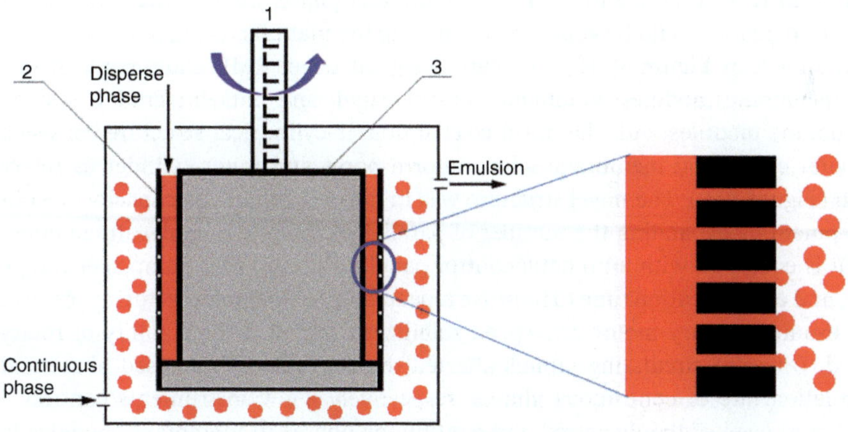

Figure 10.9 Principle of rotary membrane emulsification equipment. (1) Rotary power wheel; (2) outer cylindrical membrane with uniform pore size; and (3) inner rotary cylinder.

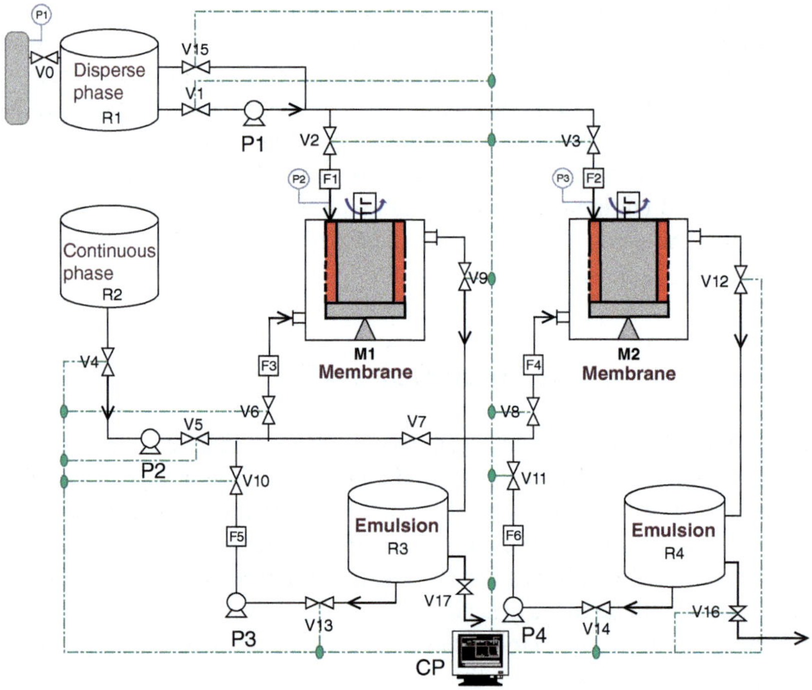

Figure 10.10 Process flow diagram of rotary membrane emulsification equipment. R: tank; V: valve; P: pump; FI: flow instrument; CP: control panel.

force, rotation speed of rotary power wheel should be constant during the whole preparation process; and (iii) to control feeding amount of the dispersed phase and avoid reverse permeation of the continuous phase, appropriate flow rate of dispersed phase should be kept constant during the manufacturing process.

As shown in Figure 10.10, our team designed novel RME equipment, involving membrane modules, automatic control panel, and data integration system. Membrane modules with the novel coaxial double-cylindrical structure consisted of outer cylindrical membrane with uniform pores and inner cylinder as rotary shaft (Figure 10.9). The novel structure was in favor of enhancing capacity of RME equipment by expanding the number of parallel membrane modules. In addition, RME is equipped with automatic control panel for controlling pump speed, flow rate, and emulsification time to improve repeatability of the manufacturing process. For example, rotary motor and speed stabilizer controlled the stability of rotary speed. Different circulating pumps affected the feed rate of dispersed phase and circulation rate of continuous phases, respectively. Flow instruments controlled the liquid level of the dispersed and continuous phases in membrane modules to ensure the continuous production process.

For different production scenarios, our team has designed novel RME equipment involving laboratory equipment and industrial equipment applied to polysaccharide systems, as shown in Figure 10.11.

Figure 10.11 A series of novel rotary membrane emulsification equipment.

10.5 Industrialization – Case Report

Membrane emulsification technology solved the problem of poor repeatability of preparation, release behavior, and drug efficacy in vivo among batches caused by wide size distributions of microspheres prepared by traditional techniques. Membrane emulsification technology has been applied in industrialization for sustained release microspheres preparation by our group. The local anesthetic drugs are widely used for regional block, postoperative analgesia, and chronic pain control in surgery. Nevertheless, postoperative pain generally causes persistent pain for several days, a single injection of local anesthetic drug just provides a nerve block lasting only three to four hours, making it difficult to achieve desirable analgesia effect. Our team

Figure 10.12 GMP production platform for sustained release microspheres by membrane emulsification.

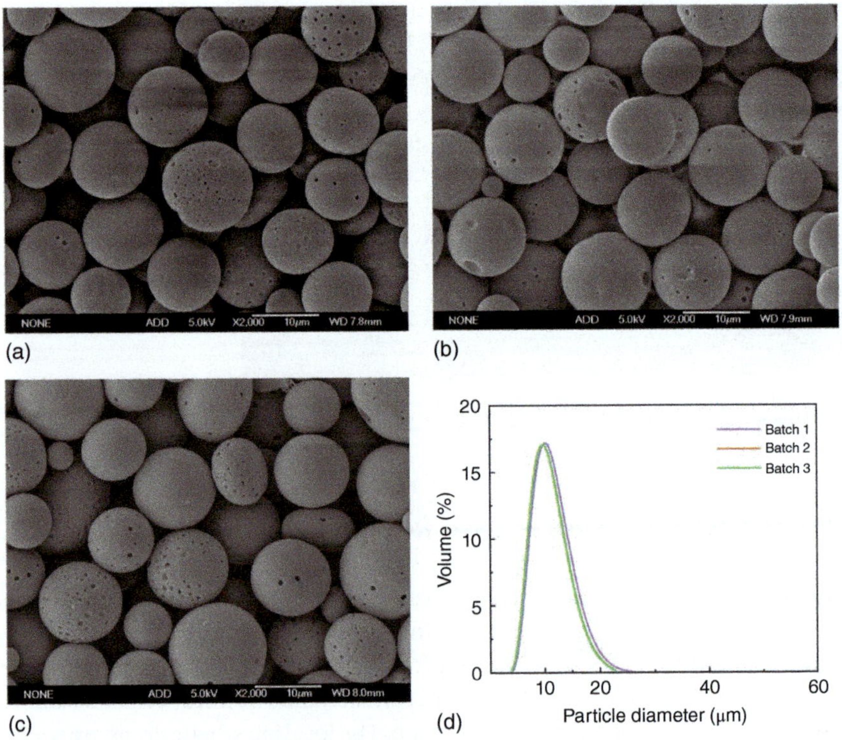

Figure 10.13 SEM micrographs (a–c) and size distributions (d) of three batches of local anesthetic drug loaded microspheres: (a) batch 1; (b) batch 2; and (c) batch 3.

cooperated with a well-known pharmaceutical company in China to develop the long-term sustained release microspheres of local anesthetic drug for injection one time for three days analgesia effect, which can improve the patient compliance and increase the bioavailability of drug. The three batches of production samples (1000 vials per each batch) were prepared by good manufacturing practice (GMP) production platform designed by our group (Figure 10.12)

As shown in Figure 10.13a–c, a few porous surfaces and regular sphericity were observed for all batches of microspheres. The size of local anesthetic drug-loaded microspheres were all around 10 μm (Figure 10.13d), which indicated that the membrane emulsification technology could guarantee the narrow size distribution of microspheres, resulting in the good reproducibility of the drug-loaded microspheres in industrialization.

10.6 Summary

Based on three subdivided membrane emulsification technology, our team has constructed an equipment platform including CME equipment, PME equipment, and RME equipment to solve the technical bottleneck of preparing high-performance emulsion and microsphere with uniform size. With the effort of about 30 years, the equipment platform has accomplished great innovative achievements for manufacturing emulsion, microsphere, and microcapsule with controllable and uniform particle size from submicron to hundreds of micrometers. Some of relevant products have been applied in biomedicine, pharmaceutical industry, and home and personal care products, such as protein separation media, sustained-release microspheres, and cell culture carriers, as described in previous chapters.

References

1 Ma, G.H., Nagai, M., and Omi, S. (1999). Effect of lauryl alcohol on morphology of uniform polystyrene-poly(methyl methacrylate) composite microspheres prepared by porous glass membrane emulsification technique. *Journal of Colloid and Interface Science* 219 (1): 110–128.

2 Ma, G.H., Nagai, M., and Omi, S. (1999). Study on preparation and morphology of uniform artificial polystyrene-poly(methyl methacrylate) composite microspheres by employing the SPG (Shirasu porous glass) membrane emulsification technique. *Journal of Colloid and Interface Science* 214 (2): 264–282.

3 Omi, S., Ma, G.H., and Nagai, M. (2000). Membrane emulsification – a versatile tool for the synthesis of polymeric microspheres. *Macromolecular Symposia* 151 (1): 319–330.

4 Hao, D.X., Gong, F.L., Hu, G.H. et al. (2008). Controlling factors on droplets uniformity in membrane emulsification: experiment and modeling analysis. *Industrial & Engineering Chemistry Research* 47 (17): 6418–6425.

5 Zhou, Q.Z., Wang, L.Y., Ma, G.H. et al. (2007). Preparation of uniform-sized agarose beads by microporous membrane emulsification technique. *Journal of Colloid and Interface Science* 311 (1): 118–127.

6 Zhang, Y.L., Lian, G.P., Zhu, S.P. et al. (2008). Investigation on the uniformity and stability of sunflower oil/water emulsions prepared by a Shirasu porous glass membrane. *Industrial & Engineering Chemistry Research* 47 (17): 6412–6417.

7 Wang, L.Y., Ma, G.H., and Su, Z.G. (2005). Preparation of uniform sized chitosan microspheres by membrane emulsification technique and application as a carrier of protein drug. *Journal of Controlled Release* 106 (1): 62–75.

8 Yuan, M.M., Na, X.M., Liu, J.X. et al. (2022). Preparation, characterization and in vivo efficacy evaluation of ropivacaine O/W emulsion by premix membrane emulsification. *Colloids and Surfaces a-Physicochemical and Engineering Aspects* 640: 128316.

9 Lv, P.P., Wei, W., Gong, F.L. et al. (2009). Preparation of uniformly sized chitosan nanospheres by a premix membrane emulsification technique. *Industrial & Engineering Chemistry Research* 48 (19): 8819–8828.

10 Wei, Q., Wei, W., Lai, B. et al. (2008). Uniform-sized PLA nanoparticles: preparation by premix membrane emulsification. *International Journal of Pharmaceutics* 359 (1–2): 294–297.

11 Tian, Q.R., Zhou, W.Q., Cai, Q. et al. (2021). Concepts, processing, and recent developments in encapsulating essential oils. *Chinese Journal of Chemical Engineering* 30: 255–271.

12 Li, X.Q., Li, Q., Gong, F.L. et al. (2015). Preparation of large-sized highly uniform agarose beads by novel rotating membrane emulsification. *Journal of Membrane Science* 476: 30–39.

13 Li, X.Q., Gong, F.L., and Ma, G.H. (2014). Preparation of uniform konjac glucomannan gel microspheres by rotating membrane emulsification. *The Chinese Journal of Process Engineering* 14 (2): 296–303.

14 Hu, Y.N., Gong, F.L., Yan, X.R. et al. (2022). Effects of membrane pore activation on microporous membrane emulsification process and emulsion droplet formation. *Colloid and Polymer Science* 300 (4): 297–307.

15 Lin, C.Z. and Guo, L.J. (2006). Numerical simulation of drop deformation and breakup in shear flow. *Journal of Engineering Thermophysics* 27 (1): 79–81.

Index

Novel Membrane Emulsification: Principles, Preparation, Processes, and Bioapplications, First Edition. Guanghui Ma.

r

s